Abstract Analytic Number Theory

JOHN KNOPFMACHER

University of the Witwatersrand
Johannesburg

DOVER PUBLICATIONS, INC., NEW YORK

Published in Canada by General Publishing Company, Ltd., 30 Lesmill Road, Don Mills, Toronto, Ontario.
Published in the United Kingdom by Constable and Company, Ltd., 10 Orange Street, London WC2H 7EG.

This Dover edition, first published in 1990, is a corrected and enlarged republication of the work originally published by North-Holland Publishing Company, Amsterdam, and American Elsevier Publishing Company, Inc., New York in 1975 as Volume 12 in the North-Holland Mathematical Library. The Preface to the Dover Edition and the Additional Bibliography are new to this edition.

Manufactured in the United States of America
Dover Publications, Inc., 31 East 2nd Street, Mineola, N.Y. 11501

Library of Congress Cataloging-in-Publication Data

Knopfmacher, John.
Abstract analytic number theory / John Knopfmacher.
p. cm.
Reprint. Originally published: Amsterdam : North Holland Pub. Co. ; New York : American Elsevier, 1975.
Includes bibliographical references.
ISBN 0-486-66344-2
1. Number theory. I. Title.
QA241.K66 1990
512′.7—dc20 90-2836
CIP

PREFACE TO THE DOVER EDITION

This edition contains corrections of some printing errors, and fills minor gaps in Proposition 2.3 of Chapter 5, and Lemma 2.5 of Chapter 6. It also contains a supplement to the original bibliography.

Johannesburg, 1989 JOHN KNOPFMACHER

PREFACE TO THE FIRST EDITION

A summary of the aims of this monograph, of the kinds of topics it deals with, and of mathematical prerequisites for reading the text may be found in the Introduction. My purpose in this preface is simply to express appreciation to some of the people who have shown interest in the book or its subject matter during the course of its preparation.

Firstly, I should like to thank Professor Dr. Wolfgang Schwarz for his very strong interest in and helpfulness concerning the original manuscript, and for very kindly arranging a visiting professorship for me at the Johann Wolfgang Goethe-Universität, Frankfurt am Main, West Germany, for a period during 1973. During that time I had the pleasant opportunity of lecturing to Professor Schwarz and colleagues of his — particularly Drs. J. Duttlinger and K.-H. Indlekofer — on many of the topics treated below. I thank them warmly for all their friendly discussions and comments, as well as for their generous hospitality during my visit.

Secondly, and just as whole-heartedly, my thanks go to local colleagues — particularly James Ridley and Allan Sinclair — for many enthusiastic conversations about this subject, and for numerous suggestions.

Johannesburg, 1974 JOHN KNOPFMACHER

CONTENTS

INTRODUCTION

The purpose of this introduction is to give a rapid survey, without much specific detail, of the aims of this book and the kinds of topics that it covers.

Abstract analytic number theory seems to have arisen first as a generalization of the classical number theory of the positive rational integers, with special emphasis on the derivation of generalizations of the famous

Prime Number Theorem. *$\pi(x) \sim x/\log x$ as $x \to \infty$, where $\pi(x)$ denotes the total number of positive rational primes $p \leqq x$.*

Such investigations may have been motivated partly by the search for an "elementary" proof of the Prime Number Theorem, which was not found until the latter part of the decade 1940—1950, and perhaps partly by the fact that they could be used to simultaneously cover at least part of Landau's classical *Prime Ideal Theorem*, which extends the Prime Number Theorem to ideals in an algebraic number field. In this way, with ordinary rational integers or ideals in algebraic number fields as essentially the only significant examples, and with the major interest centering around the "stability" of classical theorems (i.e. the extent to which they remain true under weakened abstract hypotheses), one might perhaps view the abstract theory as largely an internally motivated variation of part of classical analytic number theory.

Recently, however, it has been noted that the ideas and techniques of the above and other parts of classical analytic number theory have a bearing on and applications to a wide variety of mathematical systems quite distinct from the multiplicative semigroups formed by the positive integers or the ideals in a given number field K.

In the first place, it had been apparent for many decades that there are a large number of classes of mathematical systems, particularly ones arising in abstract algebra, which have elementary "unique factorization" properties analogous to those of the positive integers 1, 2, 3, ... For example, this is basically the content of theorems of the Krull–Schmidt type, which are known to be valid for many categories of mathematical objects. However, until very recently, the analogy mentioned was usually taken to both begin and end with the trivial algebraic concept of "unique factorization".

In the case of certain of these newer systems, though, there is an additional property which seems to make them both more "arithmetical" in a sense, and more amenable to further techniques of classical number theory. This property stems from the existence of a function measuring the "size" of an individual object (usually the object's cardinality or dimension, or "degree" in some sense), with the crucial attribute that there are not "too many" inequivalent objects (i.e., only a finite number of inequivalent ones whose "size" does not exceed any chosen bound). The first new system of this type which was noted and investigated by the methods of classical multiplicative analytic number theory was the very familiar category $\mathscr{A}$ of all *finite abelian groups.*

Let $G_{\mathscr{A}}$ denote the semigroup formed by the set of all isomorphism classes of groups in $\mathscr{A}$ under the product operation induced by the direct product of groups. It was discovered by E. Cohen that, with the "size" or "norm" function on $G_{\mathscr{A}}$ induced by the cardinality of any group in a given isomorphism class, one can carry over to $G_{\mathscr{A}}$ not only the simple unique factorization properties implied by the "Fundamental Theorem on Finite Abelian Groups" but also many asymptotic conclusions of the analytic theory of numbers. Apart from any interest provided by the existence of such analogies in a specific and natural non-classical arithmetical system, such results also yield asymptotic enumeration theorems regarding the "average" numbers of isomorphism classes of finite abelian groups with certain natural types of properties. Theorems of the latter kind could also be considered to be of intrinsic interest, regardless of any abstract arithmetical interpretation. In fact, Cohen's work depends heavily on an earlier enumeration theorem of Erdős and Szekeres, which provides an asymptotic mean-value for the total number $a(n)$ of all isomorphism classes of groups of order n in $\mathscr{A}$, as $n \rightarrow \infty$. (Cohen's results do not seem to include the formulation of a "Prime Abelian Group Theorem" analogous to the Prime Number Theorem, but such a theorem is fairly easy to deduce from the Prime Number Theorem or from its abstract version discussed below; for $\mathscr{A}$ this provides an asymptotic enumeration result concerning the indecomposable finite abelian groups.)

Some years later, work of the present author showed that $G_{\mathscr{A}}$ is only the first of infinitely many non-classical but perfectly natural arithmetical systems, which may be investigated fruitfully with the aid of methods of ordinary analytic number theory. Surprisingly perhaps, it turned out that in many cases the previously-mentioned abstract theory of numbers provides a most appropriate tool, which not only brings out the analogies of the new systems with ordinary multiplicative number theory but can also be used to derive some deep asymptotic enumeration theorems as well — principally as corollaries of an *Abstract Prime Number Theorem* subject to certain hypotheses. Further development of the previous abstract theory then also leads to other enumeration theorems in various special cases of interest. In this connection, however, it should be emphasized that these enumeration theorems only become corollaries of the abstract discussion after certain (usually non-trivial) theorems have been proved, which show that the appropriate abstract hypotheses are actually valid for the particular systems considered; a similar comment applies to Landau's Prime Ideal Theorem mentioned earlier.

Most of the non-classical types of arithmetical systems referred to here arise by considering the isomorphism classes of objects in certain particular categories of interesting mathematical objects. In a variety of cases, however, the asymptotic behaviour of the number of isomorphism classes of objects of a certain type and "size" is different from that suggested by analogy with ordinary number theory. By a remarkable coincidence, many systems of this kind turn out to be amenable to the techniques and results of a quite distinct branch of classical number theory – the branch known as *additive* analytic number theory. Thus, although the earlier abstract theory is then no longer applicable, another field of number theory — which was initiated for very different purposes by Hardy and Ramanujan around 1917 — turns out to provide tools of precisely the type now required. Some of the resulting "arithmetical" conclusions contrast strongly with the corresponding facts valid for the positive integers and their closer analogues; in particular there now arises an *Additive Abstract Prime Number Theorem* involving asymptotic statements of a different kind from those of the previous theorems in this direction.

The purpose of this book is to provide a detailed introduction to the topics outlined above. In this respect, it does not attempt to give an exhaustive account of all or the sharpest known results in this area in every case. However, by dealing carefully with some of the more fundamental concepts and theorems, it is hoped that the treatment will both convince the reader of the reasonably wide scope of the ideas discussed and of their interest for the particular systems considered in detail, and be accessible to readers with only a moderate mathematical background. Roughly three years' experience of university mathematics should be perfectly ample for most purposes below, although in some cases the text will refer to concepts and theorems involving topics (like those of algebraic number theory or the theory of Lie groups and symmetric Riemannian manifolds, for example) which may not be familiar to a reader with only this background. However, such references to topics of a more advanced kind occur mainly in the description of natural examples that provide motivation for the later abstract discussion. The treatment of such concrete examples (which makes up most of Chapter 1) is intended to stress the rather common occurrence of "arithmetical" systems in many branches of mathematics — a fact which, in this book, is taken as basic motivation for the abstract theory. Nevertheless, it is *not* necessary for immediate purposes that the reader be thoroughly familiar with all such specific systems. (Actually, the outline descriptions of parts of relatively advanced subjects that are included at certain stages in this book should usually suffice for immediate needs. In the case of the examples of Chapter 1 especially, we suggest that *to begin with the reader should merely glance fairly rapidly at the descriptions given*, then refer to these again when references are made in later chapters, and only consider detailed treatments of such background topics when this is convenient or a specific need arises.)

There is one particular field with which a moderate familiarity on the part of the reader would perhaps be useful for present purposes. This is classical multiplicative analytic number theory — a subject which can easily be introduced in its initial stages during the early years of university mathematics, although it is not always included. (The levels in analytic number theory reached in the books of Hardy and Wright [1] and Chandrasekharan [1], for example, should be more than adequate for a reader of the present text.) Even here, although some previous knowledge in this direction would probably be helpful, it is worth remarking that in principle such familiarity is not strictly essential. For, in the first place, it will be trivially obvious that certain abstract hypotheses considered below are valid for the ordinary positive integers, and, secondly, that part of the abstract theory which covers the rational integers as a special case is often only slightly more difficult at the level of useful generality than the restricted theory for the classical case alone. Thus, in theory, the present treatment could also be used as an introduction to classical analytic number theory, although it is not suggested that it should be so used — except perhaps by readers who are already fairly sophisticated in other branches of mathematics. (Even such readers might find it illuminating to refer on occasion to a more classical type of text, such as one of those mentioned above, say.)

The plan of the book is as follows: The general arithmetical systems that have been referred to give rise in the first place to the concept of an *arithmetical semigroup*. This is defined in Chapter 1. After a discussion of various natural examples of arithmetical semigroups, the rest of Part I is concerned mainly with algebraic properties of such semigroups and complex-valued "arithmetical" functions on them, and with the application of some of the conclusions to questions of enumeration. In particular, Part I deals with the algebraic enumeration of the isomorphism classes of objects in various specific mathematical categories of interest. Here, the term "algebraic enumeration" is used to indicate that the numbers involved are shown to be in principle computable from identities for the coefficients of certain explicit formal generating functions or series. Questions of asymptotic enumeration of the kinds mentioned earlier often become more amenable when such algebraic relations are already known, but in themselves the results of algebraic enumeration usually provide no insight into the asymptotic conclusions that may be valid. It is for this purpose that methods of analytic number theory often become relevant — even for a mathematician who might wish to solve a particular problem of asymptotic enumeration without having a special interest in number theory *per se*. Such asymptotic questions are included amomgst the topics treated in the second and third parts of the book, on the basis of some of the algebraic results of Part I.

Part II of the book deals with arithmetical semigroups having analytical properties of a kind similar to those of the classical semigroups of all positive integers or non-zero integral ideals in a given algebraic number field K. A portion of the discussion is concerned with establishing the validity of a certain "Axiom A" for the particular semigroups associated with various interesting categories of objects. This in itself amounts to the proof of a number of asymptotic enumeration theorems regarding the average numbers of isomorphism classes of objects of large cardinal in those categories. Apart from this, the relevant chapters investigate consequences of the abstract Axiom A, including the first Abstract Prime Number Theorem discussed above.

Part III contains two chapters, of which the first develops the analytical theory of *additive* arithmetical semigroups of a certain type. These semigroups are the ones implicitly referred to above in connection with the classical work of Hardy and Ramanujan in the field of additive analytic number theory. The final chapter treats *arithmetical formations*, which are systems whose theory allows both a generalization and a refinement of the theory of abstract arithmetical semigroups. Put roughly, an arithmetical formation consists of an arithmetical semigroup together with an equivalence relation on the semigroup which partitions it into classes that in many ways are analogous to arithmetical progressions of positive integers. The two chapters of Part III each contain proofs of further "abstract prime number theorems", of kinds appropriate to the types of system discussed in either case.* Once again, asymptotic enumeration theorems play rôles which are often of intrinsic interest regardless of the abstract number-theoretical background. In addition, some of these theorems are useful in establishing the validity of relevant abstract axioms for special systems of interest, while others occur as corollaries of the abstract discussion itself (after the axioms have been verified).

Most of the material below has not appeared previously in book form, and some of it is at most only implicit in the relevant research articles covering the field. In writing the text, it has seemed advisable not to attempt to give credit or assign priority to the particular authors whose work may be related to given theorems, except in a few special cases (usually ones in which the named authors had proved the particular results for the ordinary integers). A procedure of this kind is probably always open to debate, and it is in no way implied that the names of relevant authors do not deserve to be attached to theorems in other instances. (In certain cases, such as that of the Abstract Prime Number Theorem based on Axiom A, a special difficulty arises: the number of relevant authors is very large and difficult to evaluate. For example, see the lengthy history of the classical Prime Number Theorem described in Landau's *Handbuch* [10].)

In order to at least partly counter-balance the present approach to the history of the subject, a fairly extensive bibliography is provided, and at the end of each chapter there is a selection of references to research papers or books with some bearing on, or interest related to, the contents of the chapter. It is worth remarking that the bibliographical references which occur in the text are usually intended to point the way to developments or topics of related interest, that are not covered fully in this book; their presence at those places is not intended to convey any deeper significance. In addition to containing directly relevant material, the bibliography also

*A quite extensive class of natural arithmetical semigroups with still different asymptotic properties is treated by the author [16].

includes a selection of further references to topics of related interest which are not mentioned in the text: e.g. the "zeta functions" of rational simple algebras, of algebraic varieties, and of diffeomorphisms of manifolds.

For the expert on analytic number theory in particular, it may be emphasized again that the present book is concerned mainly with outlining the general scope of certain ideas and methods of analytic number theory within a variety of different mathematical fields besides classical number theory; there is no attempt at deriving the "best possible" or sharpest known asymptotic estimates in every case. (For information about sharper estimates in the analytic theory of the positive integers, we refer in particular to the quite recent books of Chandrasekharan [2], Huxley [1], Prachar [2] and Schwarz [4]; for information about the further analytic theory of algebraic number fields, particular reference may be made to the books of Goldstein [1], Landau [9], Lang [2] and Narkiewicz [1].)

A question that an expert might perhaps ask concerns whether the present development of abstract analytic number theory is sufficiently "ripe" for the publication of an account in book form. In this connection, the author believes that the present development of the subject is certainly ripe enough for the writing of an *introduction* to the field, and this is what this book attempts to provide. Examination of the text and articles listed in the bibliography will show that there are many aspects of this field that remain open to further investigation, and hopefully the subject will continue to develop rapidly in the near future. (A selection of unsolved questions is included at the end of the book.)

Comment on notation. A reference in the text of the type "Proposition 4.3.7" indicates Proposition 3.7 of Chapter 4, § 3; a reference of the type "Proposition 2.6" indicates Proposition 2.6 of § 2 of the chapter in which the reference occurs.

PART I

ARITHMETICAL SEMIGROUPS AND ALGEBRAIC ENUMERATION PROBLEMS

Part I is concerned with the algebraic properties of arithmetical semigroups and arithmetical functions on such semigroups, and with the application of conclusions about these properties to questions of enumeration. In particular, it deals with the enumeration of the isomorphism classes of objects in various specific categories of interest. This type of enumeration is of algebraic nature, involving the coefficients of formal generating functions or series of certain kinds. Questions of asymptotic enumeration, dealing particularly with the asymptotic behaviours of the numbers of isomorphism classes of objects of large cardinal or dimension in various natural categories, are included amongst the topics investigated in Parts II and III on the foundation of the algebraic results of Part I.

CHAPTER 1

ARITHMETICAL SEMIGROUPS

This chapter discusses a selection of natural examples of concrete mathematical systems that have "arithmetical" properties which are closely akin to those of the positive integers 1, 2, 3, ... in certain ways. The examples given split into two types. Firstly, there are "algebraic" systems arising from the study of rings that include or generalize the semigroup of positive integers itself. Secondly, we consider a selection of examples which arise from specific categories of mathematical objects for which a theorem of the Krull–Schmidt type is valid. It is the extension of methods of classical analytic number theory, so as to encompass such examples associated with explicit categories, which provides the abstract treatment of the later chapters with its most convincing justification, perhaps.

§ 1. Integral domains and arithmetical semigroups

We begin with a formal definition of the elementary but fundamental concept to be studied below.

Let G denote a commutative semigroup with identity element 1, relative to a multiplication operation denoted by juxtaposition. Suppose that G has a finite or countably infinite subset P (whose elements are called the *primes* of G) such that every element $a \neq 1$ in G has a unique factorization of the form

$$a = p_1^{\alpha_1} p_2^{\alpha_2} \ldots p_r^{\alpha_r},$$

where the p_i are distinct elements of P, the α_i are positive integers, r may be arbitrary, and uniqueness is understood to be only up to the order of the factors indicated. Such a semigroup G will be called an **arithmetical semigroup** if in addition there exists a real-valued *norm* mapping $|\ |$ on G such that:

(i) $|1| = 1$, $|p| > 1$ for $p \in P$,

(ii) $|ab| = |a|\,|b|$ for all $a, b \in G$,

(iii) the total number $N_G(x)$ of elements $a \in G$ of norm $|a| \leqq x$ is finite, for each real $x > 0$.

It is not difficult to verify that the conditions (i)–(iii) are equivalent to conditions (i) and (ii) together with

(iii)′ the total number $\pi_G(x)$ of elements $p \in P$ of norm $|p| \leqq x$ is finite, for each real $x > 0$.

Before turning to the discussion of natural examples of arithmetical semigroups, it may be emphasized that it is especially the finiteness conditions (iii), (iii)′ that are crucial in our use of the adjective 'arithmetical', and in the later consideration of methods of analytic number theory in order to obtain information about special semigroups of interest.

1.1. Example. The proto-type of all arithmetical semigroups is of course the multiplicative semigroup $G_{\mathbf{Z}}$ of all *positive integers* $\{1, 2, 3, \ldots\}$, with its subset $P_{\mathbf{Z}}$ of all rational primes $\{2, 3, 5, 7, \ldots\}$. Here we may define the norm of an integer n to be $|n| = n$, so that the number $N_{\mathbf{Z}}(x) = N_{G_{\mathbf{Z}}}(x) = [x]$, the greatest integer not exceeding x.

Although the function $N_{\mathbf{Z}}(x)$ would be too trivial to mention if one were not interested in a wider arithmetical theory, the corresponding function $\pi(x) = \pi_{G_{\mathbf{Z}}}(x)$ remains mysterious to this day. The asymptotic behaviour of $\pi(x)$ for large x forms the content of the Prime Number Theorem, which states that

$$\pi(x) \sim x/\log x \quad \text{as } x \to \infty.$$

A suitably generalized form of this theorem will be proved in Part II.

1.2. Example: Euclidean domains. A simple way in which unique factorization arises in elementary abstract algebra is by means of the study of Euclidean domains, or more generally principal ideal domains.

If D denotes an integral domain, two elements $a, b \in D$ are called *associated* if and only if $a = bu$ for some unit $u \in D$. The relation of being associated is an equivalence relation on D, and it is easy to see that the resulting set G_D of all *associate classes* $\bar{a}$ of non-zero elements $a \in D$ forms a commutative semigroup with identity under the multiplication operation $\bar{a} \cdot \bar{b} = \overline{ab}$. In the case when D is a principal ideal domain, the content of the *Unique Factorization Theorem* for D is that the elements $\bar{a} \neq \bar{1}$ of G_D admit unique factorization into powers of the classes $\bar{p}$ of prime elements $p \in D$.

If, in addition, D is a Euclidean domain with norm function $|\ |$, one may define a norm on G_D satisfying conditions (i) and (ii) above by letting

$|\bar{a}|=|a|$. In certain interesting cases this norm satisfies condition (iii) above, and G_D forms an arithmetical semigroup. The following are illustrations.

Firstly, if D is the ring $\mathbf{Z}$ of all *rational integers*, then it is clear that G_D may be identified essentially with the semigroup $G_{\mathbf{Z}}$ discussed above. A different example arises when D denotes the ring $\mathbf{Z}[\sqrt{-1}]$ of all *Gaussian integers* $m+n\sqrt{-1}$ $(m,n\in\mathbf{Z})$. It is familiar that this ring forms a Euclidean domain if one assigns the norm m^2+n^2 to the number $m+n\sqrt{-1}$. Since $\mathbf{Z}[\sqrt{-1}]$ has only the four units 1, -1, $\sqrt{-1}$, $-\sqrt{-1}$, one sees that

$$N_{\mathbf{Z}[\sqrt{-1}]}(x) = N_{G_{\mathbf{Z}[\sqrt{-1}]}}(x) = \tfrac{1}{4}\sum_{n\leqq x} r(n) < \infty,$$

where $r(n)$ denotes the total number of lattice points (a, b) (i.e., points (a, b) with integer components a, b) on the circle $x^2+y^2=n$ $(n\geqq 1)$ in the Euclidean plane $\mathbf{R}^2$. Thus $G_{\mathbf{Z}[\sqrt{-1}]}$ forms an arithmetical semigroup.

For the sake of general interest, we note that the primes of $G_{\mathbf{Z}[\sqrt{-1}]}$ are the associate classes in $\mathbf{Z}[\sqrt{-1}]$ of the numbers

(i) $1+\sqrt{-1}$,

(ii) all rational primes $p\equiv 3 \pmod 4$,

(iii) all factors $a+b\sqrt{-1}$ of rational primes $p\equiv 1 \pmod 4$. For a proof of this statement, see Hardy and Wright [1], say.

Another well-known Euclidean domain is the domain $\mathbf{Z}[\sqrt{2}]$ of all real numbers of the form $m+n\sqrt{2}$ $(m,n\in\mathbf{Z})$, in which $m+n\sqrt{2}$ is assigned the norm $|m^2-2n^2|$. This domain has infinitely many units; in fact, they are all the numbers of the form $\pm(1+\sqrt{2})^n$ $(n\in\mathbf{Z})$. Nevertheless, it can be proved that $N_{G_{\mathbf{Z}[\sqrt{2}]}}(x)$ is finite for each $x>0$, and hence that $G_{\mathbf{Z}[\sqrt{2}]}$ forms an arithmetical semigroup also. (See the book of Cohn [1], which in fact makes a study of general quadratic number domains. For the statement about units above, and the following remark, see for example Hardy and Wright [1].)

The primes of $G_{\mathbf{Z}[\sqrt{2}]}$ have a classification similar to the one mentioned for $G_{\mathbf{Z}[\sqrt{-1}]}$ above: they are the associate classes in $\mathbf{Z}[\sqrt{2}]$ of the numbers

(i) $\sqrt{2}$,

(ii) all rational primes $p\equiv \pm 3 \pmod 8$,

(iii) all factors $a+b\sqrt{2}$ of rational primes $p\equiv \pm 1 \pmod 8$.

Lastly, consider a polynomial ring $F[t]$ in an indeterminate t over a field F. This is a familiar example of a Euclidean domain, in which one usually defines $|f|=2^{\deg f}$ for $0\neq f\in F[t]$. In the case when F is a finite Galois field $\mathrm{GF}(q)$, it is easy to see that $G_{F[t]}$ forms an arithmetical semigroup. In fact, one notes that

$$N_{G_{F[t]}}(x) = \sum \{q^n\colon\ 0 \leqq n \leqq \log x/\log 2\}.$$

1.3. Example: Ideals in an algebraic number field. Although we shall *not* presuppose a previous knowledge of algebraic number theory or of some of the concepts discussed in the next section, as indicated in the Introduction, we do feel it highly desirable to at least briefly outline some facts about the existence and general nature of various interesting examples of natural arithmetical semigroups in different fields of mathematics. The reader not familiar with some of the terms employed may perhaps still find it useful to glance generally at some of the examples motivating a study of abstract arithmetical semigroups, leaving a deeper consideration of individual systems to some later stage. In particular, the next few paragraphs are intended to be viewed in the spirit of the foregoing remarks.

Let K denote an algebraic number field, i.e., a finite extension of the *rational* field $\mathbf{Q}$, and let D denote the ring of all algebraic integers contained in K. For example, D coincides respectively with the previously considered rings $\mathbf{Z}$, $\mathbf{Z}[\sqrt{-1}]$ or $\mathbf{Z}[\sqrt{2}]$ according as K denotes $\mathbf{Q}$, $\mathbf{Q}(\sqrt{-1})$ or $\mathbf{Q}(\sqrt{2})$, where $\mathbf{Q}(\sqrt{m})$ denotes the set of all complex numbers of the form $a+b\sqrt{m}$ $(a,b\in\mathbf{Q})$. In the general case of an arbitrary algebraic number field K, the ring D is an integral domain which need not necessarily be a unique factorization domain. However, Dedekind showed that unique factorization can be "restored" in a sense, if one considers the set G_K of all non-zero ideals in D (also called the *integral* ideals in K).

It is easy to see that G_K forms a commutative semigroup with D itself as identity element, under the usual multiplication of ideals in a commutative ring. The sense in which unique factorization in D may be "restored" is given by the statement that every proper ideal I can be factorized uniquely into a product of powers of *prime* ideals. (For a domain D of the present type, a prime ideal is the same as a maximal ideal in D, and is irreducible (indecomposable) with respect to ideal multiplication.)

It is an interesting fact that the quotient ring D/I is finite in cardinal for each ideal $I\in G_K$, and that the definition $|I|=\text{card}\,(D/I)$ provides a norm function on G_K satisfying the conditions (i) to (iii) for an arithmetical semigroup. Moreover, the prime ideals in D all arise by factorizing the principal ideals (p) for all rational primes p. Thus $(p)=P_1P_2\ldots P_m$, say, where P_i denotes a prime ideal, and since it can be shown that (p) has norm $p^{[K:\mathbf{Q}]}$, it follows from the multiplicative property (ii) of the norm that $\pi_K(x) = \pi_{G_K}(x)$ is finite for each $x > 0$. (For an elementary discussion of ideals in an algebraic number field, see for example Pollard [1] or LeVeque [1].)

The asymptotic behaviour of the function $\pi_K(x)$ for large x forms the content of Landau's **Prime Ideal Theorem:**

$$\pi_K(x) \sim x/\log x \quad \textit{as } x \to \infty;$$

this will be discussed again later on.

Before turning to the next section, it may be noted that, if the above domain D happens to be a principal ideal domain, then the previously-defined semigroup G_D coincides essentially (i.e., is isomorphic) with the semigroup G_K of ideals. Further, for special domains such as $\mathbf{Z}$, $\mathbf{Z}[\sqrt{-1}]$ and $\mathbf{Z}[\sqrt{2}]$, the natural isomorphism $\bar{a} \leftrightarrow (a)$ preserves norms. Thus the corresponding special arithmetical semigroups for these domains can all be subsumed under the study of the general arithmetical semigroups G_K for arbitrary algebraic number fields K.

§ 2. Categories satisfying theorems of the Krull–Schmidt type

The following discussion is concerned mainly with certain specific classes of mathematical objects, such as groups, rings, topological spaces, and so on, together with the standard "direct product" operations and isomorphism relations appropriate to those classes. It is convenient, though admittedly not quite precise, to temporarily ignore the corresponding morphisms and refer to such classes of objects as *categories*. For example, in talking of the category $\mathscr{G}$ of all finite groups, one often thinks intuitively of the groups (or their isomorphism classes) alone, even though the importance of general homomorphisms is undeniable. Also, although abstract category theory will not be assumed or used here, the author personally believes that certain concepts, such as those of 'morphism' or 'functor' for example, may in due course become important in extending the type of arithmetical theory considered below. It is for these reasons that the term 'category' will be retained here, despite the slight abuse of language or lack of immediate necessity.

Now consider some category $\mathfrak{C}$ which admits a "direct product" or "sum" operation $\times$ on its objects. Suppose that this operation $\times$ preserves $\mathfrak{C}$-isomorphisms, that it is commutative and associative up to $\mathfrak{C}$-isomorphism, and that $\mathfrak{C}$ contains a "zero" object 0 (unique up to $\mathfrak{C}$-isomorphism) such that $A \times 0 \cong A$ for all objects A in $\mathfrak{C}$. Then suppose that a theorem of the *Krull–Schmidt* type is valid for $\mathfrak{C}$, i.e., suppose that every object $A \not\cong 0$ can be expressed as a finite $\times$-product $A \cong P_1 \times P_2 \times \ldots \times P_m$ of objects $P_i \not\cong 0$ that are indecomposable with respect to $\times$, in a way that is unique up to permutation of terms and $\mathfrak{C}$-isomorphism. In most natural situations at least, one may reformulate these conditions on $\mathfrak{C}$ by stating that the various isomorphism classes $\bar{A}$ of objects A in $\mathfrak{C}$ form a set $G_{\mathfrak{C}}$ that is

(i) a commutative semigroup with identity with respect to the multiplication operation $\bar{A} \times \bar{B} = \overline{A \times B}$,

(ii) a semigroup with the unique factorization property with respect to the isomorphism classes of the *indecomposable* objects in $\mathfrak{C}$.

For this reason, one may call the $\mathfrak{C}$-isomorphism classes $\bar{P}$ of indecomposable objects P the **primes** of $\mathfrak{C}$ or $G_{\mathfrak{C}}$.

It has always been obvious that categories with properties of the above kind are in some senses formally remniscent of the multiplicative semigroup $G_{\mathbf{Z}}$ of positive integers. However, less attention appears to have been paid to the fact that, in many interesting cases (some of which are illustrated below), the category $\mathfrak{C}$ also admits a real-valued *norm* function $|\ |$ on objects which is invariant under $\mathfrak{C}$-isomorphism and has the following properties:

(i) $|0| = 1$, $|P| > 1$ for every indecomposable object P,

(ii) $|A \times B| = |A||B|$ for all objects A, B,

(iii) the total number $N_{\mathfrak{C}}(x)$ of $\mathfrak{C}$-isomorphism classes of objects A of norm $|A| \leq x$ is finite, for each real $x > 0$.

Obviously, in such circumstances, the definition $|\bar{A}| = |A|$ provides a norm function on $G_{\mathfrak{C}}$ satisfying the required conditions for an arithmetical semigroup. For these reasons, a category $\mathfrak{C}$ with such further properties may be called an **arithmetical category**.

Here again it is the finiteness condition (iii) that is taken as especially significant in introducing the adjective 'arithmetical'. It is worth noting that the functions $N_{\mathfrak{C}}(x) = N_{G_{\mathfrak{C}}}(x)$ and $\pi_{\mathfrak{C}}(x) = \pi_{G_{\mathfrak{C}}}(x)$ may well be taken as fundamentally important in the study of $\mathfrak{C}$ whether or not one is interested in the more "arithmetical" aspects of $\mathfrak{C}$ or $G_{\mathfrak{C}}$. For, these are basic enumeration functions related to the enumeration of the isomorphism classes of (i) arbitrary objects, (ii) indecomposable objects, in $\mathfrak{C}$, respectively. It is an interesting fact that, even if one is not greatly concerned with abstract number theory *per se*, general methods of analytic number theory can often be extremely useful in investigating the asymptotic behaviour of $N_{\mathfrak{C}}(x)$ or $\pi_{\mathfrak{C}}(x)$ as $x \to \infty$, or in merely obtaining formal algebraic formulae relating to these functions.

The following examples show that norm functions on categories arise in various ways.

2.1. Example: Finite abelian groups. One of the simplest non-trivial examples of an arithmetical category is provided by the category $\mathscr{A}$ of all *finite abelian groups*, together with the usual direct product operation and the norm function $|A| = \text{card}\,(A)$. Here the Krull–Schmidt theorem reduces to the well known Fundamental Theorem on finite abelian groups, the indecomposable objects of this kind being simply the various *cyclic* groups $\mathbf{Z}_{p^r}$ of *prime-power* order p^r.

An interesting subcategory of $\mathscr{A}$ which also forms an arithmetical category is the category $\mathscr{A}(p)$ of all finite abelian *p-groups*, p a fixed rational prime. Here the primes of the category are the isomorphism classes of the cyclic groups $\mathbf{Z}_{p^r}$, where p is the fixed rational prime in question.

2.2. Example: Semisimple finite rings. Let $\mathfrak{S}$ denote the category of all *semisimple associative rings of finite cardinal.* (A convenient reference for our discussion of associative rings and algebras is the book of Herstein [1].) Every ring of finite cardinal is certainly an Artinian ring. Hence it follows from the Wedderburn–Artin theory of semisimple Artinian rings that the Krull–Schmidt theorem is valid for the category $\mathfrak{S}$, the corresponding primes being the isomorphism classes of all *simple* finite rings. By the theory of simple Artinian rings, every simple finite ring S can be expressed as the ring $M_n(L)$ of all $n \times n$ matrices with entries in some division ring L, for some positive integer n. Here the division ring L must obviously be finite in cardinal, and so by Wedderburn's theorem on finite division rings the ring L must be a field. Thus L is a finite Galois field $\text{GF}(p^r)$ for some prime-power p^r, and card $(S) = p^{rn^2}$. In particular, the above remarks show that $\mathfrak{S}$ is an arithmetical category relative to direct sum of rings and the norm function $|R| = \text{card}\,(R)$.

Two interesting arithmetical subcategories of $\mathfrak{S}$ are the category $\mathfrak{S}_c$ of all *commutative* rings in $\mathfrak{S}$, and the category $\mathfrak{S}(p)$ of all rings in $\mathfrak{S}$ that have cardinal some power of p (semisimple finite *p-rings*), where p is a fixed rational prime. The above discussion of simple finite rings shows that the primes of the category $\mathfrak{S}_c$ are the isomorphism classes of the various Galois fields $\text{GF}(p^r)$, while the primes of $\mathfrak{S}(p)$ are the classes of the various rings $M_n(L)$, where L is any one of the Galois fields $\mathrm{GF}(p^r)$ with p the fixed rational prime in question.

2.3. Example: Semisimple finite-dimensional algebras. Let $\mathfrak{S}(F)$ denote the category of all *semisimple finite-dimensional associative algebras* over a given field F. It follows from the classical Wedderburn theory of such algebras that the Krull–Schmidt theorem is valid for this category, the indecomposable algebras being the *simple* finite-dimensional algebras over F. The latter algebras can be described as the various total matrix algebras $M_n(L)$, where L runs through all finite-dimensional division algebras over F. In certain cases, such as those illustrated below, one can define a norm function on $\mathfrak{S}(F)$ such that $\mathfrak{S}(F)$ becomes an arithmetical category (with direct sum as product operation) by letting $|A|=2^{\dim A}$ (or $c^{\dim A}$, for any fixed $c>1$).

For example, if F is an *algebraically closed* field, then every finite-dimensional division algebra over F is isomorphic to F. Therefore in this case $\mathfrak{S}(F)$ is an arithmetical category whose primes are the isomorphism classes of the total matrix algebras $M_n(F)$ $(n=1, 2, \ldots)$.

Again, if F is the field $\mathbf{R}$ of all *real numbers*, or more generally any *real closed* field, then the only finite-dimensional division algebras over F are essentially F itself, the field $F(\sqrt{-1})$ or the 4-dimensional standard quaternion algebra $F(\mathrm{i}, \mathrm{j}, \mathrm{k})$ over F. (The field $\mathbf{R}(\sqrt{-1})$ is of course the field $\mathbf{C}$ of all *complex* numbers.) Thus here $\mathfrak{S}(F)$ is also an arithmetical category and its primes are represented by the algebras $M_n(L)$, where $L=F$, $F(\sqrt{-1})$ or $F(\mathrm{i}, \mathrm{j}, \mathrm{k})$ $(n=1, 2, \ldots)$.

Another example arises when F is any finite *Galois* field $\mathrm{GF}(q)$. In this case, the finite-dimensional division algebras must be finite in cardinal, and so with the aid of Wedderburn's theorem on finite division rings one sees that the division algebras in question must be the various finite extensions $\mathrm{GF}(q^m)$ of $\mathrm{GF}(q)$. The total matrix algebras over these different extension fields then represent the various primes for the present arithmetical category $\mathfrak{S}(F)$.

2.4. Example: Compact Lie groups and semisimple Lie algebras. Let $\mathfrak{L}_s(F)$ denote the category of all *semisimple finite-dimensional Lie algebras* over a given field F. (For the theory of such algebras, see Jacobson [2].) The structure theory for algebras in $\mathfrak{L}_s(F)$ shows that the Krull–Schmidt theorem holds, the indecomposable Lie algebras being those that are *simple*. In certain cases, $\mathfrak{L}_s(F)$ becomes an arithmetical category if one uses the direct sum operation on Lie algebras and defines a norm function by letting

$|A|=c^{\dim A}$ for a fixed $c>1$. In particular, this is the case if F is an *algebraically closed field of characteristic zero.* In that case, with a finite number of exceptions, the simple finite-dimensional Lie algebras are represented by four infinite classes of algebras. In common notation, these are listed in the manner:

(i) A_n $(n\geqq 1)$, $\dim A_n=n^2+2n$;
(ii) B_n $(n\geqq 2)$, $\dim B_n=2n^2+n$;
(iii) C_n $(n\geqq 3)$, $\dim C_n=2n^2+n$;
(iv) D_n $(n\geqq 4)$, $\dim D_n=2n^2-n$.

By considering the category $\mathfrak{L}_s(\mathbf{C})$, where $\mathbf{C}$ denotes the field of complex numbers, the theory of Lie groups and their associated Lie algebras shows that the category $\mathfrak{G}_c$ of all *compact simply-connected Lie groups* is an arithmetical category under Lie group direct product and norm function $|H|=c^{\dim H}$, where $\dim H$ denotes the manifold dimension of the Lie group H. Further, if one ignores the (indecomposable) circle group S^1, there is a dimension-preserving 1–1 correspondence between the primes of $\mathfrak{G}_c$ and those of $\mathfrak{L}_s(\mathbf{C})$. (For further information concerning these Lie groups, see Wolf [1] or Helgason [1], say.)

2.5. Example: Symmetric Riemannian manifolds. Let $\mathfrak{R}_{sc}$ denote the category of all *compact simply-connected globally symmetric Riemannian manifolds.* The structure theory of such manifolds goes back to É. Cartan (see the books of Helgason [1] and Wolf [1]), and implies that $\mathfrak{R}_{sc}$ forms an arithmetical category under Riemannian product of manifolds and the norm function $|M|=c^{\dim M}$ $(c>1$ fixed), where $\dim M$ denotes the manifold dimension of M. Further, the indecomposable objects in this category are the compact simply-connected *irreducible* symmetric spaces, which may be listed (up to isometry) in the following manner:

	Space	*Range*	*Dimension*
(i)	$SU(n+1)$	$n\geqq 1$	$n(n+2)$
(ii)	$Spin(2n+1)$	$n\geqq 2$	$n(2n+1)$
(iii)	$Sp(n)$	$n\geqq 3$	$n(2n+1)$
(iv)	$Spin(2n)$	$n\geqq 4$	$n(2n-1)$
(v)	$SU(n)/SO(n)$	$n>1$	$\frac{1}{2}(n-1)(n+2)$
(vi)	$SU(2n)/Sp(n)$	$n>1$	$(n-1)(2n+1)$
(vii)	$SU(p+q)/S(U_p\times U_q)$	$p+q\geqq 2$	$2pq$
(viii)	$SO(p+q)/SO(p)\times SO(q)$	$p+q>4$	pq

	Space	*Range*	*Dimension*
(ix)	$SO(2n)/U(n)$	$n>2$	$n(n-1)$
(x)	$Sp(n)/U(n)$	$n>2$	$n(n+1)$
(xi)	$Sp(p+q)/Sp(p)\times Sp(q)$	$p+q>2$	$4pq$
(xii)	A finite number of exceptional spaces.		

2.6. Example: Pseudo-metrizable finite topological spaces. As an illustration of a situation in which the abstract "direct product" does *not* arise from the cartesian product of sets and the norm function arises differently from the previous cases, consider the category $\mathfrak{P}$ of all *pseudo-metrizable topological spaces of finite cardinal;* here the T_1-axiom is *not* assumed. Clearly, a space X lies in $\mathfrak{P}$ if and only if each connected component C of X lies in $\mathfrak{P}$, and since there are only a finite number of such components C these must be both open and closed. Thus X is the topological sum (disjoint union) of its connected components. It follows that $\mathfrak{P}$ becomes an arithmetical category if one takes the topological sum as "product" and defines $|X|=2^{\mathrm{card}(X)}$. The primes of $\mathfrak{P}$ are the homeomorphism classes of the various *connected* spaces in $\mathfrak{P}$, and it can be shown (see the author [1]) that up to homeomorphism there is exactly one connected space C_n of cardinal n in $\mathfrak{P}$, for $n=1, 2, \ldots$.

2.7. Example: Finite modules over a ring of algebraic integers. As a final explicit example, consider the category $\mathfrak{F}=\mathfrak{F}_D$ of all *finitely-generated torsion modules* over the ring D of all algebraic integers in some given algebraic number field K. The ring D is a domain of the type treated in Chapter 6 of Jacobson [1], and so one can deduce from the module theory given there that $\mathfrak{F}$ satisfies the Krull–Schmidt theorem, and that the indecomposable modules in $\mathfrak{F}$ are all isomorphic to *cyclic* modules of the form D/P^r, where P is a *prime* ideal in D. (In the particular case in which D is a principal ideal domain, this statement follows from the more familiar theory of modules over principal ideal domains. For example, that is the case when D is one of the domains $\mathbf{Z}$, $\mathbf{Z}[\sqrt{-1}]$, $\mathbf{Z}[\sqrt{2}]$ considered earlier.)

By the discussion of Example 1.3 above, it follows that D/P^r has finite cardinal $|P|^r$. Hence, every module in $\mathfrak{F}$ has finite cardinal and, if one defines $|M|=\mathrm{card}\,(M)$ for M in $\mathfrak{F}$ and uses the direct sum of modules, then it follows that $\pi_{\mathfrak{F}}(x)$ is finite for each real $x>0$ and that $\mathfrak{F}$ is an arithmetical category. In fact, $\mathfrak{F}$ may be described as the category of all *modules of finite cardinal* over D; thus, for $D=\mathbf{Z}$, $\mathfrak{F}$ reduces to the category $\mathscr{A}$ of all finite abelian groups. The primes of $\mathfrak{F}$ in the general case are then the various isomorphism classes of cyclic modules D/P^r for P a prime ideal in D and $r=1, 2, \ldots$.

A reader familiar with the Krull–Schmidt theorem for general algebraic systems will no doubt by this stage have noticed many other examples of arithmetical categories. Amongst other interesting examples, one might perhaps also mention:

(i) the category $\mathscr{G}$ of all *finite groups*;

(ii) the category $\mathscr{G}(p)$ of all finite *p-groups*, for a fixed rational prime p;

(iii) the category $\mathscr{R}$ of all *finite rings*;

(iv) the category $\mathscr{R}(p)$ of all finite rings of cardinal some power of p (finite *p-rings*), for a fixed rational prime p;

(v) the category $\mathscr{N}$ of all *nilpotent* finite rings;

(vi) the category $\mathscr{N}(p)$ of all *nilpotent* finite p-rings, for a fixed rational prime p;

(vii) the category $\mathfrak{A}(F)$ of all *finite-dimensional associative algebras* over a finite Galois field F;

(viii) the category $\mathfrak{N}(F)$ of all *nilpotent* algebras in $\mathfrak{A}(F)$;

(ix) the category of all *finite-dimensional representation modules* of a given finite group H over some fixed field F of characteristic prime to card (H);

(x) the category $\mathfrak{T}$ of all *finite topological spaces*;

(xi) the category (class) of all *finite graphs* of a given interesting species; et cetera.

(Regarding (xi), one may note that morphisms other than isomorphisms are not often considered in the theory of graphs.)

Clearly, there are still many other examples of great interest. However, in selecting the explicit examples 2.1 to 2.7, our concern has been especially with illustrating situations in which it is possible to provide a fairly complete description of the primes for the categories in question. For, it is in such cases that the methods considered later are particularly effective; although many of the later results of Part I of this book are certainly applicable to the above examples (i) to (xi), the information that such results provide tends to be greater the greater one's knowledge of the primes (or their norms) in particular situations. Also, although it is conceivable that results from Parts II or III are valid for the category $\mathscr{G}$ or $\mathscr{G}(p)$ (by way of illustration), the present lack of information about indecomposable finite groups makes it difficult to reach a definite decision on this point, at the moment.

Although the above comments about primes are usually appropriate, there are instances in which fairly precise asymptotic information can be obtained for functions, like $N_{\mathfrak{C}}(x)$ when $\mathfrak{C}$ is a given arithmetical category, despite virtual ignorance of the primes. For examples of this type of phenomenon, see Sims [1] regarding the category $\mathscr{G}(p)$, Kruse and Price [1, 2] concerning $\mathfrak{A}(F)$ and $\mathfrak{N}(F)$, the author [5, 7] in connection with $\mathscr{N}(p)$, $\mathfrak{N}(F)$ and various related categories, and Wright [1, 2] regarding certain classes of finite graphs. The arithmetical semigroups G that may be associated with the classes of graphs considered by Wright have the interesting property that $\pi_G(x) \sim N_G(x)$ as $x \to \infty$ (i.e., "almost all" elements are prime); such semigroups are not covered by Parts II and III, but a discussion may be found in the author's paper [13].

As a technical point, readers familiar with algebraic K-theory may observe that, for an arithmetical category $\mathfrak{C}$, the group of fractions of $G_{\mathfrak{C}}$ is precisely the *Grothendieck* group $K_0(\mathfrak{C}, \times)$. Although it is often very useful to work with this group, for present purposes, it will be more convenient to consider $G_{\mathfrak{C}}$. Similarly, it is here more convenient to work with the semigroup G_K of all non-zero integral ideals in a given algebraic number field K than with the group of *fractional* ideals of K, although in other contexts this might not be the case.

Lastly, it is interesting to note that certain of the examples discussed above are clearly **arithmetically equivalent** in the sense that there are norm-preserving isomorphisms between the arithmetical semigroups in question. These and other examples suggest that it might eventually prove interesting to investigate more general **arithmetical morphisms** of arithmetical semigroups. In this connection, certain viewpoints of algebraic K-theory (and concepts such as *morphism* (in a category), *functor*, and *group of fractions*) might perhaps turn out to be particularly relevant. However, the question will not be studied further in this book.

Selected bibliography for Chapter 1

Section 1: Cohn [1], Hardy and Wright [1], Knopfmacher [1], Narkiewicz [1], Pollard [1], Wegmann [1].

Section 2: Albert [1], Bass [1], E. Cohen [1], Helgason [1], Herstein [1], Jacobson [1, 2], Jónsson and Tarski [1], Kaplansky [1], Knopfmacher [1, 5, 7], Kurosh [1], Pareigis [1], Tits [1], Wolf [1].

CHAPTER 2

ARITHMETICAL FUNCTIONS

This chapter begins with a study of general arithmetical functions, i.e., complex-valued functions on a given arithmetical semigroup G. In order to facilitate later calculations with special arithmetical functions of interest, and in particular the solution of certain enumeration questions, it is convenient to represent arithmetical functions by means of formal Dirichlet series and then develop an elementary theory of formal analysis for such series. For present purposes, the main "analytic" results needed are a few basic ones about formal infinite sums and products. A reader prepared to take such results for granted could ignore some of the initial proofs and so proceed more rapidly to the more interesting applications of the formal analysis that appear in the later sections of this chapter and in the next chapter.

§ 1. The Dirichlet algebra of an arithmetical semigroup

Let G denote an arbitrary arithmetical semigroup. By an *arithmetical function* on G we understand any complex-valued function defined over G. The set of all arithmetical functions on G will be denoted by Dir (G). This set can be made into a complex vector space (of infinite dimension) by means of the point-wise operations

$$(f+g)(a) = f(a)+g(a) \qquad [f,g \in \mathrm{Dir}\,(G),\ a \in G],$$
$$(\alpha \cdot f)(a) = \alpha f(a) \qquad [\alpha \text{ complex},\ f \in \mathrm{Dir}\,(G),\ a \in G].$$

Further, this vector space becomes an associative algebra, which we shall call the **Dirichlet algebra** of G, under the *convolution* multiplication operation $*$ defined by

$$(f*g)(a) = \sum_{bc=a} f(b)g(c) \qquad [f,g \in \mathrm{Dir}\,(G),\ a \in G].$$

It is not difficult to verify the various statements involved in the assertion that Dir (G) forms an algebra under the above operations, as well as the fact that Dir (G) forms a commutative algebra, with an identity element 1 (which sends $1 \in G$ to the number 1, and every other element $a \in G$ to 0). For calculation purposes, it is often convenient to represent functions $f \in \mathrm{Dir}(G)$ in the form of *formal Dirichlet series* over G:

$$f = f(z) = \sum_{a \in G} f(a) a^{-z},$$

where z denotes a type of "indeterminate" (but not one in the usual algebraic sense of the term).

Note that the definition of an arithmetical semigroup G implies that G is countable. For the sake of definiteness, one could if desired express G in the form

$$G = \{1 = a_1, a_2, a_3, \ldots\},$$

where $|a_i| \leqq |a_{i+1}|$, and write

$$f(z) = \sum_{i=1}^{\infty} f(a_i) a_i^{-z}.$$

However, even if one prefers to think of the notation in this way, it is usually less cumbersome to treat the elements of G themselves as indices. In using the notation of formal series, multiplication will be denoted by juxtaposition, so that

$$f(z) + g(z) = \sum_{a \in G} [f(a) + g(a)] a^{-z},$$

$$\alpha \cdot f(z) = \sum_{a \in G} [\alpha f(a)] a^{-z},$$

$$f(z) g(z) = \sum_{a \in G} (f * g)(a) a^{-z},$$

where $f, g \in \mathrm{Dir}(G)$, and α is complex.

In studying Dir (G), it is sometimes useful to consider the **norm** function $\| \ \|$ defined as follows: First define the **order** $\langle f \rangle$ of a function $f \in \mathrm{Dir}(G)$ by

$$\langle f \rangle = \begin{cases} \min \{|a| : f(a) \neq 0\} & \text{if } 0 \neq f \in \mathrm{Dir}(G), \\ \infty & \text{if } 0 = f \in \mathrm{Dir}(G), \end{cases}$$

where the symbol ∞ will be treated in the usual formal way, so that for example $1/\infty = 0$, and $x < \infty$ for every real x. Then define

$$\|f\| = \frac{1}{\langle f \rangle}.$$

1.1. Proposition. *The norm function* $\| \ \|$ *is a "non-archimedean valuation" on* Dir (G), *i.e., it has the following properties:*

(i) $\|f\| \geqq 0$, and $\|f\| = 0$ *if and only if* $f = 0$;

(ii) $\|f+g\| \leqq \max\{\|f\|, \|g\|\}$ *for all* $f, g \in \mathrm{Dir}(G)$;

(iii) $\|f(z)g(z)\| = \|f\| \|g\|$ *for all* $f, g \in \mathrm{Dir}(G)$.

Proof. Properties (i) and (ii) are easy to verify, especially if one first notes the inequality $\langle f+g \rangle \geqq \min(\langle f \rangle, \langle g \rangle)$. Similarly, it follows easily from properties of the norm $| \ |$ on G and the definition of multiplication that $\langle f(z)g(z) \rangle \geqq \langle f \rangle \langle g \rangle$.

In order to establish equality in this case, it will be sufficient to consider any $f \neq 0$ and $g \neq 0$ in Dir (G), and let $a_1, \ldots, a_r, b_1, \ldots, b_s$ denote all the elements of G such that $|a_i| = \langle f \rangle$ and $|b_j| = \langle g \rangle$. Then $a_1^{-z}, \ldots, a_r^{-z}$, $b_1^{-z}, \ldots, b_s^{-z}$ are elements of the subalgebra $[p_1^{-z}, \ldots, p_m^{-z}]$ of Dir (G) generated by the elements p_i^{-z} corresponding to all the prime divisors $p_1, \ldots, p_m$ of $a_1, \ldots, a_r, b_1, \ldots, b_s$ in G. (Note that, by the chosen convention for representing arithmetical functions by formal Dirichlet series, a^{-z} $(a \in G)$ denotes that function h such that $h(a) = 1$ and $h(b) = 0$ for all $b \neq a$ in G. Also note that, in dealing with elements of G, one may define terms like *divisor* in the obvious way, by saying that the elements b, c are divisors of $a \in G$ if and only if $a = bc$ in G; one then sometimes writes $b|a$, $c|a$, $c = a/b$, $b = a/c$.)

Now, because the elements of G admit unique factorization into powers of the primes $p \in P$, it follows from the definition of the algebra operations in Dir (G) that $p_1^{-z}, \ldots, p_m^{-z}$ are algebraically independent over the subalgebra C generated by 1. The latter subalgebra is clearly isomorphic to the field $\mathbf{C}$ of all complex numbers, and so it follows that $[p_1^{-z}, \ldots, p_m^{-z}]$ is isomorphic to a polynomial algebra $\mathbf{C}[x_1, \ldots, x_m]$ in m algebraically independent indeterminates $x_1, \ldots, x_m$ over $\mathbf{C}$. Since $\mathbf{C}[x_1, \ldots, x_m]$ is an integral domain, it follows that the non-zero elements $\sum_{i=1}^{r} f(a_i) a_i^{-z}$ and $\sum_{j=1}^{s} g(b_j) b_j^{-z}$ of $[p_1^{-z}, \ldots, p_m^{-z}]$ must have a non-zero product. This implies that $f(z)g(z) \neq 0$ and that $\langle f(z)g(z) \rangle = \langle f \rangle \langle g \rangle$. The proposition therefore follows. □

1.2. Corollary. *The algebra* Dir (G) *is an integral domain.*

Although we shall make no use of the fact, it may be interesting to point out that Dir (G) *is actually a unique factorization domain;* see the author [9]. The following is a simple characterization of the units in Dir (G):

1.3. Proposition. A *function* $f \in \mathrm{Dir}\,(G)$ *is a unit (i.e., is invertible with respect to multiplication) if and only if* $f(1) \neq 0$ *(i.e.,* $\|f\| = 1$*).*

Proof. If $f \in \mathrm{Dir}\,(G)$ is a unit, then $f(z)g(z) = 1$ for some function g. By the definition of multiplication in Dir (G), this implies that $f(1)g(1) = 1$, and hence $f(1) \neq 0$.

Conversely, if $f(1) \neq 0$, one may recursively define complex numbers $g(b)$ $(b \in G)$ such that the resulting function g satisfies $f(z)g(z) = 1$, as follows:

(i) let $g(1) = f(1)^{-1}$;

(ii) if $g(b)$ has already been defined for all $b \in G$ such that $|b| < |a|$, where $|a| > 1$, define $g(a)$ to be the unique solution of the equation

$$f(1)g(a) + \sum \{f(b)g(c):\ bc = a,\ c \neq a\} = 0.$$

This definition shows that $(f * g)(a) = 0$ for $a \neq 1$, and hence that $f(z)g(z) = 1$. □

§ 2. Infinite sums and products

One of our main reasons for introducing the norm $\|\ \|$ on Dir (G) is that it defines a metric

$$\varrho(f,g) = \|f - g\| \qquad [f, g \in \mathrm{Dir}\,(G)],$$

which helps in the theoretical discussion of certain concepts of formal analysis in Dir (G). In particular, this "analysis" emanates from a concept of 'pseudo-convergence' of infinite sums or products in Dir (G), which may be defined either in a purely algebraic way or by means of metric convergence relative to ϱ:

Let $f_1, f_2, \ldots \in \mathrm{Dir}\,(G)$ be a sequence of arithmetical functions such that, for each $a \in G$, $f_n(a) \neq 0$ for at most a finite number of indices n. In such a case, $\sum_{n=1}^{\infty} f_n$ will be called a **pseudo-convergent** series, and its **sum** will be defined by

$$\sum_{n=1}^{\infty} f_n = \sum_{n=1}^{\infty} f_n(z) = \sum_{a \in G} \left(\sum_{n=1}^{\infty} f_n(a) \right) a^{-z}.$$

(The sums $\sum_{n=1}^{\infty} f_n(a)$ are essentially finite sums.) Obviously, the 'sum' defined is independent of the order of the terms f_n.

2.1. Proposition. *Let $f_1, f_2, \ldots \in \mathrm{Dir}(G)$. Then the following statements are equivalent:*

(i) $\sum_{n=1}^{\infty} f_n$ *is a pseudo-convergent series;*

(ii) $\sum_{n=1}^{\infty} f_n$ *is a convergent series relative to* ϱ*;*

(iii) $\|f_n\| \to 0$ *as* $n \to \infty$*;*

(iv) $\langle f_n \rangle \to \infty$ *as* $n \to \infty$.

If $\sum_{n=1}^{\infty} f_n$ *is a pseudo-convergent series, then its sum coincides with* $\lim_{N \to \infty} \sum_{n=1}^{N} f_n$ *(relative to* ϱ*).*

Proof. Convergence relative to the metric ϱ is defined in the obvious way, as in ordinary analysis. Hence, if $\sum_{n=1}^{\infty} f_n$ converges relative to ϱ, then

$$\|f_N\| \leqq \left\| \sum_{n=1}^{N} f_n - f \right\| + \left\| f - \sum_{n=1}^{N-1} f_n \right\| \to 0 \quad \text{as } N \to \infty,$$

where $f = \lim_{N \to \infty} \sum_{n=1}^{N} f_n$. Thus the condition (ii) implies (iii), and (iii) is obviously equivalent to (iv).

Now suppose that $\langle f_n \rangle \to \infty$ as $n \to \infty$. Then, given any $x > 0$, there is an integer N_x such that $\langle f_n \rangle > x$ for all $n \geqq N_x$. Thus $f_n(a) = 0$ for all $a \in G$ with $|a| \leqq x$, when $n \geqq N_x$. Hence, for any given $a \in G$, $f_n(a) = 0$ for all $n \geqq N_{|a|}$, which implies that $f_n(a) \neq 0$ for at most a finite number of indices n. Therefore $\sum_{n=1}^{\infty} f_n$ is pseudo-convergent.

Conversely, if $\sum_{n=1}^{\infty} f_n$ is a pseudo-convergent series, then, for any given $a \in G$, there is an integer N_a such that $f_n(a) = 0$ for all $n \geqq N_a$. By using condition (iii) in the definition of an arithmetical semigroup, which states that the total number $N_G(x)$ of elements $a \in G$ of norm at most x is finite for each $x > 0$, one sees that

$$N_x' = \max\{N_a \colon |a| \leqq x\} < \infty.$$

Hence $f_n(a) = 0$ for all $a \in G$ with $|a| \leqq x$ whenever $n \geqq N_x'$. Thus, for any $x > 0$, there is an integer N_x' such that $\langle f_n \rangle > x$ for all $n \geqq N_x'$. Therefore statement (iv) holds.

In order to complete the proof that statements (i) to (iv) are equivalent, we now show that statement (iii) implies (ii). Therefore suppose that $\|f_n\| \to 0$ as $n \to \infty$. Then, for any $k > 0$,

$$\left\| \sum_{n=1}^{N+k} f_n - \sum_{n=1}^{N} f_n \right\| \leqq \max(\|f_{N+1}\|, \ldots, \|f_{N+k}\|) \to 0 \quad \text{as } N \to \infty.$$

Thus $F_1, F_2, \ldots$ is a Cauchy sequence relative to the metric ϱ, where $F_N = \sum_{n=1}^{N} f_n$. By the next lemma, which is also of some independent interest, it follows that F_N must tend to some limit F as $N\to\infty$. Therefore $\sum_{n=1}^{\infty} f_n$ is convergent relative to the metric ϱ.

2.2. Lemma. *The Dirichlet algebra* Dir (G) *is complete relative to the metric* ϱ.

Proof. Let $F_1, F_2, \ldots \in \mathrm{Dir}\,(G)$ denote an arbitrary Cauchy sequence relative to the metric ϱ. Then, given any $x>0$, there is an integer N_x such that $\langle F_n - F_m \rangle > x$ for all $m,n \geqq N_x$. Therefore $F_n(a) = F_m(a)$ for all $a\in G$ with $|a| \leqq x$, if $m,n \geqq N_x$. Thus, for any given $a \in G$, the sequence $F_1(a)$, $F_2(a)$, $F_3(a)$, ... eventually reaches some constant value $F(a)$, say. So $F_n(a) = F(a)$ whenever $n \geqq N_{|a|}$.

By the finiteness of the numbers $N_G(x)$ for $x>0$, it follows that

$$M_x = \max\,\{N_{|a|} : |a| \leqq x\} < \infty.$$

Therefore $F_n(a) = F(a)$ for all $a \in G$ with $|a| \leqq x$, whenever $n \geqq M_x$. Thus $\langle F_n - F \rangle > x$ for all $n \geqq M_x$, and so $F_n \to F$ as $n\to\infty$. This shows that the metric space formed by Dir (G) relative to ϱ is complete. □

The final assertion of Proposition 2.1 is straightforward, and will be left as an exercise. □

In order to define infinite products in Dir (G), the simplest procedure appears to be to call a product $\prod_{n=1}^{\infty} f_n$ $(f_n \in \mathrm{Dir}\,(G))$ **pseudo-convergent** if and only if $\prod_{n=1}^{N} f_n$ tends to some limit F as $N\to\infty$, relative to the metric ϱ (i.e., the product is convergent relative to ϱ). Then F may be called the **value** of the infinite product.

For some purposes, it might be convenient to exclude the value 0, as in ordinary analysis, but for present considerations this is immaterial because the later discussion will be concerned exclusively with products of the following type.

2.3. Proposition. *Let* $f_1, f_2, \ldots \in \mathrm{Dir}\,(G)$ *be arithmetical functions such that, for each* $n = 1, 2, \ldots$,

$$f_n(z) = 1 + \bar{f}_n(z),$$

where $\|\bar{f}_n\| < 1$ *and* $\sum_{n=1}^{\infty} \bar{f}_n$ *is a pseudo-convergent series. Then* $\prod_{n=1}^{\infty} f_n$ *is a pseudo-convergent product, and its value is*

$$1 + \sum_{\substack{a\in G \\ a\neq 1}} \Big(\sum_{\substack{b_1 b_2 \ldots b_r = a \\ r\geqq 1,\ \text{distinct } n_i}} f_{n_1}(b_1) f_{n_2}(b_2) \ldots f_{n_r}(b_r) \Big)\, a^{-z}.$$

Proof. First note that the asserted value for the infinite product is well-defined in view of the stated assumptions on $f_1, f_2, \ldots$, because those assumptions reduce the sum in parentheses to a finite sum. (It is understood that each $b_i \neq 1$ if $r > 1$.) Then observe that, for any given $x > 0$, the finiteness of these sums for each $a \in G$ with $|a| \leqq x$ (in particular) together with the finiteness of $N_G(x)$ implies that there is some integer M_x' such that, if $F_N = \prod_{n=1}^{N} f_n$ and F is the asserted "value" for the infinite product, then $F_N(a) = F(a)$ for all $a \in G$ with $|a| \leqq x$ whenever $N \geqq M_x'$. In other words, $F(a)$ is precisely the value at a of a suitable finite product F_M with the property that $F_N(a) = F_M(a)$ for all $N \geqq M = M_a$, and one may choose $M_x' = \max\{M_a : |a| \leqq x\}$.

It follows that $\langle F_N - F \rangle > x$ for $N \geqq M_x'$, and so $F_N \to F$ as $N \to \infty$. This proves the proposition. □

The above proposition is useful in that it shows that, for such an infinite product, the 'value' is both independent of the order of the terms f_n and may be calculated in the obvious desired way. This and the similar fact about pseudo-convergent series will allow us later on to prove various formulae involving special arithmetical functions of interest by purely formal algebraic arguments of a uniform type, rather than by special inductive arguments designed separately for each individual case, or by methods of ordinary analysis which do not extend to all arithmetical semigroups. (At the appropriate later stage in this chapter, the reader may find it interesting to compare the present methods with those given in Chapter XVII of Hardy and Wright [1] for the semigroup $G_{\mathbf{Z}}$ of positive integers; the latter methods make use of ordinary analysis, but Hardy and Wright also make some remarks about the possible use of formal series in deriving certain identities. In the case of arbitrary arithmetical semigroups, where convergence behaviour in the ordinary sense may differ from that over $G_{\mathbf{Z}}$, standard analysis is less readily applicable.)

The following proposition assembles a few elementary facts about pseudo-convergent sums and products, sufficient for most purposes below.

2.4. Proposition. (i) *Let* $\sum_{n=1}^{\infty} f_n$ *be a pseudo-convergent series in* Dir (G) *and let* $g \in$ Dir (G). *Then* $\sum_{n=1}^{\infty} f_n(z) g(z)$ *is a pseudo-convergent series and*

$$\sum_{n=1}^{\infty} f_n(z) g(z) = \left(\sum_{n=1}^{\infty} f_n(z) \right) g(z).$$

(ii) *Let* $f \in$ Dir (G) *be a unit, so that* f *may be expressed in the form* $f(z) = \alpha[1 + \bar{f}(z)]$, *where* $\alpha = f(1) \neq 0$ *and* $\|\bar{f}\| < 1$. *Then*

$$[f(z)]^{-1} = \alpha^{-1} \sum_{r=0}^{\infty} (-1)^r [\bar{f}(z)]^r.$$

(iii) *Let* $\prod_{n=1}^{\infty} f_n$ *and* $\prod_{n=1}^{\infty} g_n$ *be pseudo-convergent products in* Dir (G). *Then* $\prod_{n=1}^{\infty} f_n(z)g_n(z)$ *is a pseudo-convergent product and*

$$\prod_{n=1}^{\infty} f_n(z)g_n(z) = \left\{\prod_{n=1}^{\infty} f_n(z)\right\}\left\{\prod_{n=1}^{\infty} g_n(z)\right\}.$$

(iv) *Let* $\prod_{n=1}^{\infty} f_n$ *be a pseudo-convergent product in* Dir (G) *such that* $f_n(1)=1$ *for each* $n=1, 2, \ldots$. *Then* $\prod_{n=1}^{\infty} [f_n(z)]^{-1}$ *is a pseudo-convergent product and*

$$\prod_{n=1}^{\infty} [f_n(z)]^{-1} = \left\{\prod_{n=1}^{\infty} f_n(z)\right\}^{-1}.$$

The proof of this proposition is straightforward, especially if one first notes the next lemma, which may be proved by elementary arguments of a standard type; the details are left as an exercise.

2.5. Lemma. (i) *Let* $f_1, f_2, \ldots$ *and* $g_1, g_2, \ldots$ *be convergent sequences in* Dir (G), *and let* $f=\lim_{n\to\infty} f_n$, $g=\lim_{n\to\infty} g_n$. *Then the following limits exist in* Dir (G) *and have the values indicated:*

$$\lim_{n\to\infty} (f_n+g_n) = f+g, \qquad \lim_{n\to\infty} f_n(z)g_n(z) = f(z)g(z).$$

(ii) *Let* $f=\lim_{n\to\infty} f_n$ *in* Dir (G), *where* $f_n(1)=1$ *for each* $n=1, 2, \ldots$. *Then* $f(1)=1$ *and*

$$[f(z)]^{-1} = \lim_{n\to\infty} [f_n(z)]^{-1}.$$

§ 3. Double series and products

In many cases below, it will be necessary to consider a notion of pseudo-convergence for double series and products. However, as in some parts of § 2, it will not be essential for the present applications to examine the most comprehensive cases or possible propositions.

Now consider a double sequence of arithmetical functions $f_{mn} \in$ Dir (G), where $m,n=1,2,\ldots$, and suppose that, for each given $a \in G$, $f_{mn}(a) \neq 0$ for at most a finite number of pairs (m, n). In such a situation, $\sum_{m,n} f_{mn}$ will be called a **pseudo-convergent double series,** and its **sum** will be defined by

$$\sum_{m,n} f_{mn} = \sum_{m,n} f_{mn}(z) = \sum_{a\in G} \Big(\sum_{m,n} f_{mn}(a)\Big)\, a^{-z}.$$

(Note that the sum in parentheses is essentially a finite sum.)

3.1. Proposition. *Let $f_{mn} \in \operatorname{Dir}(G)$ be a double sequence of arithmetical functions. Then $\sum_{m,n} f_{mn}$ is a pseudo-convergent double series if and only if for each pair (m, n) both $\sum_{r=1}^{\infty} f_{mr}$ and $\sum_{r=1}^{\infty} f_{rn}$ are pseudo-convergent series. If that is the case, then*

$$\sum_{m,n} f_{mn} = \lim_{\min(M,N)\to\infty} \sum_{m=1}^{M} \sum_{n=1}^{N} f_{mn}$$

$$= \sum_{m=1}^{\infty} \left(\sum_{n=1}^{\infty} f_{mn} \right) = \sum_{n=1}^{\infty} \left(\sum_{m=1}^{\infty} f_{mn} \right).$$

Conversely, if the above limit exists, then $\sum_{m,n} f_{mn}$ is a pseudo-convergent double series.

Proof. The first statement is an easy consequence of the definitions. Next suppose that $\sum_{m,n} f_{mn}$ is pseudo-convergent, and let F denote its sum. For any given $a \in G$, there is an integer M_a such that $f_{mn}(a)=0$ whenever $m, n \geqq M_a$. Let

$$F_{MN} = \sum_{m=1}^{M} \sum_{n=1}^{N} f_{mn}$$

and let

$$M_x' = \max \{M_a : |a| \leqq x\},$$

given $x>0$. Then $f_{mn}(a)=0$ for all $a \in G$ with $|a| \leqq x$, whenever $m, n \geqq M_x'$. Therefore, by the definition of F, $F_{MN}(a)=F(a)$ for all $a \in G$ with $|a| \leqq x$, whenever $M, N \geqq M_x'$. This implies that

$$F = \lim_{\min(M,N)\to\infty} F_{MN}.$$

Conversely, suppose that such a limit does exist and denote it by F. Then, given any $x>0$, there is an integer M_x such that $F_{MN}(a)=F(a)$ for all $a \in G$ with $|a| \leqq x$, whenever $m,n \geqq M_x$. Therefore, for a fixed element $a \in G$, $F_{MN}(a)=F(a)$ whenever $M, N \geqq M_{|a|}$. This implies that $f_{mn}(a)=0$ whenever $m, n \geqq M_{|a|}$, and hence $\sum_{m,n} f_{mn}$ is pseudo-convergent.

The assertion about repeated sums is easy to verify. □

Although pseudo-convergence may be defined in terms of natural limits relative to the metric ϱ, the algebraic definition is especially useful later on. Similarly, although it is briefer to define pseudo-convergent double products in terms of limits as below, for the applications made later on it is sufficient to study products of the type considered in the next proposition.

Firstly, let $f_{mn} \in \mathrm{Dir}\,(G)$, where $m, n = 1, 2, \ldots$. We call $\prod_{m,n} f_{mn}$ a **pseudo-convergent double product** if and only if $F_{MN} = \prod_{m=1}^{M} \prod_{n=1}^{N} f_{mn}$ tends to some limit F as $\min(M, N) \to \infty$. Then F will be called the **value** of the double product.

3.2. Proposition. *Let $f_{mn} \in \mathrm{Dir}\,(G)$ $(m, n = 1, 2, \ldots)$ be functions such that, for each pair (m, n),*

$$f_{mn}(z) = 1 + \bar{f}_{mn}(z)$$

where $\|\bar{f}_{mn}\| < 1$ and $\sum_{m,n} \bar{f}_{mn}$ is pseudo-convergent. Then $\prod_{m,n} f_{mn}$ is pseudo-convergent and its value is

$$1 + \sum_{\substack{a \in G \\ a \neq 1}} \left(\sum_{\substack{b_1 b_2 \ldots b_r = a,\ r \geqq 1, \\ \text{distinct pairs } (m_i,\, n_i)}} \bar{f}_{m_1 n_1}(b_1) \bar{f}_{m_2 n_2}(b_2) \ldots \bar{f}_{m_r n_r}(b_r) \right) a^{-z} =$$

$$= \prod_{m=1}^{\infty} \left(\prod_{n=1}^{\infty} f_{mn} \right) = \prod_{n=1}^{\infty} \left(\prod_{m=1}^{\infty} f_{mn} \right),$$

where the single products are all pseudo-convergent.

The proof of this proposition is similar to that of Proposition 2.3, which may also be used to verify the assertions regarding the repeated products; the details are left as an exercise.

Although one could of course define certain other concepts of elementary analysis in the context of Dir (G), such as those of 'absolute' convergence or the 'elementary' functions (exponential, logarithmic, trigonometric, and so on) for example, details will be omitted until the need arises. However, it is useful for some of the later discussion to observe in particular that the above discussion of double series and products may easily be extended to a similar treatment of **triple** series and products. In addition, we note that there is a useful **formal differentation** operation on Dir (G) defined by

$$f'(z) = -\sum_{a \in G} f(a)(\log |a|)\, a^{-z}.$$

It is easy to verify that the rule $f \to f'$ defines a derivation on Dir (G), i.e., it is linear and satisfies the usual formula for the derivative of a product. Further, it is a continuous mapping relative to the metric ϱ, and all functions in Dir (G) are differentiable in this sense.

§ 4. Types of arithmetical functions

Let G denote an arbitrary arithmetical semigroup. In dealing with the Dirichlet algebra Dir (G), we shall not so much be interested in arbitrary arithmetical functions as in certain special functions or types of functions. In particular, we shall be concerned with the following types of functions:

Let $f \in \text{Dir}(G)$. The function f will be called **multiplicative** if and only if $f(ab)=f(a)f(b)$ whenever $a,b \in G$ have g.c.d. $(a, b)=1$ and $f(1)=1$. (The *greatest common divisor* or g.c.d. (c, d) of given elements $c,d \in G$ may be defined as a common divisor of c and d that is divisible by any other common divisor of c and d; it is easy to see that the g.c.d. always exists for elements of an arithmetical semigroup G, because of the unique factorization property of G.) If $f(1)=1$ and $f(ab)=f(a)f(b)$ for all $a,b \in G$, then f will be called **completely multiplicative**. (It is convenient to exclude the zero function $0 \in \text{Dir}(G)$ in each case.)

To a lesser extent we shall consider functions $f \in \text{Dir}(G)$ that are **additive** (i.e., $f(ab)=f(a)+f(b)$ whenever $(a, b)=1$), and functions f that are **completely additive** (i.e., $f(ab)=f(a)+f(b)$ for all $a,b \in G$). Here $f(1)=0$ in each case.

Lastly, we shall consider functions $f \in \text{Dir}(G)$ that are **prime-independent,** i.e., such that for any given positive integer n, $f(p^n)=f(q^n)$ for all primes $p, q \in P$.

4.1. Proposition. (i) *Let* $M(G)$ *denote the set of all multiplicative functions* $f \in \text{Dir}(G)$, *and let* PIM (G) *denote the set of all prime-independent functions* $f \in \text{M}(G)$. *Then both* $\text{M}(G)$ *and* PIM (G) *form subgroups of the group of units of* Dir (G).

(ii) *Let* $\text{A}(G)$ *denote the set of all additive functions* $f \in \text{Dir}(G)$, *let* CA (G) *denote the set of all completely additive functions* $f \in \text{Dir}(G)$, *and let* PIA (G) *denote the set of all prime-independent functions* $f \in \text{A}(G)$. *Then* $\text{A}(G)$, CA (G) *and* PIA (G) *all form subspaces of the complex vector space* Dir (G).

Proof. Part (ii) of the proposition is a trivial consequence of the various definitions. For part (i), consider $f \in \text{M}(G)$. Then note that

$$f(p_1^{\alpha_1} p_2^{\alpha_2} \dots p_r^{\alpha_r}) = f(p_1^{\alpha_1}) f(p_2^{\alpha_2}) \dots f(p_r^{\alpha_r})$$

for distinct primes $p_i \in P$ and positive integers α_i. If one sets

$$f_p(z) = 1 + f(p)p^{-z} + f(p^2)p^{-2z} + \dots + f(p^n)p^{-nz} + \dots,$$

it follows that $f(z)$ may be represented as the value of the pseudo-convergent product $\prod_{p\in P} f_p(z)$, i.e.,

$$f(z) = \prod_{p\in P} (1+f(p)p^{-z}+\ldots+f(p^n)p^{-nz}+\ldots).$$

(If preferred, one may think of P as being arranged in some sequence $p_1, p_2, \ldots$ with $|p_i|\leqq|p_{i+1}|$, and take pseudo-convergent sums or products over the indices i. However, this does not affect the limiting values in question, and it is usually more convenient to treat the elements $p\in P$ themselves as indices.)

Conversely, suppose that $f\in \text{Dir}\,(G)$ is an arbitrary function with a pseudo-convergent product representation of the form

$$f(z) = \prod_{p\in P} f_p(z),$$

where f_p is a function of the form

$$f_p(z) = 1+f_{p1}p^{-z}+f_{p2}p^{-2z}+\ldots+f_{pn}p^{-nz}+\ldots$$

(i.e. $f_p(1)=1$ and $f_p(a)=0$ whenever $a\neq 1$ or a power of p). Then it follows from Proposition 2.3 that f is a multiplicative function. Therefore a function f is multiplicative if and only if it has a pseudo-convergent product representation of the form $f(z)=\prod_{p\in P} f_p(z)$, where f_p is a function of the above type. Further, this representation is unique.

Now let $f=\prod_{p\in P} f_p$ and $g=\prod_{p\in P} g_p$ be multiplicative functions represented in the above way. Then, by Proposition 2.4,

$$f(z)g(z) = \prod_{p\in P} f_p(z)g_p(z), \qquad [f(z)]^{-1} = \prod_{p\in P} [f_p(z)]^{-1},$$

and, partly by the same proposition, one sees that both $f_p(z)g_p(z)$ and $[f_p(z)]^{-1}$ have the general form

$$1+c_1p^{-z}+c_2p^{-2z}+\ldots+c_np^{-nz}+\ldots.$$

Hence $f(z)g(z)$ and $[f(z)]^{-1}$ are also multiplicative. Therefore $M(G)$ forms a group.

In the case when both f and g are also prime-independent, $f_p(z)$ and $g_p(z)$ are power series in p^{-z} whose coefficients are independent of $p\in P$. Hence the previous coefficients $c_1, c_2, \ldots$ are then also independent of p in each case. Therefore PIM (G) also forms a group. □

4.2. Corollary. (i) *A function* $f \in \text{Dir}(G)$ *is multiplicative if and only if it has pseudo-convergent product representation of the form* $f = \prod_{p \in P} f_p$, *where* f_p *is a function of the form*

$$f_p(z) = 1 + c_{p1}p^{-z} + c_{p2}p^{-2z} + \ldots + c_{pn}p^{-nz} + \ldots .$$

In that case, the representation is unique.

(ii) *A function* $f \in \text{Dir}(G)$ *is completely multiplicative if and only if it has a pseudo-convergent product representation as above, where* f_p *is a function of the form*

$$f_p(z) = (1 - c_p p^{-z})^{-1}.$$

In that case, the representation is unique.

Proof. Part (i) follows from the above proof, while, if f is a completely multiplicative function, then

$$c_{pn} = f(p^n) = [f(p)]^n,$$

so that

$$f_p(z) = 1 + c_p p^{-z} + c_p^2 p^{-2z} + \ldots + c_p^n p^{-nz} + \ldots = (1 - c_p p^{-z})^{-1},$$

where

$$c_p = c_{p1}. \qquad \square$$

Now let $\mathbf{C}[[t]]$ denote the algebra of all formal power series

$$a_0 + a_1 t + a_2 t + \ldots + a_n t^n + \ldots$$

with complex coefficients a_i, in a transcendental element (indeterminate) t over the field $\mathbf{C}$ of complex numbers. If $f \in \text{PIM} = \text{PIM}(G)$ and $p \in P$, then f is completely determined by the coefficients c_i in the expansion

$$f_p(z) = 1 + c_1 p^{-z} + c_2 p^{-2z} + \ldots + c_n p^{-nz} + \ldots .$$

Therefore the rule that assigns to f the power series

$$\hat{f} = 1 + c_1 t + c_2 t^2 + \ldots + c_n t^n + \ldots$$

defines a 1–1 mapping $\hat{}: \text{PIM} \to \mathbf{C}[[t]]$, and this mapping is actually a homomorphism of the group PIM into the group of units of $\mathbf{C}[[t]]$. By the earlier discussion, the image of this group monomorphism is the group of all formal power series with constant term 1.

Now $\mathbf{C}[[t]]$ has a standard topology (the (t)-*adic* topology) which may be defined by the following metric σ: First, define the *order* $o(f)$ of a non-zero power series

$$f = a_0 + a_1 t + \ldots + a_n t^n + \ldots$$

to be the smallest index i such that $a_i \neq 0$. Then define

$$\sigma(f, g) = \begin{cases} 2^{-o(f-g)} & \text{if } f \neq g, \\ 0 & \text{if } f = g. \end{cases}$$

It is well known (see for example Zariski and Samuel [1]) that this metric makes $\mathbf{C}[[t]]$ into a complete topological ring.

If one uses the subspace topologies or, equivalently, the restrictions of the metrics ϱ and σ, he may now verify the continuity statements contained in the following proposition.

4.3. Proposition. *The above mapping* $\hat{}$: PIM $\to \mathbf{C}[[t]]$ *defines a topological group isomorphism of* PIM *with the group of all power series with constant term* 1. □

It will be seen later, that this proposition conveniently allows one to transfer various calculations involving **PIM-functions** (i.e. elements of PIM) to simpler ones within $\mathbf{C}[[t]]$. Further, it will appear shortly that PIM-functions are particularly common amongst special arithmetical functions that occur naturally.

§ 5. The zeta and Möbius functions

On grounds of simplicity, one might expect the constant function ζ_G such that $\zeta_G(a)=1$ for all $a \in G$ to have some significance in the study of the arithmetical semigroup G. In fact, it turns out that ζ_G and its inverse μ_G in Dir (G) are of basic importance for this purpose. The function ζ_G is called the **zeta** function, and its inverse μ_G is known as the **Möbius** function, on G.

In the series notation, we have

$$\zeta_G(z) = \sum_{a \in G} a^{-z}.$$

Since ζ_G is a PIM-function, Corollary 4.2 implies that it satisfies the **Euler product formula**

$$\zeta_G(z) = \prod_{p \in G} (1 - p^{-z})^{-1}.$$

To a large extent, the importance and usefulness of ζ_G will be seen to stem from this formula, which may be regarded as a formal analytic translation of the unique factorization property of G. In particular, the Euler product formula provides the first step in solving certain enumeration problems for an arithmetical category $\mathfrak{C}$.

5.1. Proposition. *The Möbius function $\mu=\mu_G$ may be evaluated as follows:*

$$\mu(a) = \begin{cases} 1 & \text{if } a = 1 \text{ in } G, \\ (-1)^r & \text{if } a = p_1 p_2 \dots p_r \text{ (distinct } p_i \in P), \\ 0 & \text{otherwise.} \end{cases}$$

Proof. By the Euler product formula and Proposition 2.4,

$$\mu(z) = \mu_G(z) = \prod_{p \in P} (1-p^{-z}).$$

Hence, by Proposition 2.3,

$$\mu(z) = 1 + \sum_{\substack{p_i \in P,\, r \geqq 1 \\ \alpha_i \geqq 1}} \mu_{\alpha_1}\mu_{\alpha_2} \dots \mu_{\alpha_r} (p_1^{\alpha_1} p_2^{\alpha_2} \dots p_r^{\alpha_r})^{-z},$$

where $\mu_\alpha = -1$ if $\alpha=1$ and $\mu_\alpha=0$ otherwise. The result therefore follows. □

Now let χ denote a completely multiplicative function on G. For certain later purposes, given any $f \in \mathrm{Dir}\,(G)$, we shall wish to consider the associated series

$$f(z, \chi) = \sum_{a \in G} \chi(a) f(a) a^{-z} = f_\chi(z),$$

where $f_\chi(a) = \chi(a) f(a)$ for $a \in G$. Thus, in particular,

$$\zeta_G(z, \chi) = \chi(z) = \sum_{a \in G} \chi(a) a^{-z},$$

and, by Corollary 4.2,

$$\zeta_G(z, \chi) = \prod_{p \in P} (1 - \chi(p) p^{-z})^{-1}.$$

It is easy to verify that the rule $f(z) \to f(z, \chi)$ (or $f \to f_\chi$) defines a continuous algebra endomorphism of the Dirichlet algebra Dir (G). Further, if $\chi(a) \neq 0$ for all $a \in G$, then this endomorphism is a topological algebra automorphism of Dir (G).

In the general case, the rule $f \to f_\chi$ also induces group endomorphisms of the group of units of Dir (G) and of $\mathrm{M}(G)$; again, these are automorphisms if $\chi(a) \neq 0$ for all $a \in G$. Thus, if $f \in \mathrm{Dir}\,(G)$ is a unit and $g(z) = [f(z)]^{-1}$, then $g(z, \chi) = [f(z, \chi)]^{-1}$. In particular,

$$\mu(z, \chi) = [\zeta_G(z, \chi)]^{-1}.$$

5.2. Proposition (Möbius Inversion Principle). *Let χ denote a completely multiplicative function on G, and let $f, g \in \operatorname{Dir}(G)$. Then*

$$f(a) = \sum_{d|a} \chi(d) g(a/d) \quad \textit{for all } a \in G$$

if and only if

$$g(a) = \sum_{d|a} \chi(d)\mu(d) f(a/d) \quad \textit{for all } a \in G.$$

In particular,

$$f(a) = \sum_{d|a} g(a/d) \quad \textit{for all } a \in G$$

if and only if

$$g(a) = \sum_{d|a} \mu(d) f(a/d) \quad \textit{for all } a \in G.$$

Proof. Note that

$$\begin{aligned} f(a) = \sum_{d|a} \chi(d) g(a/d) \text{ (all } a) &\Leftrightarrow f = \chi * g \\ &\Leftrightarrow f(z) = \chi(z) g(z) \\ &\Leftrightarrow f(z)\, \zeta_G(z, \chi) g(z) \\ &\Leftrightarrow g(z) = [\zeta_G(z, \chi)]^{-1} f(z) \\ &\Leftrightarrow g(z) = \mu(z, \chi) f(z) \\ &\Leftrightarrow g = \mu_\chi * f \\ &\Leftrightarrow g(a) = \sum_{d|a} \mu_\chi(d) f(a/d) \text{ (all } a) \\ &\Leftrightarrow g(a) = \sum_{d|a} \chi(d)\mu(d) f(a/d) \text{ (all } a). \quad \square \end{aligned}$$

5.3. Corollary.

$$\sum_{d|a} \mu(d) = \begin{cases} 1 & \text{if } a = 1, \\ 0 & \text{if } a \neq 1. \end{cases}$$

Proof. Let $\delta(1) = 1$ and $\delta(a) = 0$ for $a \neq 1$, so that $\delta(z)$ is the identity element $1 \in \operatorname{Dir}(G)$. Then

$$\mu(a) = \sum_{d|a} \mu(d) \delta(a/d) \quad \text{for all } a \in G.$$

Therefore the Möbius inversion principle (with $f = \delta$, $g = \mu$) implies that

$$\delta(a) = \sum_{d|a} \mu(a/d) = \sum_{d|a} \mu(d) \quad \text{for all } a \in G.$$

The corollary follows. (Alternatively, we could argue directly from the defining equation $\mu(z)\zeta_G(z)=1$.) □

The following is a slightly different type of inversion principle.

5.4. Proposition. *Let χ denote a completely multiplicative function in* Dir (G), *and let F, H denote complex-valued functions of a positive real variable x. Then*

$$F(x) = \sum_{\substack{a\in G \\ |a|\leq x}} \chi(a)H(x/|a|) \quad \textit{for all } x > 0$$

if and only if

$$H(x) = \sum_{\substack{a\in G \\ |a|\leq x}} \chi(a)\mu(a)F(x/|a|) \quad \textit{for all } x > 0.$$

Proof. Assuming the first equation for each $x>0$,

$$\begin{aligned}\sum_{|a|\leq x} \chi(a)\mu(a)F(x/|a|) &= \sum_{|a|\leq x} \chi(a)\mu(a) \sum_{|b|\leq x/|a|} \chi(b)H(x/|ab|) \\ &= \sum_{|ab|\leq x} \mu(a)\chi(ab)H(x/|ab|) \\ &= \sum_{|c|\leq x} \chi(c)H(x/|c|) \sum_{ab=c} \mu(a) \\ &= \chi(1)H(x/|1|) = H(x),\end{aligned}$$

with the aid of Corollary 5.3. The converse argument is similar. □

§ 6. Further natural arithmetical functions

Amongst various arithmetical functions that one might look at naturally in studying an arithmetical semigroup G there are the following:

(i) the **divisor** function d such that $d(a)$ is the total number of divisors of $a\in G$;

(ii) the **generalized divisor** function d_k $(k\geq 1)$ such that $d_k(a)$ is the total number of ordered k-tuples $(b_1, \ldots, b_k)$ with $b_1 b_2 \ldots b_k = a\in G$; then $d_1=\zeta_G$ and $d_2=d$;

(iii) the **unitary-divisor** function d_* such that $d_*(a)$ is the total number of *unitary* divisors of $a\in G$ (i.e., divisors d such that the g.c.d. $(d, a/d)=1$);

(iv) the **prime-divisor** functions ω, Ω and β such that

$$\omega(1) = \Omega(1) = 0, \qquad \beta(1) = 1,$$

and, for $a=p_1^{\alpha_1}p^{\alpha_2}\dots p_r^{\alpha_r}$, where the $p_i \in P$ are distinct and $\alpha_i > 0$,

$$\omega(a) = r, \qquad \Omega(a) = \alpha_1 + \dots + \alpha_r, \qquad \beta(a) = \alpha_1\alpha_2\dots\alpha_r;$$

(v) the characteristic function q_k of the set G_k of all k-**free** elements in G (i.e., elements a with no divisors of the form $b^k \neq 1$); thus

$$q_k(a) = \begin{cases} 1 & \text{if } a \in G \text{ is } k\text{-free,} \\ 0 & \text{otherwise;} \end{cases}$$

(vi) the **Liouville** function λ, and the function λ_k, such that

$$\lambda(a) = (-1)^{\Omega(a)},$$

while

$$\lambda_k(a) = \begin{cases} \lambda(a) & \text{if } a \text{ is } k\text{-free,} \\ 0 & \text{otherwise;} \end{cases}$$

then $\lambda_1(z) = q_1(z) = 1$, while λ_2 is the Möbius function μ;

(vii) the **Euler** function φ such that

$$\varphi(a) = \sum_{\substack{|b| \leqq |a| \\ (b,a)=1}} 1,$$

i.e., $\varphi(a)$ is the total number of elements b with $|b| \leqq |a|$ such that b and a are coprime: $(b, a) = 1$;

(viii) the **von Mangoldt** function Λ such that

$$\Lambda(a) = \begin{cases} \log|p| & \text{if } a \text{ is a power } p^r \text{ for some } p \in P,\ r > 0, \\ 0 & \text{otherwise.} \end{cases}$$

Although some of these functions may look rather artificial, their significance or uses will become clearer later on. For example, the von Mangoldt function is of great technical importance in the proof of the abstract prime number theorem based on Axiom A, while the function λ is useful in investigating those elements $a \in G$ with $\Omega(a)$ even, or odd, respectively. In some cases, as in the next chapter, natural arithmetical functions on particular semigroups also come to the fore in connection with various enumeration questions for an arithmetical category.

It may be noticed that many of the functions defined above, and in the next chapter, are PIM-functions. Further, they bear a close relationship to the zeta function, which in the case of the above functions is shown by the next theorem.

Before stating the theorem, consider the operation of 'substitution' $z \to mz$ in Dir (G), where m is a positive integer: Given $f \in \text{Dir}(G)$, this operation is defined by letting

$$f(mz) = \sum_{a \in G} f(a)(a^m)^{-z} = \sum_{a \in G} f(a) a^{-mz},$$

i.e., $f(mz)$ is that function $h(z)$ such that

$$h(a) = \begin{cases} f(b) & \text{if } a = b^m \text{ for some } b \in G, \\ 0 & \text{otherwise.} \end{cases}$$

It is not difficult to verify that the rule $f(z) \to f(mz)$ defines a 1–1 continuous algebra endomorphism of Dir (G).

6.1. Theorem. *In terms of the above notation, the following formulae are valid:*

(i) $d_k(z) = [\zeta_G(z)]^k$; *in particular,* $d(z) = [\zeta_G(z)]^2$.

(ii) $d^2(z) = [\zeta_G(z)]^4/\zeta_G(2z)$, *where* d^2 *denotes the point-wise square:* $d^2(a) = [d(a)]^2$ *for* $a \in G$.

(iii) $d_*(z) = [\zeta_G(z)]^2/\zeta_G(2z)$.

(iv) $\beta(z) = \zeta_G(z)\zeta_G(2z)\zeta_G(3z)/\zeta_G(6z)$.

(v) $q_k(k) = \zeta_G(z)/\zeta_G(kz)$.

(vi) $\lambda(z) = \zeta_G(2z)/\zeta_G(z)$.

(vii) $\lambda_k(z) = \zeta_G(2z)/\zeta_G(z)\zeta_G(kz)$ *if* k *is even,*

$= \zeta_G(2z)\zeta_G(kz)/\zeta_G(z)\zeta_G(2kz)$ *if* k *is odd.*

(viii) $\varphi(z) = \left(\sum_{a \in G} N_G(|a|)a^{-z}\right)/\zeta_G(z)$.

(ix) $\Lambda(z) = -\zeta_G{}'(z)\zeta_G(z)$.

Proof. We note that the first seven formulae involve PIM-functions. Therefore it is possible to make use of the monomorphism $\hat{}: \text{PIM} \to \mathbf{C}[[t]]$ defined in § 4, in the following way: Suppose that f is a PIM-function and that it has already been proved that $\hat{f}$ is a finite product of the form

$$\hat{f} = \prod_i (1 - t^{m_i})^{-k_i},$$

where the m_i and k_i are integers and $m_i > 0$. Then note that, since the substitution operation $z \to mz$ defines a continuous algebra endomorphism of

Dir (G),

$$\zeta_G(mz) = \prod_{p \in P} (1-p^{-mz})^{-1} = \zeta_m(z),$$

say. Therefore

$$\hat{\zeta}_m = (1-t^m)^{-1},$$

and it follows that

$$\hat{f} = \prod_i \hat{\zeta}_m^{\,k_i}.$$

Hence

$$f(z) = \prod_i [\zeta_G(m_i z)]^{k_i}.$$

(For later purposes, it is worth observing that, by the topological property of ^ mentioned in Proposition 4.3, the same argument is valid if the product for f is not finite but still convergent relative to the (t)-adic topology. In that case, the final formula for $f(z)$ remains true as a pseudo-convergent product expression. In actual fact, the numbers k_i could even be arbitrary complex numbers. For, one may define $(1-t^m)^{-k}$ and $[\zeta_G(mz)]^k$ by means of the (respectively, convergent or pseudo-convergent) 'binomial series' expansion formulae when k is any complex number. Since these exponentiation operations (for suitable types of power series or functions) may be verified to be continuous relative to the topologies on $\mathbf{C}[[t]]$ or Dir (G), it follows that the above argument remains valid once again.)

The required statements (i) to (vii) will now follow once suitable formulae have been established for the various series $\hat{f}$; for statement (i), however, it is simpler to proceed by induction. For, if $k>1$, then

$$d_k(a) = \sum_{b_1 b_2 \dots b_k = a} 1 = \sum_{bc=a} \sum_{b_1 b_2 \dots b_{k-1} = c} 1$$

$$= \sum_{bc=a} d_{k-1}(c) = \sum_{bc=a} \zeta_G(b) d_{k-1}(c).$$

Therefore $d_k = \zeta_G * d_{k-1}$, i.e.,

$$d_k(z) = \zeta_G(z) d_{k-1}(z).$$

(ii) It is easy to see that $d(p^r) = r+1$ for $p \in P$. Therefore

$$\hat{d}^2 = 1 + 4t + 9t^2 + \dots + (r+1)^2 t^r + \dots.$$

On the other hand,

$$\left\{\frac{[\zeta_G(z)]^4}{\zeta_G(2z)}\right\}^{\wedge} = \frac{1-t^2}{(1-t)^4} = \frac{1+t}{(1-t)^3}$$

$$= (1+t)[1+3t+6t^2+\ldots+\tfrac{1}{2}(r+1)(r+2)t^r+\ldots]$$

$$= 1+4t+9t^2+\ldots+(r+1)^2t^r+\ldots$$

$$= \hat{d^2}.$$

(iii) The only unitary divisors of p^r ($p\in P$, $r>0$) are 1 and p. Therefore $d_*(p^r)=2$ and

$$\hat{d}_* = 1+2(t+t^2+\ldots+t^r+\ldots) = 1+2t(1-t)^{-1}$$

$$= \frac{1+t}{1-t} = \frac{1-t^2}{(1-t)^2}$$

$$= \left\{\frac{[\zeta_G(z)]^2}{\zeta_G(2z)}\right\}^{\wedge}.$$

(iv) We have

$$\left\{\frac{\zeta_G(z)\zeta_G(2z)\zeta_G(3z)}{\zeta_G(6z)}\right\}^{\wedge} = \frac{1-t^6}{(1-t)(1-t^2)(1-t^3)}$$

$$= \frac{1+t^3}{(1-t)^2(1+t)} = 1+\frac{t}{(1-t)^2}$$

$$= 1+t+2t^2+3t^3+\ldots+rt^r+\ldots$$

$$= \hat{\beta},$$

since $\beta(p^r)=r$ for $p\in P$, $r>0$.

(v) From the definition of q_k,

$$\hat{q}_k = 1+t+t^2+\ldots+t^{k-1}$$

$$= \frac{1-t^k}{1-t} = \left\{\frac{\zeta_G(z)}{\zeta_G(kz)}\right\}^{\wedge}.$$

(vi) By definition, $\lambda(p^r)=(-1)^r$ for $p\in P$. Therefore

$$\hat{\lambda} = 1-t+t^2-t^3+\ldots+(-1)^rt^r+\ldots = (1+t)^{-1}$$

$$= \frac{1-t}{1-t^2} = \left\{\frac{\zeta_G(2z)}{\zeta_G(z)}\right\}^{\wedge}.$$

(vii) We have

$$\begin{aligned}\hat{\lambda}_k &= 1-t+t^2-\ldots+(-1)^{k-1}t^{k-1}\\ &= \frac{(1-t^2)(1-t+t^2-\ldots+(-1)^{k-1}t^{k-1})}{1-t^2}\\ &= \frac{(1-t)[1+(-1)^{k-1}t^k]}{1-t^2}\\ &= \begin{cases}\dfrac{(1-t)(1-t^k)}{1-t^2} & \text{if } k \text{ is even,}\\ \dfrac{(1-t)(1-t^{2k})}{(1-t^2)(1-t^k)} & \text{if } k \text{ is odd.}\end{cases}\end{aligned}$$

The formulae follow.

In order to prove the final formulae, consider

6.2. Lemma. *For any* $a \in G$;

(i) $\varphi(a) = \sum_{d|a} \mu(d) N_G(|a/d|)$;

(ii) $N_G(|a|) = \sum_{d|a} \varphi(d)$;

(iii) $\Lambda(a) = \sum_{d|a} \mu(d) \log|a/d| = -\sum_{d|a} \mu(d) \log|d|$;

(iv) $\log|a| = \sum_{d|a} \Lambda(d)$.

Proof. By the Möbius inversion principle, (i) is equivalent to (ii), and the first part of (iii) is equivalent to (iv), while the second equation of (iii) follows from the first with the aid of Corollary 5.3.

Now note that

$$\varphi(a) = \sum_{\substack{|b| \leq |a| \\ (a,b)=1}} 1 = \sum_{|b|=|a|} \delta(a,b),$$

where

$$\delta(a,b) = \begin{cases} 1 & \text{if } (a,b) = 1, \\ 0 & \text{otherwise.}\end{cases}$$

Therefore, by Corollary 5.3,

$$\begin{aligned}\varphi(a) &= \sum_{|b| \leq |a|} \sum_{d|(a,b)} \mu(d) = \sum_{\substack{|b| \leq |a| \\ d|a,\, d|b}} \mu(d)\\ &= \sum_{d|a} \mu(d) \sum_{\substack{|b| \leq |a| \\ d|b}} 1 = \sum_{d|a} \mu(d) \sum_{|cd| \leq |a|} 1\\ &= \sum_{d|a} \mu(d) \sum_{|c| \leq |a/d|} 1 = \sum_{d|a} \mu(d) N_G(|a/d|).\end{aligned}$$

This proves (i). For (iv), consider an element $a=p_1^{\alpha_1}p_2^{\alpha_2}\dots p_r^{\alpha_r}$, where the $p_i \in P$ are distinct and $\alpha_i > 0$. Then

$$\sum_{d|a} \Lambda(d) = \sum_{p_i^s|a} \log|p_i| = \sum_{i=1}^{r} \sum_{s=1}^{\alpha_i} \log|p_i| = \log|a|. \qquad \square$$

This lemma shows that $\varphi = \mu * f$, where $f(a) = N_G(|a|)$, while $\Lambda = \mu * g$, where $g(a) = \log|a|$. Hence the last formulae of Theorem 6.1 follow. □

Now let χ denote a completely multiplicative function on G and m a positive integer. Given $f \in \mathrm{Dir}(G)$, we *define*

$$f(mz, \chi) = f_\chi(mz).$$

Then, using this convention, if $f_m(z) = f(mz)$, we have

$$f_m(z, \chi) = \sum_{a \in G} \chi(a^m) f(a) a^{-mz} = \sum_{a \in G} \chi^m(a) f(a) a^{-mz}$$
$$= f_{\chi^m}(mz) = f(mz, \chi^m),$$

where g^m denotes the point-wise m^{th} power:

$$g^m(a) = [g(a)]^m \quad \text{for } g \in \mathrm{Dir}(G),\ a \in G.$$

By using this and the fact that the rule $f \to f_\chi$ defines a continuous algebra endomorphism of Dir (G), one may deduce the next proposition, which has uses later.

6.3. Proposition. *Suppose that $f \in \mathrm{Dir}(G)$ can be represented as a pseudo-convergent product of the form*

$$f(z) = \prod_i [\zeta_G(m_i z)]^{k_i},$$

where the k_i are complex numbers and the m_i are positive integers. Then, for any completely multiplicative function $\chi \in \mathrm{Dir}(G)$,

$$f(z, \chi) = \prod_i [\zeta_G(m_i z, \chi^{m_i})]^{k_i}. \qquad \square$$

(Actually, for most natural purposes it is sufficient to consider products of this type in which the k_i are rational integers.)

By combining this proposition with Theorem 6.1, we obtain

6.4. Corollary. *Let χ denote a completely multiplicative function on G. Then:*

(i) *For any positive integer m,*

$$d_k(mz, \chi^m) = [\zeta_G(mz, \chi^m)]^k.$$

(ii) $d^2(z, \chi)=[\zeta_G(z, \chi)]^4/\zeta_G(2z, \chi^2)$.
(iii) $d_*(z, \chi)=[\zeta_G(z, \chi)]^2/\zeta_G(2z, \chi^2)$.
(iv) $\beta(z, \chi)=\zeta_G(z, \chi)\zeta_G(2z, \chi^2)\zeta_G(3z, \chi^3)/\zeta_G(6z, \chi^6)$.
(v) $q_k(z, \chi)=\zeta_G(z, \chi)/\zeta_G(kz, \chi^k)$.
(vi) $\lambda(z, \chi)=\zeta_G(2z, \chi^2)/\zeta_G(z, \chi)$.
(vii) $\lambda_k(z, \chi)=\zeta_G(2z, \chi^2)/\zeta_G(z, \chi)\zeta_G(kz, \chi^k)$ *if k is even,*
$=\zeta_G(2z, \chi^2)\zeta_G(kz, \chi^k)/\zeta_G(z, \chi)\zeta_G(2kz, \chi^{2k})$ *if k is odd.*
(viii) $\varphi(z, \chi)=(\sum_{a\in G}\chi(a)N_G(|a|)a^{-z})/\zeta_G(z, \chi)$.
(ix) $\Lambda(z, \chi) = -\zeta_G'(z, \chi)/\zeta_G(z, \chi)$. □

We end this section by noting some further interpretations of the unitary-divisor function d_* and the prime-divisor function β:

Consider an element $a=p_1^{\alpha_1}p_2^{\alpha_2}\dots p_r^{\alpha_r}\in G$, where the $p_i\in P$ are distinct and $\alpha_i>0$. The **core**, or 'greatest square-free divisor', of a is defined to be the element $a_*=p_1p_2\dots p_r$; the element a is called **square-full** if and only if a_*^2 divides a. Then note that a unitary divisor c of a is either 1 or a product of distinct powers $p_i^{\alpha_i}$. Therefore the unitary divisors of a are in 1–1 correspondence with all the divisors of the core a_* of a. Hence

$$d_*(a) = d(a_*) = 2^{\omega(a)}.$$

On the other hand, the square-full divisors of a are all elements of the form $d=p_1^{\beta_1}p_2^{\beta_2}\dots p_r^{\beta_r}$, where (if one introduces the convention that $1_*=1$ so that 1 is regarded as square-full) each $\beta_i=0$ or $2\leq\beta_i\leq\alpha_i$. This shows that the total number of square-full divisors of a is $\alpha_1\alpha_2\dots\alpha_r=\beta(a)$.

The above remarks imply:

6.5. Proposition. (i) *For any* $a\in G$,

$$d_*(a) = d(a_*) = 2^{\omega(a)}.$$

(ii) *For any* $a\in G$, $\beta(a)$ *is equal to the total number of square-full divisors of* a. □

§ 7. ζ-formulae

In the preceding section, there was some discussion of functions $f\in \mathrm{Dir}\,(G)$ such that f could be represented as a pseudo-convergent product of the form

$$f(z) = \prod_i [\zeta_G(m_i z)]^{k_i},$$

where the k_i are complex numbers and the m_i are positive integers.

In the first place, examples appeared for which this product is finite and the k_i are all rational integers. In such cases, we shall say that the function f possesses a **finite ζ-formula.** In the next chapter, some examples will be given of natural functions f having a representation as above, in which the k_i are rational integers but the product is not necessarily finite. In such a case, we shall simply say that f possesses a **ζ-formula.** Finally, if f admits a general representation of the above form, we shall say that f possesses a **generalized** ζ-formula.

The significance of the last concept is indicated by the next theorem.

7.1. Theorem. *A function* $f \in \text{Dir}(G)$ *is a* PIM-*function if and only if* f *possesses a generalized ζ-formula.*

Proof. In § 6 and in the above definitions, the products $\prod_i$ were written as though the indices i were ranging over a single sequence of values. In fact, there would be no essential change to § 6 if the products in question were replaced by double or triple products, say. The same applies to the above definitions, but in fact we shall show that every PIM-function possesses a generalized ζ-formula in which the indices i range over a single strictly increasing sequence of positive integers and the integers m_i increase strictly also.

Now suppose that f possesses a generalized ζ-formula. Then note that every function of the form $[\zeta_G(mz)]^k$ is a PIM-function, because $\zeta_G(mz) \in \text{PIM}$ and it was remarked in § 6 that exponentiation (of suitable types of functions) is continuous in Dir (G), so that exponentiation carries over to the factors g_p in a product $g = \prod_{p \in P} g_p$. One may then use Proposition 2.4(iii) in order to deduce that $f \in \text{PIM}$, or else one may observe directly that a finite product of functions of the form $[\zeta_G(mz)]^k$ must lie in PIM since PIM is a group, and then deduce the general case from the next lemma (which is also of independent interest).

7.2. Lemma. *The subsets* M(G), PIM, A(G), CA(G) *and* PIA(G) *defined in Proposition* 4.1, *as well as the subset* CM(G) *of all completely multiplicative functions, are all closed in the topology of* Dir (G) *defined by the metric* ϱ.

Proof. The arguments are all similar, and so we shall consider only the group PIM. For this purpose, suppose that f denotes some limit point of PIM in Dir (G). Then, given any $x > 0$, there is at least one $g \in \text{PIM}$ such that $\langle f - g \rangle > x$.

Now let $a,b \in G$ be coprime. Then, if $x=|ab|$, there exists a PIM-function g such that $\langle f-g \rangle > x$. Therefore

$$f(1) = g(1) = 1,$$
$$f(ab) = g(ab) = g(a)g(b) = f(a)f(b).$$

Hence f is multiplicative. Similarly, if $p,q \in P$ and $x=\max(|p|^n, |q|^n)$, then there exists a PIM-function g such that $f(p^n)=g(p^n)$ and $f(q^n)=g(q^n)$. Therefore $f(p^n)=f(q^n)$. Hence f is prime-independent. □

For the converse of Theorem 7.1, now let f denote an arbitrary PIM-function, and suppose that

$$\hat{f} = 1 + k_1 t^{m_1} + \sum_{r>m_1} c_r t^r.$$

In that case,

$$(1-t^{m_1})^{k_1} \hat{f} = 1 + k_2 t^{m_2} + \sum_{r>m_2} c_r' t^r = \hat{f}_2,$$

say, where $m_2 > m_1$ and $\hat{f}_2$ is the image under $\hat{}$ of some PIM-function f_2 satisfying

$$f(z) = [\zeta_G(m_1 z)]^{k_1} f_2(z).$$

Then apply the same procedure to f_2, in order to obtain a PIM-function f_3 satisfying

$$f_2(z) = [\zeta_G(m_2 z)]^{k_2} f_3(z).$$

By carrying on in this way, one obtains a sequence of PIM-functions $f, f_2, f_3, \ldots$ such that $(\hat{f}_n - 1) \to 0$ as $n \to \infty$, relative to the (t)-adic topology in $\mathbf{C}[[t]]$. The topological property of the mapping $\hat{}$ stated in Proposition 4.3 therefore implies that $f_n \to 1$ as $n \to \infty$, relative to the metric ϱ on Dir (G). If

$$F_N(z) = [\zeta_G(m_1 z)]^{k_1} \ldots [\zeta_G(m_N z)]^{k_N},$$

it follows now that $F_N \to f$ as $N \to \infty$. Therefore f possesses a generalized ζ-formula. □

7.3. Corollary. *A function* $f \in \mathrm{Dir}(G)$ *possesses a* ζ*-formula if and only if* $f \in \mathrm{PIM}$ *and* f *takes only rational integer values.*

Proof. If f is a PIM-function taking only rational integer values, then the numbers k_i above will now be integers. Conversely, if f does possess a ζ-formula, then this implies that f takes only rational integer values, since that is the case for all integral powers of $\zeta_G(mz)$. □

Now consider the case of *finite* ζ-formulae, which arose naturally in § 6. Formulae (i) and (ii) of Theorem 6.1 provide finite ζ-formulae for the divisor function d and its point-wise square d^2. Also, every point-wise power d^k of d possesses a ζ-formula, by Corollary 7.3. It therefore seems interesting to ask whether any of the additional powers d^k possess finite ζ-formulae. One might ask similar questions about other natural arithmetical functions. One method of approaching such problems stems from the next proposition and its corollary below.

In order to state the proposition, first consider the subring $\mathbf{Z}[[t]]$ of $\mathbf{C}[[t]]$ consisting of all power series with rational integer coefficients. Then consider the nth *cyclotomic* polynomial

$$\Phi_n = \prod (t-\omega),$$

where the product ranges over all $\varphi(n)$ primitive n^{th} roots of unity ω. It is well known (see for example Lang [1]) that the cyclotomic polynomials have rational integer coefficients, are irreducible over the field $\mathbf{Q}$ of rational numbers, and satisfy the *inversion formulae*

$$t^n - 1 = \prod_{d|n} \Phi_d, \qquad \Phi_n = \prod_{d|n} (t^d - 1)^{\mu(n/d)},$$

where μ now denotes the Möbius function on the semigroup $G_{\mathbf{Z}}$. In particular, the last equation together with Corollary 5.3 shows that the constant term of Φ_n is

$$\prod_{d|n} (-1)^{\mu(n/d)} = \prod_{d|n} (-1^{\mu(d)} = (-1)^{\Sigma_{d|n}\mu(d)}$$

$$= \begin{cases} -1 & \text{if } n = 1, \\ 1 & \text{otherwise.} \end{cases}$$

Since we are interested in power series having constant term 1, in the following definitions we adopt the convention that the first cyclotomic polynomial shall now be taken to be $-\Phi_1 = 1-t$. Then, for the sake of brevity, we shall call a power series $g \in \mathbf{Z}[[t]]$ a **cyclotomic rational** if and only if g can be expressed as a finite product of cyclotomic polynomials or inverses of these; if g is a product of cyclotomic polynomials alone, we shall call it a **cyclotomic integer.**

7.4. Proposition. *A function $f \in \text{Dir}(G)$ possesses a finite ζ-formula if and only if $f \in \text{PIM}$ and the associated power series $\hat{f}$ is a cyclotomic rational.*

Proof. Suppose that f has a finite ζ-formula of the form

$$f(z) = \prod_i [\zeta_G(m_i z)]^{k_i}.$$

Then $f \in$ PIM and

$$\hat{f} = \prod_i (1-t^{m_i})^{-k_i} = \prod_i (-\Phi_1)^{-k_i} \prod_{1 \neq d | m_i} \Phi_d{}^{-k_i}.$$

Therefore $\hat{f}$ is a cyclotomic rational.

Conversely, if $f \in$ PIM and f is a cyclotomic rational with

$$\hat{f} = (-\Phi_1)^{r_1} \prod_{n_i \neq 1} \Phi_{n_i}{}^{r_i},$$

then the formula for Φ_n quoted above implies that

$$\hat{f} = (1-t)^{r_1} \prod_{n_i \neq 1} \prod_{d|n_i} [(-1)(1-t^d)]^{\mu(n_i/d) r_i}$$
$$= (1-t)^{r_1} \prod_{n_i \neq 1} \prod_{d|n_i} (1-t^d)^{\mu(n_i/d) r_i},$$

since $\prod_{d|n} (-1)^{\mu(n/d)} = 1$ if $n \neq 1$. It follows that

$$f(z) = [\zeta_G(z)]^{-r_1} \prod_{n_i \neq 1} \prod_{d|n_i} [(\zeta_G(dz)]^{-\mu(n_i/d) r_i}.$$

Hence f possesses a finite ζ-formula. □

7.5. Corollary. *If $f \in$ PIM and $\hat{f}$ is a polynomial, then f possesses a finite ζ-formula if and only if $\hat{f}$ is a cyclotomic integer.*

Proof. Let g denote a polynomial that is a cyclotomic rational. Then recall that the cyclotomic polynomials are irreducible, and note that we may write $g = h_1/h_2$, where h_1 and h_2 are coprime cyclotomic integers. This implies that $h_1 = g h_2$ and, since the polynomial ring $\mathbf{Z}[t]$ is a unique factorization domain, it follows that g must be a cyclotomic integer also. This proves the corollary. □

Now let f denote an arbitrary PIM-function, and consider its point-wise k^{th} power f^k: $f^k(a) = [f(a)]^k$ for $a \in G$. In order to be able to prove that f^k does not possess a finite ζ-formula, in certain interesting special cases, consider the assumption that

$$\hat{f} = 1 + \sum_{r>1} a_r t^r,$$

where $a_r = a(r)$ is a polynomial of degree s in r. In that case, let

$$g_k(z) = [\zeta_G(z)]^{-(sk+1)} f^k(z),$$

so that

$$\hat{g}_k = (1-t)^{sk+1}[f^k]\hat{}.$$

Then the coefficient of t^r in $\hat{g}_k$ may be regarded as the $(sk+1)^{\text{th}}$ repeated "difference" of the coefficient $a_r{}^k$ of t^r in $[f^k]\hat{}$. On the stated assumption, $a_r{}^k=[a(r)]^k$ is a polynomial of degree sk in r, and so the "difference" mentioned (the coefficient of t^r in $\hat{g}_k$) must be zero whenever $r>sk+1$. Therefore $\hat{g}_k$ is a polynomial of degree at most $sk+1$ in t. Therefore, on the assumption about a_r, Corollary 7.5 implies that f^k *possesses a finite* ζ*-formula if and only if* $\hat{g}_k$ *is a cyclotomic integer.*

In order to make use of this criterion we note the following simple lemma, which certainly applies to all cyclotomic integers.

7.6. Lemma. *Let* $g=1+c_1t+\ldots+c_nt^n$ *be a polynomial with complex coefficients whose zeros all lie on the unit circle. Then* $|c_r|\leqq\binom{n}{r}$, *the* r^{th} *binomial coefficient.*

Proof. Let

$$g = c_n \prod_{i=1}^{n} (t+w_i),$$

where the w_i are complex numbers. Then $c_r=c_n\sigma_{n-r}$, where σ_i denotes the ith elementary symmetric polynomial in n variables evaluated at $w_1, \ldots, w_n$. Also,

$$1 = c_n\sigma_n = c_n w_1 w_2 \ldots w_n,$$

and so $|c_n|=1$. Since $|\sigma_i|\leqq\binom{n}{i}$ by the assumption on the zeros, it follows that

$$|c_r| \leqq \binom{n}{n-r} = \binom{n}{r}. \qquad \square$$

7.7. Example. Let $f=d_*$, the unitary-divisor function. Here the coefficient $a_r=d_*(p^r)=2$ $(p\in P)$. Therefore the special assumption about a_r is satisfied, with $s=0$. The above definition of g_k then leads to

$$\hat{g}_k = 1+(2^k-1)t.$$

This polynomial is a cyclotomic integer if and only if $k=1$, when $\hat{g}_1=\Phi_2$. It follows that $d_*{}^k$ has a finite ζ-formula if and only if $k=1$ (the case already considered in Theorem 6.1).

7.8. Example. Now let f denote the divisor function d. For this function, $a_r=r+1$ which is a polynomial of degree 1 in r. In the general situation above, in which $a_r = a(r)$ is a polynomial of degree s in r, it was noted that $\hat{g}_k$ is a polynomial of degree at most $sk+1$ in t. However, if $a(0)=1$, it may be verified that $\hat{g}_k$ can have degree at most sk in t. In the present special case, it follows that $\hat{g}_k$ has degree at most k in t, and the definition of g_k shows that

$$\hat{g}_k = 1 + (2^k - k - 1)t + \dots .$$

Therefore the first coefficient

$$2^k - k - 1 > k = \binom{k}{1} \quad \text{when } k > 2.$$

By Lemma 7.6, it follows that $\hat{g}_k$ is not a cyclotomic integer, and hence that d^k does not possess a finite ζ-formula, when $k>2$. This answers our earlier question about d^k and explains why Ramanujan, who discovered the finite ζ-formula for d^2 when $G = G_{\mathbf{Z}}$, produced no formulae for higher powers of d.

7.9. Example. Finally, let $f=\beta$, the prime-divisor function defined in § 6. Now $a_r = r$, so that the special assumption is satisfied with $s=1$. Therefore the corresponding polynomial $\hat{g}_k$ has degree at most $k+1$ in t, and a calculation shows that

$$\hat{g}_k = 1 - kt + [2^k + \tfrac{1}{2}(k+1)(k-2)]t^2 + \dots .$$

The coefficient of t^2 is greater than $\binom{k+1}{2}$ when $k>1$. Therefore $\hat{g}_k$ is not a cyclotomic integer for $k>1$, by Lemma 7.6. Thus β^k does not possess a finite ζ-formula when $k>1$.

We conclude this section with a more general proposition, whose statement is illustrated by the above examples.

7.10. Proposition. *Let $f \in$ PIM, and suppose that*

$$\hat{f} = 1 + \sum_{r \geq 1} a_r t^r,$$

where $a_r = a(r)$ is a polynomial of degree s in r whose coefficients are rational integers. If $a(r)$ is not identically 1, -1 *or* 0, *then at most a finite number of the point-wise powers f^k possess finite ζ-formulae.*

Proof. Let n denote the least positive integer r such that $|a(r)|>1$, and let $w = |a(n)|$. In terms of the earlier notation, if $sk+1 \geqq n$, then the coefficient of t^n in $\hat{g}_k$ is

$$\pm w^k + \sum_{r=1}^{n-1} (-1)^r \binom{sk+1}{r} [a(n-r)]^k + (-1)^n \binom{sk+1}{n}.$$

CHAPTER 3

ENUMERATION PROBLEMS

In many cases, particularly in the study of arithmetical semigroups associated with categories, one may interpret the ζ-formulae studied in the previous chapter as the answers to certain types of enumeration questions. In the context of arithmetical categories, those questions often also have an intrinsic character essentially independent of the present number-theoretical formulation. A major problem to be considered in this chapter is that of enumerating the isomorphism classes of objects in a given arithmetical category. It will be seen below that the Euler product formula for the zeta function provides a basic stepping-stone towards the solution of this problem, and that it leads to complete algebraic solutions for quite a number of specific categories of interest.

§ 1. A special algebra homomorphism

The reader with some interest in algebra may have noticed quite early that much of our discussion of the Dirichlet algebra Dir (G) of an arithmetical semigroup G would be largely unaffected if complex-valued functions were replaced by functions $f: G \to \boldsymbol{k}$, where $\boldsymbol{k}$ is a general field with a real valuation $| \ |$ (say). This would especially be the case if $\boldsymbol{k}$ were a field of characteristic 0 and one could factorize the norm mapping on G through $\boldsymbol{k}$ and its valuation $| \ |$. In this way, one would obtain a *generalized* Dirichlet algebra Dir $(G, \boldsymbol{k})$, whose properties might perhaps be worth studying in detail.

In this book, it will not be necessary to go further into the above possibility, but instead it will be useful to consider the possibility of assuming slightly less about the initial semigroup G. For present purposes, it will be sufficient to withdraw the requirement of unique factorization and to consider a semigroup H of the following type: Let H denote a commutative semigroup with identity element 1, with the property that every element $a \in H$ has only a finite number of divisors in H, and such that there

Since $|a(r)|\leqq 1$ for $r<n$, the absolute value of the preceding sum is not less than

$$w^k - \sum_{r=1}^{n} \binom{sk+1}{r} > \binom{sk+1}{n}$$

for k sufficiently large, because $w^k \geqq 2^k$ and $\binom{sk+1}{n} + \sum_{r=1}^{n} \binom{sk+1}{r}$ is a polynomial in k. Therefore Lemma 7.6 implies that $\hat{g}_k$ is not a cyclotomic integer for k sufficiently large. This establishes the conclusion. (It is easy to see that the proposition is false when $a(r)$ is identically 1 or -1; also $f(z)=1$ if $a(r)$ is identically zero.) □

In addition to asking questions concerning the existence of generalized ζ-formulae of different kinds, one might ask whether or not the numbers m_i and k_i in a given formula

$$f(z) = \prod_i [\zeta_G(m_i z)]^{k_i}$$

are *unique*. In fact, such uniqueness is fairly easy to establish with the aid of the mapping ˆ, and so will be left as an exercise. (I thank J. N. Ridley for this remark.)

Selected bibliography for Chapter 2

Sections 1–3: Ahern [2], Belgy [1], Carlitz [5], Cashwell and Everett [1, 2], E. Cohen [7], Doubilet, Rota and Stanley [1], Hardy and Wright [1], Horadam [5], Knopfmacher [9], Lu [1], Rota [1], Shapiro [4], Scheid [1–3].

Section 4: Knopfmacher and Ridley [1].

Sections 5 *and* 6: E. Cohen [17], Hardy and Wright [1], Horadam [1, 2, 4], Knopfmacher [6, 9], Knopfmacher and Ridley [1], Suryanarayana and Sita Rama Chandra Rao [1], Wegmann [1].

Section 7: Knopfmacher and Ridley [1].

exists a real-valued *norm* mapping $|\ |$ on H satisfying the conditions:

(i) $|1|=1$, and $|a|>1$ for $a\neq 1\in H$,

(ii) $|ab|=|a|\,|b|$ for all $a, b\in H$,

(iii) the total number $N_H(x)$ of elements $a\in H$ of norm $|a|\leqq x$ is finite, for each real $x>0$.

Under these assumptions, let Dir (H) denote the set of all complex-valued functions on H. One may then imitate the earlier treatment of Dir (G) when G is an arithmetical semigroup, and in the first place make Dir (H) into a complex algebra in the obvious similar fashion. Elements of Dir (H) may be represented as formal Dirichlet series over H, and one may define a norm $\|\ \|$ on Dir (H) in the same general way as in the discussion of Dir (G). In that case, this 'Dirichlet algebra' for H will retain many of the properties of Dir (G), although it may perhaps not be an integral domain and $\|\ \|$ may only be a pseudo-valuation (i.e., have the properties of $\|\ \|$ on Dir (G) given in Proposition 2.1.1, except that $\|f(z)g(z)\|\leqq \leqq\|f\|\,\|g\|$ without equality necessarily holding in all cases).

In particular, these principles may be applied to the semigroup

$$|G| = \operatorname{Im}\{|\ |: G\to \mathbf{R}\},$$

where G is some given arithmetical semigroup. Then there is a continuous identity-preserving algebra homomorphism

$$\tilde{}: \operatorname{Dir}(G)\to \operatorname{Dir}(|G|)$$

defined by

$$\tilde{f}(z) = \sum_{q\in|G|}\Big(\sum_{|a|=q} f(a)\Big)q^{-z} = \sum_{a\in G} f(a)\,|a|^{-z},$$

given $f\in$ Dir (G). (The norm on $|G|$ is taken to be the identity mapping.)

It should be noted that the present Dirichlet series over $|G|$ are still formal series, but it is through these that one may make a convenient transition to ordinary series of analysis when desired at a later stage. For the moment, without making such a transition, it is clear that the earlier-mentioned problem of enumerating the isomorphism classes in an arithmetical category is intimately related with the associated **enumerating function**

$$\zeta_G(z) = \sum_{a\in G}|a|^{-z} = \sum_{q\in|G|} G(q)q^{-z}$$

where $G(q)$ denotes the total number of elements $a\in G$ of norm $|a|=q$.

Now observe that, since $\tilde{}$ is a continuous identity-preserving algebra homomorphism, it will preserve pseudo-convergent sums and products. (Note that the concept of *pseudo-convergence* may be carried over to

Dir (H), and so to Dir ($|G|$), in the obvious way.) Roughly speaking, this remark may be paraphrased so as to state that the homomorphism "preserves formulae". In particular, the Euler product formula for $\zeta_G(z)$ implies that

$$\tilde{\zeta}_G(z) = \prod_{p\in P} (1-|p|^{-z})^{-1} = \prod_{q\in|G|_0} (1-q^{-z})^{-P(q)},$$

where $|G|_0=|G|\setminus\{1\}$, and $P(q)$ is the number of primes of norm q in G. Given a knowledge of the norms $|p|$ of the primes $p\in P$, this formula provides a mechanical algebraic procedure for calculating the successive numbers $G(q)$ as $q\in|G|$ increases monotonically. In other words, this formula, which (if confusion seems unlikely) we may also refer to as the *Euler product formula*, in principle provides a complete algebraic solution to the enumeration problem discussed above.

Before illustrating this statement by means of examples, we now introduce some further terminology and notation, which seems especially appropriate and convenient for certain types of arithmetical semigroups:

If one re-examines the examples listed in Chapter 1, it will be noticed that in some cases the norm functions are not entirely "natural", in that they arise by exponentiating certain more immediate functions with slightly different properties (e.g. 'dimension' or 'degree' functions). This suggests the formal concept of an **additive arithmetical semigroup,** which we define to be a semigroup G with the same unique factorization properties as in Chapter 1, § 1, but with the norm mapping $|\ |$ now replaced by a real-valued *degree* mapping ∂ on G such that

(i) $\partial(1)=0$, $\partial(p)>0$ for $p\in P$,

(ii) $\partial(ab)=\partial(a)+\partial(b)$ for all $a,b\in G$,

(iii) the total number $N_G^{\#}(x)$ of elements $a\in G$ of degree $\partial(a)\leq x$ is finite, for each real $x>0$.

Obviously, this concept is only formally different from that of an arithmetical semigroup, since one may clearly interchange norm and degree mappings by the rules

$$|a| = c^{\partial(a)}, \qquad \partial(a) = \log_c |a| \qquad (a\in G),$$

where $c>1$ is fixed. However, as indicated above, it is nevertheless sometimes more natural or convenient to use the new terminology. It is clear, and we shall take as understood, how one may similarly define an **additive arithmetical category** $\mathfrak{C}$ and its associated additive arithmetical semigroup $G_{\mathfrak{C}}$.

Now let G denote an additive arithmetical semigroup with degree mapping ∂. By a remark at the beginning of Chapter 1, § 1, it is clear that the above conditions (i) to (iii) are equivalent to (i) and (ii) together with

(iii)′ the total number $\pi_G{}^{\#}(x)$ of primes $p \in P$ of degree $\partial(p) \leqq x$ is finite, for each $x > 0$.

Now consider the enumerating function $\tilde{\zeta}_G(z)$ for G that arises by associating with G the norm function

$$|a| = c^{\partial(a)} \qquad (a \in G),$$

where $c > 1$ is fixed by some suitable choice. If confusion seems unlikely, we may also refer to $\tilde{\zeta}_G(z)$ as the *zeta function* of G (or of the arithmetical category $\mathfrak{C}$ to which G is associated, if appropriate) and sometimes later even omit the tilde ˜. In the present situation, we then have

$$\tilde{\zeta}_G(z) = \sum_{q \in |G|} G(q) q^{-z} = \sum_{u \in \partial(G)} G^{\#}(u) c^{-uz},$$

where $c^{-uz} = (c^u)^{-z}$ and $G^{\#}(u)$ denotes the total number of elements $a \in G$ such that $\partial(a) = u$. Further, the Euler product formula (in the new sense) may be written as

$$\tilde{\zeta}_G(z) = \prod_{p \in P} (1 - c^{-\partial(p) z})^{-1} = \prod_{u \in \partial(G)_0} (1 - c^{-uz})^{-P^{\#}(u)},$$

where $P^{\#}(u)$ denotes the total number of primes $p \in P$ such that $\partial(p) = u$, and $\partial(G)_0 = \partial(G) \setminus \{0\}$.

Now write

$$y^u = c^{-uz} = (c^u)^{-z} \text{ for } u \in \partial(G),$$

so that in Dir ($|G|$) we have

$$y^u y^v = (c^u)^{-z}(c^v)^{-z} = (c^{u+v})^{-z} = y^{u+v}.$$

Thus y^u behaves symbolically like a "u^{th} power", and we may write

$$\tilde{\zeta}_G(z) = \sum_{u \in \partial(G)} G^{\#}(u) y^u = Z_G(y),$$

say. The Euler product formula then becomes

$$Z_G(y) = \prod_{u \in \partial(G)_0} (1 - y^u)^{-P^{\#}(u)}.$$

A frequent simplification. Before the reader becomes too irritated or discouraged by symbolic "u^{th} powers" and formal sums and products defined over obscure subsets $|G|$ or $\partial(G)$ of the field of real numbers, we hasten to emphasize that in a large proportion of naturally interesting special cases (such as those stemming from or included in Chapter 1) the natural norm or degree functions take only rational integer values. (If the degree function takes only rational integer values, one can always choose c so that the corresponding norm function takes only rational integer values.) In cases of this kind, there is no need to use generalized Dirichlet algebras of the form Dir (H).

For, in the first place, if G has an integer-valued norm mapping $|\ |$, then one may work directly with the homomorphism

$$\tilde{}: \operatorname{Dir}(G) \to \operatorname{Dir}(G_{\mathbf{Z}})$$

given by

$$\tilde{f}(z) = \sum_{n=1}^{\infty} \Big(\sum_{|a|=n} f(a) \Big) n^{-z} \qquad [f \in \operatorname{Dir}(G)].$$

The enumerating function (zeta function) may now be written as

$$\zeta_G(z) = \sum_{n=1}^{\infty} G(n) n^{-z}$$

(with the tilde $\tilde{}$ omitted), and the Euler product formula has the form

$$\zeta_G(z) = \prod_{m=1}^{\infty} (1 - m^{-z})^{-P(m)}.$$

Secondly, if G has an integer-valued degree mapping ∂, one may use a corresponding integer-valued norm of the form

$$|a| = k^{\partial(a)} \quad \text{for } a \in G,$$

where $k > 1$ is an integer. If one substitutes $y = k^{-z}$ in Dir $(G_{\mathbf{Z}})$, one then obtains an 'enumerating power series' or **generating function**

$$Z_G(y) = \sum_{n=0}^{\infty} G^{\#}(n) y^n,$$

and the Euler product formula becomes

$$Z_G(y) = \prod_{m=1}^{\infty} (1 - y^m)^{-P^{\#}(m)}.$$

Over here, one may view y as an indeterminate in the ordinary algebraic sense; for, y is algebraically independent over the subalgebra C of Dir $(G_{\mathbf{Z}})$ generated by 1, and so the set of all power series $\sum_{n=0}^{\infty} c_n y^n$ (with complex c_n) forms a subalgebra of Dir $(G_{\mathbf{Z}})$ which is isomorphic to the formal power series algebra $\mathbf{C}[[t]]$ discussed earlier.

Bearing the convenience of these simplifications in mind, one may wonder how necessary it is to consider norm or degree mappings that are not integer-valued. One reason for considering the more general case is that one may then derive and use a certain "Normalization Principle" in the context of Part II of this book. Another reason lies in the fact that certain analytical techniques for obtaining asymptotic enumeration formulae (discussed in Part III below) seem to be more conveniently phrased in terms of general real-valued norms or degrees. (One specific situation, in which it becomes useful if not essential to consider degree mappings that are not integer-valued, is considered in the author's paper [12]; this covers certain natural arithmetical categories with integer-valued norm mappings, which nevertheless seem most easily dealt with in terms of associated degree mappings that need not necessarily take only integral values.)

§ 2. Enumeration and zeta functions in special cases

At this stage it may be interesting to re-examine the main examples of Chapter 1, and see what information the above general discussion yields when applied to such special but important cases.

2.1. Example: The Riemann zeta function. For the semigroup $G_{\mathbf{Z}}$ of positive integers, $|G_{\mathbf{Z}}|=G_{\mathbf{Z}}$ and $\tilde{}$ is the identity automorphism of Dir $(G_{\mathbf{Z}})$. The zeta function of $G_{\mathbf{Z}}$ is then

$$\zeta(z) = \sum_{n=1}^{\infty} n^{-z},$$

and is called the **Riemann** zeta function. (In using this terminology, and in referring to the other special zeta functions below, one often thinks of $\zeta(z)$ and the other functions as functions of a complex variable z (or y, if one is considering $Z_G(y)$), and questions of ordinary convergence then become relevant. In this chapter, however, we shall continue to work only with formal series.) For the Riemann zeta function, we have the classical Euler product formula

$$\zeta(z) = \prod \{(1-p^{-z})^{-1}\text{: rational primes } p\}.$$

2.2. Example: Euclidean domains. For the arithmetical semigroups of associate classes of non-zero elements of the domains $\mathbf{Z}[\sqrt{-1}]$ and $\mathbf{Z}[\sqrt{2}]$ discussed in Chapter 1, the enumerating functions are special cases of the Dedekind zeta function treated below. However, the information on prime elements in these domains listed in Chapter 1, § 1, allows us to write down the Euler products explicitly in terms of products over rational primes.

Thus, for the Gaussian integers $\mathbf{Z}[\sqrt{-1}]$ we have

$$\zeta_{\mathbf{Z}[\sqrt{-1}]}(z) = (1-2^{-z})^{-1} \prod \{(1-p^{-2z})^{-1}\text{: rational primes } p \equiv 3 \pmod 4\}$$
$$\times \prod \{(1-p^{-z})^{-2}\text{: rational primes } p \equiv 1 \pmod 4\}.$$

(Here, and throughout this section, the tilde over $\tilde{\zeta}_G(z)$ is omitted.)

Similarly, for the domain $\mathbf{Z}[\sqrt{2}]$ one obtains

$$\zeta_{\mathbf{Z}[\sqrt{2}]}(z) = (1-2^{-z})^{-1} \prod \{(1-p^{-2z})^{-1}\text{: rational primes } p \equiv \pm 3 \pmod 8\}$$
$$\times \prod \{(1-p^{-z})^{-2}\text{: rational primes } p \equiv \pm 1 \pmod 8\}.$$

A very simple but nevertheless quite interesting example is provided by the semigroup of associate classes of non-zero elements of a *Galois polynomial ring* $\mathrm{GF}[q, t]$, i.e. the polynomial ring $F[t]$ in an indeterminate t over the finite Galois field $F = \mathrm{GF}(q)$ with q elements. This is an additive arithmetical semigroup $G = G[q, t]$ with degree mapping $\partial(\bar{f}) = \deg f$. It is easy to see that there are exactly q^n elements of degree n in $G[q, t]$. Hence $G[q, t]$ has the generating function

$$Z_{G[q,t]}(y) = \sum_{n=0}^{\infty} q^n y^n = (1-qy)^{-1}.$$

Consider the corresponding Euler product formula

$$Z_{G[q,t]}(y) = \prod_{m=1}^{\infty} (1-y^m)^{-P^{\#}(m)}.$$

We shall now show how this formula leads to an explicit equation for $P^{\#}(m)$, and hence to the following proposition.

2.3. Proposition. *The total number of monic irreducible polynomials of degree m in* $\mathrm{GF}[q, t]$ *is*

$$P^{\#}(m) = m^{-1} \sum_{d|m} \mu(d) q^{m/d},$$

where μ is the Möbius function on the positive integers.

Proof. By the Möbius inversion principle, the proposition will follow once we have established the equation

$$q^m = \sum_{d|m} d P^{\#}(d).$$

One way to do this is to take 'logarithms' on both sides of the Euler product formula and then compare coefficients of the resulting series. In order to do this, we could first use the usual 'logarithmic series' formula to define $\log(1+h)$ whenever h is a power series with $o(h)>0$, where the order $o(h)$ is defined as in Chapter 2, § 4. After verifying that log is continuous relative to the (y)-adic topology for series of the form $1+h$, and that

$$\log\{(1+h_1)(1+h_2)\} = \log(1+h_1)+\log(1+h_2),$$

one could then proceed with the details indicated. However, familiarity with the algebra of differentiation may perhaps make it slightly simpler to use 'logarithmic derivatives' instead:

Firstly, if

$$f = \sum_{n=0}^{\infty} c_n y^n,$$

define its *formal derivative* by the usual formula

$$f' = \sum_{n=1}^{\infty} nc_n y^{n-1}.$$

One can verify that the rule $f \to f'$ is continuous relative to the (y)-adic topology and is a derivation, i.e., satisfies the usual rules for differentiation. If f is a unit, we define its **logarithmic derivative** to be $\mathrm{D_L}(f)=f'/f$. Then $\mathrm{D_L}$ is a continuous operation relative to the (y)-adic topology (since formal differentiation, inversion and multiplication are continuous), and

$$\mathrm{D_L}(fg) = \mathrm{D_L}(f)+\mathrm{D_L}(g)$$

if f, g are both units. (If desired, one could define and treat logarithms and logarithmic derivatives of arithmetical functions over a general arithmetical semigroup, in a very similar way.)

Now take logarithmic derivatives on both sides of the above Euler product formula. Then

$$\mathrm{D_L}(Z_{G[q,t]}(y)) = q(1-qy)^{-1} = \sum_{n=0}^{\infty} q^{n+1}y^n,$$

while on the other hand the continuity and the additive property of $\mathrm{D_L}$ give

$$\begin{aligned}\mathrm{D_L}(Z_{G[q,t]}(y)) &= \sum_{m=1}^{\infty} mP^{\#}(m)y^{m-1}(1-y^m)^{-1}\\ &= \sum_{m=1}^{\infty} mP^{\#}(m)y^{m-1}(1+y^m+y^{2m}+\ldots).\end{aligned}$$

By comparing the coefficients of y^{n-1} one then obtains

$$q^n = \sum_{\substack{m,r\\ m-1+rm=n-1}} mP^{\#}(m) = \sum_{m(r+1)=n} mP^{\#}(m) = \sum_{d|n} dP^{\#}(d),$$

since the preceding sum allows $r=0$. This proves the proposition. □

2.4. Example: The Dedekind zeta function. Let G_K denote the arithmetical semigroup of all non-zero integral ideals in a given algebraic number field K. The enumerating function for G_K is

$$\zeta_K(z) = \sum_{I \in G_K} |I|^{-z} = \sum_{n=1}^{\infty} K(n) n^{-z},$$

where $K(n)$ denotes the total number of ideals of norm n in G_K. The function $\zeta_K(z)$ is known as the **Dedekind** zeta function of K.

The facts about prime ideals in G_K mentioned in Chapter 1 show that the Euler product formula for G_K has the form

$$\zeta_K(z) = \prod \{[(1-p^{-\alpha_1 z}) \dots (1-p^{-\alpha_k z})]^{-1}\text{: rational primes } p\},$$

where $p^{\alpha_1}, \dots, p^{\alpha_k}$ are the norms of the distinct prime ideals $P_1, \dots, P_k$ in the decomposition $(p)=P_1^{e_1} \dots P_k^{e_k}$. (Since (p) has norm $p^{[K:\mathbf{Q}]}$, it follows that $e_1\alpha_1 + \dots + e_k\alpha_k = [K:\mathbf{Q}]$.) By direct considerations, or by means of this formula and Corollary 2.4.2, one notes that $K(n)$ is a multiplicative function of $n \in G_{\mathbf{Z}}$.

2.5. Example: Finite abelian groups. For the category $\mathscr{A}$ of all finite abelian groups, the zeta function may be written as

$$\zeta_{\mathscr{A}}(z) = \sum_{n=1}^{\infty} a(n) n^{-z},$$

where $a(n)$ denotes the total number of isomorphism classes of abelian groups of order n. The discussion of primes in $\mathscr{A}$ given in Chapter 1 shows that here the Euler product may be written as a pseudo-convergent double product:

$$\begin{aligned}\zeta_{\mathscr{A}}(z) &= \prod \{(1-p^{-rz})^{-1}\text{: rational primes } p,\ r \geqq 1\}\\ &= \prod_{r=1}^{\infty} \prod_{p} (1-p^{-rz})^{-1}\\ &= \prod_{r=1}^{\infty} \zeta(rz),\end{aligned}$$

by the Euler product formula for the Riemann zeta function. Again, these formulae or direct considerations show that $a(n)$ is a multiplicative function in Dir $(G_{\mathbf{Z}})$. Further, the last equation provides an explicit infinite ζ-formula for $a(n)$ as an element of Dir $(G_{\mathbf{Z}})$.

Now consider the subcategory $\mathscr{A}(p)$ of all finite abelian p-groups, where p is a fixed rational prime. Since p is now fixed, for some purposes it is natural to regard $\mathscr{A}(p)$ as an *additive* arithmetical category, with degree mapping defined by

$$\partial(A) = \log_p \operatorname{card}(A).$$

In that case, $\mathscr{A}(p)$ has exactly one prime of degree r for each $r=1, 2, \ldots$. Therefore the Euler product formula implies that $\mathscr{A}(p)$ has the generating function

$$Z_{\mathscr{A}(p)}(y) = \prod_{r=1}^{\infty} (1-y^r)^{-1} = \sum_{n=0}^{\infty} \boldsymbol{p}(n) y^n,$$

where $\boldsymbol{p}(n)=a(p^n)$ is the total number of isomorphism classes of abelian groups of degree n in the above sense. In fact, for $n>0$, $\boldsymbol{p}(n)$ is a function of the positive integer n familiar to number theorists as the **partition** function, which may be defined arithmetically as the total number of ways of partitioning n into a sum of positive integers (where the order of the summands is disregarded). For example, $\boldsymbol{p}(5)=7$ since

$$\begin{aligned} 5 &= 1+4 = 1+1+3 = 1+1+1+2 = 1+1+1+1+1 \\ &= 2+3 = 2+2+1. \end{aligned}$$

It will be seen later that the study of additive arithmetical semigroups is often, as in this case, closely related to 'additive' number theory in the classical sense. (For more information on this and the partition function, see for example Hardy and Wright [1].)

2.5. Example: Semisimple finite rings. Let $\mathfrak{S}$ denote the category of all semisimple finite rings. The enumerating function for $\mathfrak{S}$ (or $G_{\mathfrak{S}}$) may be written as

$$\zeta_{\mathfrak{S}}(z) = \sum_{n=1}^{\infty} S(n) n^{-z},$$

where $S(n)$ denotes the total number of isomorphism classes of rings of cardinal n in $\mathfrak{S}$. In this case, the primes in the category are the isomorphism classes of the various simple finite rings, and the discussion in Chapter 1,

§ 2, shows that the corresponding Euler product may be expressed as a pseudo-convergent triple product

$$\begin{aligned}\zeta_{\mathfrak{S}}(z) &= \prod \{(1-p^{-rm^2 z})^{-1}\text{: rational primes } p,\ r \geqq 1,\ m \geqq 1\} \\ &= \prod_{\substack{r\geqq 1\\ m\geqq 1}} \prod_{p} (1-p^{-rm^2 z})^{-1} \\ &= \prod_{\substack{r\geqq 1\\ m\geqq 1}} \zeta(rm^2 z),\end{aligned}$$

by the Euler product formula for the Riemann zeta function. The function $S(n)$ of the positive integer n is multiplicative, and the last equation provides an explicit infinite ζ-formula for it as an element of Dir $(G_{\mathbf{Z}})$.

Now consider the subcategory $\mathfrak{S}(p)$ of all semisimple finite p-rings, where p is now a fixed rational prime. As with $\mathscr{A}(p)$, it is convenient to regard this category as an additive arithmetical category, with degree mapping defined by

$$\partial(R) = \log_p \text{card}\,(R).$$

Here the discussion in Chapter 1 shows that the primes of $\mathfrak{S}(p)$ may be arranged in a double sequence (P_{rm}) $(r,m=1,2,\ldots)$, where $\partial(P_{rm})=rm^2$. Hence the Euler product has the form

$$Z_{\mathfrak{S}(p)}(y) = \prod_{\substack{r\geqq 1\\ m\geqq 1}} (1-y^{rm^2})^{-1} = \sum_{n=0}^{\infty} s(n)\,y^n,$$

where $s(n)=S(p^n)$ is the total number of isomorphism classes of rings of degree n in $\mathfrak{S}(p)$, and $s(n)$ may be viewed as a natural generalized partition function of the positive integer n. For small n, it is easy to calculate $s(n)$; for example, $s(5)=8$. (Tables of values for $p(n)$ and $s(n)$ when $n\leqq 50$ are provided in Appendix 2.)

By the discussion in Chapter 1, it is not difficult to verify that the subcategory $\mathfrak{S}_c$ of all commutative rings in $\mathfrak{S}$ is 'arithmetically equivalent' to the category $\mathscr{A}$, in the sense indicated at the end of Chapter 1. Hence $\mathfrak{S}_c$ and $\mathscr{A}$ have the same enumerating function. Similarly, the subcategory $\mathfrak{S}_c(p)$ of all p-rings in $\mathfrak{S}_c$ (for a fixed rational prime p) is arithmetically equivalent to the category $\mathscr{A}(p)$, and has the same generating function.

2.6. Example: Semisimple finite-dimensional algebras. Consider the category $\mathfrak{S}(F)$ of all semisimple finite-dimensional associative algebras over a given field F. Three cases were discussed in Chapter 1, § 2, and in each of them $\mathfrak{S}(F)$ is an additive arithmetical category with the degree function

$$\partial(A) = \dim A.$$

Case (i): F is an *algebraically closed* field. In this case there is exactly one prime of degree m^2 for each $m=1, 2, \dots$. Therefore $\mathfrak{S}(F)$ has the generating function

$$Z_{\mathfrak{S}(F)}(y) = \prod_{m=1}^{\infty} (1-y^{m^2})^{-1} = \sum_{n=0}^{\infty} p^{(2)}(n) y^n,$$

where $p^{(2)}(n)$ is the total number of isomorphism classes of n-dimensional algebras in $\mathfrak{S}(F)$, and, for $n>0$, $p^{(2)}(n)$ may be interpreted arithmetically (in the classical sense) as the total number of partitions of n into a sum of squares.

Case (ii): F is a *real closed* field R. Here, by the discussion in Chapter 1, the primes split into three sequences (P_m), (Q_m) and (R_m), where

$$\partial(P_m) = m^2, \qquad \partial(Q_m) = 2m^2, \qquad \partial(R_m) = 4m^2 \qquad (m = 1, 2, \dots).$$

Therefore

$$Z_{\mathfrak{S}(R)}(y) = \prod_{m=1}^{\infty} [(1-y^{m^2})(1-y^{2m^2})(1-y^{4m^2})]^{-1}$$

$$= \sum_{n=0}^{\infty} s_R(n) y^n,$$

say, where $s_R(n)$ enumerates the isomorphism classes of n-dimensional algebras in $\mathfrak{S}(R)$.

Case (iii): F is a finite *Galois* field $\mathrm{GF}(q)$. The description of the primes for $\mathfrak{S}(F)$ in this case given in Chapter 1 shows that now the category $\mathfrak{S}(F)$ is arithmetically equivalent to the category $\mathfrak{S}(p)$ above. Hence $\mathfrak{S}(F)$ has the same generating function as $\mathfrak{S}(p)$ in this case.

2.7. Example: Compact Lie groups, semisimple Lie algebras, and symmetric Riemannian manifolds. The description given under Examples 2.4 and 2.5 in Chapter 1 shows that the categories considered there may be regarded as additive arithmetical categories. We leave the computation of the generating functions (correct up to a finite number of factors) as an exercise.

2.8. Example: Pseudo-metrizable finite topological spaces. For illustrative purposes, it is interesting to note that the description given in Chapter 1 shows that the category $\mathfrak{P}$ of these spaces forms an additive arithmetical category, if one defines the *degree* of a space X to be card (X). There is then exactly one prime of degree m for each $m=1, 2, \ldots$. Hence $\mathfrak{P}$ is arithmetically equivalent to the category $\mathscr{A}(p)$ above, and has the same generating function $\sum_{n=0}^{\infty} p(n)y^n$, where $p(n)$ is the partition function.

2.9. Example: Finite modules over a ring of algebraic integers. Finally, consider the category $\mathfrak{F}=\mathfrak{F}_D$ of all modules of finite cardinal over the ring D of all algebraic integers in a given algebraic number field K. If $a_D(n)$ denotes the total number of isomorphism classes of modules of cardinal n in $\mathfrak{F}$, then

$$\zeta_{\mathfrak{F}}(z) = \sum_{n=1}^{\infty} a_D(n) n^{-z}$$

and the description of the indecomposable modules in $\mathfrak{F}$ given in Chapter 1 shows that

$$\begin{aligned}\zeta_{\mathfrak{F}}(z) &= \prod \{(1-(|P|^r)^{-z})^{-1}\text{: prime ideals } P \text{ in } D,\ r \geqq 1\} \\ &= \prod_{r=1}^{\infty} \prod \{(1-|P|^{-rz})^{-1}\text{: prime ideals } P \text{ in } D\} \\ &= \prod_{r=1}^{\infty} \zeta_K(rz),\end{aligned}$$

where $\zeta_K(z)$ is the Dedekind zeta function discussed above. It follows that $a_D(n)$ is a multiplicative function of $n \in G_{\mathbf{Z}}$.

In studying $a_D(n)$ and the category $\mathfrak{F}$ or semigroup $G_{\mathfrak{F}}$ in general, it is sometimes convenient to factorize the homomorphism $\tilde{}$: Dir $(G_{\mathfrak{F}}) \to$ Dir $(G_{\mathbf{Z}})$ as follows: First note that there is an identity-preserving and norm-preserving homomorphism $\Phi: G_{\mathfrak{F}} \to G_K$ defined by sending the isomorphism class of D/P^r (P a prime ideal, $r \geqq 1$) to P^r, and extending this rule multiplicatively. Then Φ is actually an epimorphism, and Φ induces a continuous identity-preserving algebra homomorphism

$$\Phi_*\text{: Dir}\,(G_{\mathfrak{F}}) \to \text{Dir}\,(G_K)$$

according to the rule $f \to \Phi_*(f)$, where

$$\Phi_*(f)(z) = \sum_{I \in G_K} \Big(\sum_{\Phi(\overline{M})=I} f(\overline{M})\Big) I^{-z}.$$

(Here $\overline{M}$ denotes the isomorphism class of the module M in $\mathfrak{F}$.) Since Φ is norm-preserving, it is clear that the sum in brackets is always finite. Also, $\tilde{}$: Dir $(G_{\mathfrak{F}}) \to$ Dir $(G_{\mathbf{Z}})$ is the composition of the homomorphism Φ_* and $\tilde{}$: Dir $(G_K) \to$ Dir $(G_{\mathbf{Z}})$.

Now consider the zeta function $\zeta_{\mathfrak{F}} \in \mathrm{Dir}\,(G_{\mathfrak{F}})$. Then

$$\Phi_*(\zeta_{\mathfrak{F}})(z) = \sum_{I \in G_K} \Big(\sum_{\Phi(\bar{M})=I} 1 \Big) I^{-z} = \sum_{I \in G_K} a(I) I^{-z},$$

where $a(I)$ denotes the total number of isomorphism classes $\bar{M}$ such that $\Phi(\bar{M})=I$. Also, the Euler product formula for $\zeta_{\mathfrak{F}}$ and the fact that Φ_* is a continuous identity-preserving algebra homomorphism implies that

$$\begin{aligned} \Phi_*(\zeta_{\mathfrak{F}})(z) &= \prod \{(1-(P^r)^{-z})^{-1}\text{: prime ideals } P \text{ in } D,\ r \geqq 1\} \\ &= \prod_{r=1}^{\infty} \prod \{(1-P^{-rz})^{-1}\text{: prime ideals } P \text{ in } D\} \\ &= \prod_{r=1}^{\infty} \zeta_{G_K}(rz), \end{aligned}$$

where $\zeta_{G_K} \in \mathrm{Dir}\,(G_K)$ is the zeta function of G_K. (Recall that $\zeta_K(z) = \tilde{\zeta}_{G_K}(z)$.)

In studying the enumerative function $a_D(n)$, it is sometimes convenient to use the decomposition

$$a_D(n) = \sum_{|I|=n} a(I)$$

and first examine $a(I)$. The above formula provides an explicit infinite ζ-formula for $a(I)$ as a function of I, and shows that $a(I)$ is a PIM-function over G_K.

In view of the multiplicative properties of $a_D(n)$ and $a(I)$, it is of interest to consider the following subcategories $\mathfrak{F}(p)$ and $\mathfrak{F}(P)$ of $\mathfrak{F}$, where p is any fixed rational prime and P is any fixed prime ideal in D. We define $\mathfrak{F}(p)$ to be the subcategory of all modules of cardinal p^n for some n, and we let $\mathfrak{F}(P)$ denote the subcategory of all *P-modules* in $\mathfrak{F}$ (i.e. modules M such that $\Phi(\bar{M})$ is some power of P). If $|P|=q$, the earlier discussion shows that q is a rational prime-power and that one may regard $\mathfrak{F}(P)$ as an additive arithmetical category with the degree mapping

$$\partial(M) = \log_q \mathrm{card}\,(M).$$

Further, the category $\mathfrak{F}(P)$ then has exactly one prime of degree r, for each $r=1, 2, \ldots$. Therefore $\mathfrak{F}(P)$ is 'arithmetically equivalent' to the category $\mathscr{A}(p)$, in the sense that there exists a *degree*-preserving isomorphism between the associated semigroups; hence $\mathfrak{F}(P)$ and $\mathscr{A}(p)$ have the same generating function.

On the other hand, one may regard the category $\mathfrak{F}(p)$ as an additive arithmetical category with the degree mapping

$$\partial(M) = \log_p \mathrm{card}(M).$$

Then the generating function for $\mathfrak{F}(p)$ is

$$Z_{\mathfrak{F}(p)}(y) = \sum_{n=0}^{\infty} a_D(p^n) y^n.$$

Also, if $(p)=P_1^{e_1}\dots P_m^{e_m}$, where the P_i are distinct prime ideals in D, $e_i>0$ and $|P_i|=p^{\alpha_i}$, then the primes for the category $\mathfrak{F}(p)$ separate into m infinite sequences $(P_{1r}), \dots, (P_{mr})$ with the property that $\partial(P_{ir})=\alpha_i r$ $(r=1, 2, \dots)$. Therefore the corresponding Euler product formula is

$$Z_{\mathfrak{F}(p)}(y) = \prod_{r=1}^{\infty} [(1-y^{\alpha_1 r})\dots(1-y^{\alpha_m r})]^{-1}$$
$$= Z(y^{\alpha_1})Z(y^{\alpha_2})\dots Z(y^{\alpha_m}),$$

where

$$Z(y) = \sum_{n=0}^{\infty} p(n) y^n$$

is the generating function for the classical partition function $p(n)$. In particular, $a_D(p^n)$ may be regarded as a generalized partition function $p_{K,p}(n)$ depending on the algebraic number field K and the rational prime p. By way of comparison with the above formula, it is interesting to note that the generating function for $\mathfrak{S}(p)$ in Example 2.5 may also be written as

$$Z_{\mathfrak{S}(p)}(y) = \prod_{m=1}^{\infty} Z(y^{m^2}).$$

(Just as the categories $\mathscr{A}$ and $\mathscr{A}(p)$ may be regarded as the initial members of the families of categories $\mathfrak{F}$ and $\mathfrak{F}(p)$, which are 'parametrized' by the underlying algebraic number field K (or its ring D of algebraic integers), so the categories $\mathfrak{S}$ and $\mathfrak{S}(p)$ discussed above may also be viewed as initial cases of analogous arithmetical categories $\mathfrak{S}_D$ and $\mathfrak{S}_D(p)$ 'parametrized' by D (or K) above; for further details, see the author [7].)

§ 3. Special functions and additive arithmetical semigroups

It was remarked earlier that the study of specific arithmetical functions, and ζ-formulae (or similar formulae) for these functions, may often be interpreted as the investigation and solution of special types of enumeration problems regarding an arithmetical semigroup or category. In this section, we note how Theorem 2.6.1 combined with the homomorphism applies to functions on an additive arithmetical semigroup, and how, for some particular classes of semigroups, this leads to especially detailed and complete arithmetical information.

Firstly, let G denote an additive arithmetical semigroup with degree mapping ∂, and, for a suitable $c>1$, consider the associated norm mapping $|a|=c^{\partial(a)}$ for $a\in G$. Write

$$y^{\partial(a)} = |a|^{-z} = c^{-\partial(a)z},$$

so that, for $f\in \mathrm{Dir}\,(G)$,

$$\tilde{f}(z) = \sum_{a\in G} f(a)\,|a|^{-z} = \sum_{a\in G} f(a)\,y^{\partial(a)} = f^{\#}(y),$$

say. In keeping with our use of the symbols $f(a)$ and $f(z)$, which hopefully has not created difficulty thus far, we shall sometimes write

$$\tilde{f}(q) = \sum_{|a|=q} f(a), \qquad f^{\#}(u) = \sum_{\partial(a)=u} f(a),$$

so that

$$\tilde{f}(z) = \sum_{q\in|G|} \tilde{f}(q)\,q^{-z}, \qquad f^{\#}(y) = \sum_{u\in\partial(G)} f^{\#}(u)\,y^{u}.$$

Thus, for example, $\tilde{\zeta}_G(q)=G(q)$, $\zeta_G{}^{\#}(u)=G^{\#}(u)$, and

$$Z_G(y) = \zeta_G{}^{\#}(y) = \tilde{\zeta}_G(z).$$

By using Theorem 2.6.1 and the definition of the Möbius function, applying the homomorphism $\tilde{}$, and changing to the symbolic powers y^u, one now obtains:

3.1. Theorem. *The arithmetical functions defined in §§ 5–6 of Chapter 2 satisfy the following formulae:*

(i) $\mu^{\#}(y) = 1/Z_G(y)$.

(ii) $d_k{}^{\#}(y) = [Z_G(y)]^k$; *hence* $d^{\#}(y) = [Z_G(y)]^2$.

(iii) $d^{2\#}(y) = [Z_G(y)]^4/Z_G(y^2)$.

(iv) $d_*{}^{\#}(y) = [Z_G(y)]^2/Z_G(y^2)$.

(v) $\beta^{\#}(y) = Z_G(y)Z_G(y^2)Z_G(y^3)/Z_G(y^6)$.

(vi) $q_k{}^{\#}(y) = Z_G(y)/Z_G(y^k)$.

(vii) $\lambda^{\#}(y) = Z_G(y^2)/Z_G(y)$.

(viii) $\lambda_k{}^{\#}(y) = Z_G(y^2)/Z_G(y)Z_G(y^k)$ *if k is even,*
$= Z_G(y^2)Z_G(y^k)/Z_G(y)Z_G(y^{2k})$ *if k is odd.*

(ix) $\varphi^{\#}(y) = \left(\sum_{a\in G} N_G{}^{\#}(\partial(a))\,y^{\partial(a)}\right)/Z_G(y)$.

(x) $\Lambda^{\#}(y) = (\log c)\,yZ_G'(y)/Z_G(y)$,

where formal differentiation is defined in the obvious way.

The meaning of $f^{\#}(y^m)$ for a positive integer m is clear. In the case when G is the semigroup $G[q, t]$ of associate classes of non-zero elements in a Galois polynomial ring GF$[q, t]$, it was shown in Example 2.2 above that

$$Z_G(y) = (1-qy)^{-1}.$$

Hence Theorem 3.1 implies

3.2. Proposition. *The following formulae hold for functions on the additive arithmetical semigroup* $G[q, t]$:

(i) $\mu^{\#}(y) = 1-qy.$

(ii) $d_k^{\#}(y) = (1-qy)^{-k}$; *hence* $d^{\#}(y) = (1-qy)^{-2}.$

(iii) $d^{2\#}(y) = (1-qy)^{-4}(1-qy^2).$

(iv) $d_*^{\#}(y) = (1-qy)^{-2}(1-qy^2).$

(v) $\beta^{\#}(y) = (1-qy^6)/\{(1-qy)(1-qy^2)(1-qy^3)\}.$

(vi) $q_k^{\#}(y) = \sum_{n=0}^{k-1} q^n y^n + \sum_{n=k}^{\infty} (q^n - q^{n-k+1}) y^n.$

(vii) $\lambda^{\#}(y) = \sum_{n=0}^{\infty} q^n y^{2n} - \sum_{n=0}^{\infty} q^{n+1} y^{2n+1}.$

(viii) $\lambda_k^{\#}(y) = (1-qy)(1-qy^k)/(1-qy^2)$ *if k is even,*
$= (1-qy)(1-qy^{2k})/\{(1-qy^2)(1-qy^k)\}$ *if k is odd.*

(ix) $\varphi^{\#}(y) = \sum_{n=0}^{\infty} q^{2n} y^n.$

(x) $\Lambda^{\#}(y) = (\log c) \sum_{n=0}^{\infty} q^{n+1} y^{n+1}.$

Some of the above formulae have quite simple and interesting arithmetical interpretations: For example, formula (vi) shows that every polynomial of degree less than k is k-free (as is obvious directly), but that, for $n \geqq k > 1$, the 'density' of k-free polynomials of degree n in GF$[q, t]$ is

$$q^{-n} q_k^{\#}(n) = 1 - q^{-(k-1)}.$$

Next, for any additive arithmetical semigroup G, let $M_{\text{even}}^{\#}(u)$ and $M_{\text{odd}}^{\#}(u)$ denote the total numbers of square-free elements a of degree u such that $\omega(a)$ is even or odd, respectively. Also, let $N_{\text{even}}^{\#}(u)$ and $N_{\text{odd}}^{\#}(u)$ denote the total numbers of elements a of degree u such that $\Omega(a)$ is even or odd, respectively. Then

$$q_2^{\#}(u) = M_{\text{even}}^{\#}(u) + M_{\text{odd}}^{\#}(u), \qquad \mu^{\#}(u) = M_{\text{even}}^{\#}(u) - M_{\text{odd}}^{\#}(u),$$

while

$$G^{\#}(u) = N_{\text{even}}^{\#}(u) + N_{\text{odd}}^{\#}(u), \qquad \lambda^{\#}(u) = N_{\text{even}}^{\#}(u) - N_{\text{odd}}^{\#}(u).$$

In the special case when G is the above semigroup $G[q, t]$, formula (i) above shows that $M_{\text{even}}^{\#}(1)=0$ and $M_{\text{odd}}^{\#}(1)=q$ (which is again directly obvious), and then that

$$M_{\text{even}}^{\#}(n) = M_{\text{odd}}^{\#}(n) \quad \text{for } n > 1.$$

(Although this may seem plausible, it is not an *a priori* conclusion.) On the other hand, for $n>0$, formula (vii) and the above equations show that

$$N_{\text{even}}^{\#}(2n) = \tfrac{1}{2}(q^{2n}+q^n), \qquad N_{\text{odd}}^{\#}(2n) = \tfrac{1}{2}(q^{2n}-q^n),$$

$$N_{\text{even}}^{\#}(2n+1) = \tfrac{1}{2}(q^{2n+1}-q^{n+1}), \qquad N_{\text{odd}}^{\#}(2n+1) = \tfrac{1}{2}(q^{2n+1}+q^{n+1}).$$

In particular it follows that

$$N_{\text{even}}^{\#}(n) \begin{cases} > N_{\text{odd}}^{\#}(n) & \text{when } n \text{ is even,} \\ < N_{\text{odd}}^{\#}(n) & \text{when } n \text{ is odd.} \end{cases}$$

We conclude this section with a few remarks on additive arithmetical semigroups of a type that arises frequently in the theory of group representations, and also in the theory of vector bundles in topological K-theory, for example. These are semigroups for which the degree mappings are integer-valued (and usually stem from 'dimension' functions on categories), and which contain only *finitely* many primes.

Let G denote such an additive arithmetical semigroup. Then the Euler product formula shows that $Z_G(y)$ is a rational function of y of the form

$$Z_G(y) = \prod_{i=1}^{m} (1-y^{r_i})^{-1},$$

the r_i being positive integers. For any function $f \in \text{Dir}(G)$ that possesses a finite ζ-formula, the associated power series $f^{\#}(y)$ will then be a rational function of the form

$$f^{\#}(y) = \prod_{j=1}^{M} (1-y^{s_j})^{-k_j},$$

where the s_j and k_j are rational integers and $s_j>0$. In any such case, one can in principle obtain complete information about the values $f^{\#}(n)$ for $n=1, 2, \ldots$, by expanding the above product into a power series. (In some cases, especially if one seeks the asymptotic behaviour of $f^{\#}(n)$ as $n\to\infty$, it may help to first decompose $f^{\#}(y)$ into partial fractions $A(\alpha-y)^{-r}$, where r is a positive integer.) Therefore, given sufficient information to be able to write down the Euler product for $Z_G(y)$, one can in principle calculate the numbers $G^{\#}(n)$ and $f^{\#}(n)$ for $n=1, 2, \ldots$, when f is any of the functions occurring in Theorem 3.1 for example.

If information about the Euler product for G is not available, or only partially available, it is sometimes still possible to draw certain arithmetical conclusions. For example, consider:

3.3. Proposition. *Let G denote an additive arithmetical semigroup for which the degree mapping ∂ is integer-valued, and which contains only finitely many primes $p \in P$. Then*

$$M_{\text{even}}{}^{\#}(n) = M_{\text{odd}}{}^{\#}(n)$$

for all except perhaps a finite number of integers n.

Proof. Under the stated hypotheses, $Z_G(y)$ has an expression as a rational function of y of the form noted above. Therefore

$$\mu^{\#}(y) = \prod_{i=1}^{m} (1-y^{r_i}),$$

where the r_i are certain positive integers. This is a polynomial of degree $N=r_1+\ldots+r_m$. Therefore, for $n>N$,

$$M_{\text{even}}{}^{\#}(n) - M_{\text{odd}}{}^{\#}(n) = \mu^{\#}(n) = 0. \qquad \square$$

Selected bibliography for Chapter 3

Section 1: Knopfmacher [9].

Section 2: Bender and Goldman [1], Doubilet, Rota and Stanley [1], Harary [1], Harary and Palmer [1], Knopfmacher [1, 3, 5—9].

Section 3: Carlitz [1—4], S. D. Cohen [2, 3], Knopfmacher [16], Shader [1, 2].

PART II

ARITHMETICAL SEMIGROUPS WITH ANALYTICAL PROPERTIES OF CLASSICAL TYPE

In Part I it was seen that the zeta function and its algebraic properties are closely related to many other arithmetical functions and their algebraic properties, and to the solution of enumeration problems for arithmetical semigroups or categories (in terms of algebraic formulae that lead to direct or recursive equations for the numbers to be evaluated). Similarly, it turns out that analytic properties of the enumerating function (regarded now as a function of a complex variable) or asymptotic properties of its coefficients have a powerful influence on the analytic or asymptotic properties of other arithmetical functions, and on the solution of enumeration problems in terms of estimates for the orders of magnitude or asymptotic behaviour of the numbers concerned. The previous algebraic results are helpful in studying such asymptotic questions, but usually provide no direct information as to the answers.

Part II of this book is concerned with arithmetical semigroups and categories whose analytic and asymptotic properties are closely parallel to those of the classical semigroups consisting of the positive integers or the non-zero integral ideals in a given algebraic number field. The fact that there are natural arithmetical categories with such characteristics, taken in conjunction with known facts about the classical semigroups, provides strong motivation for the investigation of arithmetical semigroups satisfying a certain *Axiom* A stated below. For the categories in question, which include those of all finite abelian groups or all semisimple finite rings, for example, this axiom is simply an abstraction of certain asymptotic enumeration theorems concerning the total numbers of isomorphism classes of objects whose norms do not exceed a large bound.

Once Axiom A has been established for particular semigroups of interest, it becomes reasonable and in fact time-saving to conduct further investigations in an abstract setting based on this axiom. In this way, without an exceptional amount of extra effort, one may simultaneously derive many analytical properties of the classical semigroups and analogues of these for other interesting systems, such as those mentioned above. For example, the classical Prime Number Theorem and Landau's Prime Ideal Theorem are both special consequences of an *Abstract Prime Number Theorem* based on Axiom A (although Landau's theorem then depends on the theorem verifying Axiom A for the given number field). When the abstract prime number theorem is applied to suitable specific categories, it also yields asymptotic enumeration theorems for the corresponding indecomposable objects. (By way of illustration, it implies an asymptotic 'Simple Ring Theorem' upon specialization to the category $\mathfrak{S}$ of all semisimple finite rings.)

It may be interesting to remark that, although asymptotic enumeration theorems of the various kinds referred to above appear to have an intrinsic significance independent of any abstract arithmetical interpretation, abstract analytic number theory is nevertheless very useful (if not essential) in their derivation, and perhaps places them in a clearer general perspective. (Similar general comments apply to the theories developed in Part III, which are founded on different axioms — axioms that are appropriate to other natural arithmetical systems, with different asymptotic enumeration properties.)

CHAPTER 4

SEMIGROUPS SATISFYING AXIOM A

This chapter introduces Axiom A and studies some of its consequences with regard to the asymptotic behaviour of arithmetical functions. In the first place, there is a discussion of some analytic properties of the zeta function $\zeta_G(z)$ in the case when z is taken to be a complex variable. Certain of these properties of the zeta function will be especially useful in Chapter 6, for the purpose of proving the abstract prime number theorem subject to Axiom A. The rest of the present chapter is concerned largely with the study of "average values" of arithmetical functions and "densities" of various arithmetically defined sets. Although some examples of natural arithmetical semigroups that satisfy Axiom A are discussed in § 1, a treatment of various arithmetical categories satisfying the axiom and some asymptotic enumeration theorems which arise in consequence is deferred to Chapter 5.

§ 1. The basic axiom

We begin with a statement of the basic asymptotic axiom to be assumed throughout most of Part II, concerning a given arithmetical semigroup G.

Axiom A. *There exist positive constants A and δ, and a constant η with $0 \leqq \eta < \delta$, such that*

$$N_G(x) = Ax^\delta + O(x^\eta) \quad \textit{as } x \to \infty.$$

Obviously, this axiom is satisfied by the semigroup $G_{\mathbf{Z}}$ of all positive integers. For,

$$N_{\mathbf{Z}}(x) = [x] = x + O(1).$$

Another interesting example is provided by the semigroup $G_{\mathbf{Z}[\sqrt{-1}]}$ of all associate classes of non-zero Gaussian integers:

1.1. Proposition. *For the Gaussian integers one has*

$$N_{\mathbf{Z}[\sqrt{-1}]}(x) = \tfrac{1}{4}\pi x + O(x^{1/2}) \quad \textit{as } x \to \infty.$$

Proof. It was noted in Chapter 1, § 1, that

$$N_{\mathbf{Z}[\sqrt{-1}]}(x) = \tfrac{1}{4} \sum_{n \leqq x} r(n),$$

where $r(n)$ denotes the total number of lattice points (a, b) on the circle of radius $n^{1/2}$ $(n \geqq 1)$ and centre $(0, 0)$ in the Euclidean plane $\mathbf{R}^2$. Therefore, disregarding the origin $(0, 0)$, we wish to estimate the total number of lattice points on the closed disc of radius $x^{1/2}$ and centre the origin. Geometrically, this may be done by associating with each lattice point (a, b) the square of side 1 of which it is the "left-hand bottom" corner. If this is done for every lattice point on the above closed disc, the total region covered by the squares will be included in the disc of radius $x^{1/2}+2^{1/2}$ and centre $(0, 0)$, and will cover all points of the disc of radius $x^{1/2}-2^{1/2}$ and centre $(0, 0)$.

It follows that the area of the region covered by the squares, which coincides with the number of lattice points in the original disc (the origin included), has a value lying in the closed interval $[\pi(x^{1/2}-2^{1/2})^2, \pi(x^{1/2}+2^{1/2})^2]$. This implies that it is of the form $\pi x + O(x^{1/2})$ as $x \to \infty$, and the proposition follows. □

1.2. Corollary. *The arithmetical function $r(n)$ of the positive integer n possesses an asymptotic 'mean-value'*

$$\lim_{N \to \infty} N^{-1} \sum_{n=1}^{N} r(n) = \pi.$$

It can be shown that the semigroup $G_{\mathbf{Z}[\sqrt{2}]}$ of all associate classes of non-zero elements of the domain $\mathbf{Z}[\sqrt{2}]$ also satisfies Axiom A, with

$$N_{\mathbf{Z}[\sqrt{2}]}(x) = (1/\sqrt{2}) \log(1+\sqrt{2})x + O(x^{1/2}).$$

More generally, let G_K denote the semigroup of all non-zero integral ideals in a given algebraic number field K. Then a theorem going back to Weber and Landau implies that G_K satisfies Axiom A, with

$$N_K(x) = N_{G_K}(x) = A_K x + O(x^{\eta_0})$$

where A_K is a positive constant that can be described explicitly in terms of K, and $\eta_0 = 1-2(1+[K:\mathbf{Q}])^{-1}$. For example, $A_K = 1$ if K is the field $\mathbf{Q}$ of rational numbers, and Proposition 1.1 shows that $A_K = \frac{1}{4}\pi$ when K is the field $\mathbf{Q}(\sqrt{-1})$.

Although this theorem about G_K is certainly of direct relevance to Part II, its proof requires a fairly detailed knowledge of parts of algebraic number theory (which is not presupposed in this book). For this reason, the proof is omitted. When the information is required for applications of the (independent) abstract theory to G_K or to modules over the domain D of all algebraic integers in K, we suggest that the reader simply accept the validity of the theorem about $N_K(x)$ for the time being. (A proof may be found in Landau's book [9]; for the simpler case of quadratic domains, which includes $\mathbf{Z}[\sqrt{2}]$ for example, see also Cohn [1].)

The following proposition indicates one way of obtaining further arithmetical semigroups satisfying Axiom A from a given one. The discussion of additional examples is deferred to Chapter 5.

1.3. Proposition. *Given any element a of an arithmetical semigroup G, let $G\langle a\rangle$ denote the set of all elements $b \in G$ that are coprime to a. Then $G\langle a\rangle$ is also an arithmetical semigroup. If G satisfies Axiom* A *as stated above, then $G\langle a\rangle$ satisfies Axiom* A *with*

$$N_{G\langle a\rangle}(x) = A\varphi_\delta(a)\,|a|^{-\delta}\,x^\delta + O(x^\eta) \quad \textit{as } x \to \infty,$$

where

$$\varphi_w(a) = \sum_{d\,|\,a} \mu(d)\,|a/d|^w \quad (w \textit{ real}).$$

Proof. It is easy to see that $G\langle a\rangle$ coincides with the subsemigroup of G generated by 1 and all primes $p \in P$ that do not divide a. Thus, if a' has the same set of prime divisors as a, then $G\langle a'\rangle = G\langle a\rangle$. Also, if one considers the norm mapping restricted to $G\langle a\rangle$, then $G\langle a\rangle$ must be an arithmetical semigroup.

Now suppose that G satisfies Axiom A, and assume firstly that a has only one prime divisor p. (If a has no prime divisors, then $a=1$ and $G\langle a\rangle = G$.) In that case, $G\langle a\rangle$ is the set of all elements b that are not multiples of p, and

$$\begin{aligned} N_{G\langle a\rangle}(x) &= N_G(x) - \sum_{\substack{|c| \leqq x \\ p\,|\,c}} 1 = N_G(x) - \sum_{|pd| \leqq x} 1 \\ &= N_G(x) - N_G(x/|p|) \\ &= Ax^\delta + O(x^\eta) - A(x/|p|)^\delta - O((x/|p|)^\eta) \\ &= A(1 - |p|^{-\delta})x^\delta + O(x^\eta). \end{aligned}$$

It follows by induction on the number of distinct prime divisors of a, when a is a general element of G, that generally

$$N_{G\langle a\rangle}(x) = A \prod_{\substack{p\in P\\ p|a}} (1-|p|^{-\delta})x^{\delta}+\mathrm{O}(x^{\eta}).$$

The proposition therefore follows, since

$$\prod_{\substack{p\in P\\ p|a}} (1-|p|^{-\delta}) = \sum_{d|a}{}' \mu(d)\,|d|^{-\delta}. \qquad \square$$

1.4. Corollary. *Let $G_{\mathbf{Z}}\langle m\rangle$ denote the set of all positive integers coprime to a given positive integer m. Then $G_{\mathbf{Z}}\langle m\rangle$ forms an arithmetical semigroup satisfying Axiom* A, *with*

$$N_{G_{\mathbf{Z}}\langle m\rangle}(x) = m^{-1}\varphi(m)x+\mathrm{O}(1) \quad \textit{as } x\to\infty,$$

where φ denotes the Euler function on $G_{\mathbf{Z}}$.

Proof. By Lemma 2.6.2, φ_1 coincides with the Euler function on the semigroup $G_{\mathbf{Z}}$. $\square$

It is worth noting that the function φ_1 is often regarded as the appropriate generalization of the Euler function on $G_{\mathbf{Z}}$ to the semigroup G_K of all non-zero integral ideals in a given algebraic number field K. For, it can be shown that, if $I\in G_K$, then $\varphi_1(I)$ has a natural interpretation as the total number of cosets in D/I whose elements generate principal ideals coprime to I. (Here, as usual, D denotes the ring of all algebraic integers in K.) For general arithmetical semigroups satisfying Axiom A as stated above, it may perhaps be best to regard φ_δ and φ as simply two of several reasonable generalizations of the classical Euler function.

A **"normalization principle".** Throughout the rest of this book, many of the discussions will be concerned with finding asymptotic estimates as $x\to\infty$ for functions of the form

$$N(f, x) = \sum_{|a|\leq x} f(a) \qquad [f\in \mathrm{Dir}\, G)].$$

If G satisfies Axiom A, calculations may sometimes be simplified by temporarily re-norming G in terms of the norm function $|\ |^*$ such that

$$|a|^* = |a|^{\delta} \qquad (a\in G).$$

For then

$$\sum_{|a|^*\leq x} 1 = \sum_{|a|^{\delta}\leq x} 1 = N_G(x^{1/\delta}) = Ax+\mathrm{O}(x^{\eta/\delta}).$$

In other words, this process converts G into an arithmetical semigroup G^* satisfying Axiom A with $\delta=1$. After dealing with general arithmetical semigroups of the latter type (for which computations may sometimes be slightly easier), one may then make a simple transformation back to the original case.

Although it is worth bearing this theoretical simplification in mind, especially if one is only seeking a solution *in principle*, the particular questions considered below (in which explicit estimates are sought) seem to gain little from it; consequently it is not used in the following discussion.

(Some motivation for allowing arbitrary $\delta>0$ is provided by results in Chapter 5, § 1 below, and also by a study of two-sided ideals in rational simple algebras which will not be pursued here; these considerations yield natural situations in which $\delta=m^{-1}$ for $m=1, 2, 3, \ldots$. [I am indebted to Professor M. Eichler for a letter drawing my attention to certain properties of the ideals just referred to.])

§ 2. Analytical properties of the zeta function

Let G denote an arbitrary arithmetical semigroup. In discussing the enumerating function $\tilde{\zeta}_G(z)=\sum_{a\in G}|a|^{-z}$, from now on we shall usually regard it as a function of a complex variable z, and if confusion seems unlikely still refer to it as the *zeta function* of G (or the arithmetical category to which G is associated, if appropriate). Similarly, we shall often remove the tilde ˜ and still refer to the formula $\zeta_G(z)=\prod_{p\in P}(1-|p|^{-z})^{-1}$ as the *Euler product* formula, in this ordinary analytic context.

Of course, under these circumstances, it becomes necessary to investigate questions of convergence in the ordinary analytic sense. This section considers some questions relating to the analytic behaviour of $\zeta_G(z)$ as a function of a complex variable z. However, since most of the later asymptotic investigations require little or, often, no information in this direction (at least, as far as the stages aimed at in this book go), there will be no attempt at completeness in this way. In fact, many of the methods used later fall under the technical heading of "elementary" methods (i.e., they require very little in the way of deeper real or complex analysis), and for this book these are quite sufficient. In particular, although $\zeta_G(z)=\sum_{q\in|G|}G(q)q^{-z}$ may now be regarded as an ordinary Dirichlet series, it will not be necessary to assume or develop the general theory of such series over here.

The basic general result about convergence is as follows.

2.1. Proposition. *Let G denote an arbitrary arithmetical semigroup. The series $\sum_{a\in G} |a|^{-z}$, $\sum_{p\in P} |p|^{-z}$ and the product $\prod_{p\in P} (1-|p|^{-z})^{-1}$ converge for exactly the same positive real values of z (if they converge for any values). If they converge for real $z=\sigma_0>0$, then they are absolutely and uniformly convergent for all complex z with* $\text{Re}\, z \geqq \sigma_0$, *and the Euler product formula holds analytically in this region.*

Proof. In considering these series and the product, it is convenient to suppose that $G=\{a_1, a_2, \ldots\}$ and $P=\{p_1, p_2, \ldots\}$ have been arranged in sequences in such a way that $|a_i| \leqq |a_{i+1}|$ and $|p_i| \leqq |p_{i+1}|$; in fact, the statement about absolute convergence shows that any arrangement may be chosen without affecting the sums or product. Now consider the inequalities

$$\begin{aligned}\sum_{|p_n|<|p_N|} |p_n|^{-\sigma} &\leqq \sum_{|p_n|<|p_N|} |a_n|^{-\sigma} \\ &\leqq \sum_{\alpha_i \geqq 0} (|p_1|^{\alpha_1} |p_2|^{\alpha_2} \ldots |p_N|^{\alpha_N})^{-\sigma} \\ &= \prod_{n=1}^{N} (1-|p_n|^{-\sigma})^{-1} = \prod_{n=1}^{N} (1+(1-|p_n|^{-\sigma})^{-1}|p_n|^{-\sigma}) \\ &\leqq \prod_{n=1}^{N} \exp [(1-|p_n|^{-\sigma})^{-1}|p_n|^{-\sigma}] \\ &\leqq \exp \left[(1-|p_1|^{-\sigma})^{-1} \sum_{n=1}^{N} |p_n|^{-\sigma}\right],\end{aligned}$$

where $\sigma>0$ and use was made of the inequality $1+y \leqq \exp y$ $(y \geqq 0)$. The first statement about convergence therefore follows, since $|p_N| \to \infty$ as $N \to \infty$ if P is infinite. (If P is finite, the conclusion is easy.)

Next, suppose that the series and product converge for the real value $\sigma_0>0$ of z, and consider a complex number z with $\text{Re}\, z=\sigma \geqq \sigma_0$. Then

$$\left|\sum_{n=M}^{N} |a_n|^{-z}\right| \leqq \sum_{n=M}^{N} |a_n|^{-\sigma} \leqq \sum_{n=M}^{N} |a_n|^{-\sigma_0}.$$

By first letting $M=1$ and then letting M increase arbitrarily, one sees now that the convergence of $\sum_{a\in G} |a|^{-\sigma_0}$ implies the absolute and uniform convergence of $\sum_{a\in G} |a|^{-z}$ for all complex z with $\text{Re}\, z \geqq \sigma_0$. The statement about $\sum_{p\in P} |p|^{-z}$ follows in the same way, and implies the absolute convergence of the product $\prod_{p\in P} (1-|p|^{-z})$ for $\text{Re}\, z \geqq \sigma_0$. Therefore the Euler product is absolutely convergent in that region.

Now let $\operatorname{Re} z \geqq \sigma_0$, and consider

$$\prod_{n=1}^{N} (1-|p_n|^{-z})^{-1} = \sum_{\alpha_i \geqq 0} (|p_1|^{\alpha_1} |p_2|^{\alpha_2} \dots |p_N|^{\alpha_N})^{-z}$$

$$= \sum_{|a|<|p_N|} |a|^{-z} + \sum_{|a| \geqq |p_N|}{}' |a|^{-z}$$

where $\sum'$ indicates a sum running only over elements a divisible by primes p_i with $i \leqq N$. It follows that

$$\left| \prod_{n=1}^{N} (1-|p_n|^{-z})^{-1} - \sum_{|a|<|p_N|} |a|^{-z} \right| \leqq \sum_{|a| \geqq |p_N|}{}' |a|^{-\sigma_0}$$

$$< \tfrac{1}{2}\varepsilon$$

(say) for N sufficiently large. (It may be assumed that P is infinite, so that $|p_N| \to \infty$ as $N \to \infty$, since, if P is finite, there is nothing further to prove.) Therefore

$$\left| \prod_{n=1}^{N} (1-|p_n|^{-z})^{-1} - \zeta_G(z) \right|$$

$$\leqq \left| \prod_{n=1}^{N} (1-|p_n|^{-z})^{-1} - \sum_{|a|<|p_N|} |a|^{-z} \right| + \left| \zeta_G(z) - \sum_{|a|<|p_N|} |a|^{-z} \right|$$

$$< \tfrac{1}{2}\varepsilon + \tfrac{1}{2}\varepsilon$$

for $N \geqq$ some M, where M may depend on $\varepsilon > 0$ but not on z, since $\sum_{a \in G} |a|^{-z}$ is uniformly convergent for $\operatorname{Re} z \geqq \sigma_0$. Therefore the Euler product is uniformly convergent to $\zeta_G(z)$ for $\operatorname{Re} z \geqq \sigma_0$. □

2.2. Corollary. *Suppose that the series* $\sum_{a \in G} |a|^{-z}$ *converges for some real value* $\sigma_0 > 0$ *of* z. *Then the series* $\sum_{q \in |G|} G(q) q^{-z}$, $\sum_{q \in |G|} P(q) q^{-z}$ *and the product* $\prod_{q \in |G|_0} (1-q^{-z})^{-P(q)}$ *are absolutely and uniformly convergent for* $\operatorname{Re} z \geqq \sigma_0$, *and in that region*

$$\sum_{q \in |G|} P(q) q^{-z} = \sum_{p \in P} |p|^{-z},$$

$$\sum_{q \in |G|} G(q) q^{-z} = \zeta_G(z) = \prod_{q \in |G|_0} (1-q^{-z})^{-P(q)}.$$

Further, $P_G(z) = \sum_{p \in P} |p|^{-z}$ *and* $\zeta_G(z)$ *are analytic functions of* z *for* $\operatorname{Re} z \geqq \sigma_0$, *and* $\zeta_G(z) \neq 0$ *in that region.*

Proof. The fact that $P_G(z)$ and $\zeta_G(z)$ are analytic functions follows from the fact that they are sums of uniformly convergent series of analytic functions. (Hence their derivatives may also be calculated by term-by-term differentiation; for example,

$$\zeta_G'(z) = -\sum_{a\in G} (\log |a|)\,|a|^{-z} = -\sum_{q\in |G|} G(q)\,(\log q)q^{-z}.)$$

In order to see that $\zeta_G(z)\neq 0$ for $\operatorname{Re} z=\sigma\geqq\sigma_0$, consider

$$|\zeta_G(z)| = \Big|\prod_{p\in P} (1-|p|^{-z})^{-1}\Big| \geqq \Big|\prod_{p\in P} (1+|p|^{-\sigma})^{-1}\Big|$$
$$\geqq \prod_{p\in P} \exp(-|p|^{-\sigma}) = \exp\Big(-\sum_{p\in P} |p|^{-\sigma}\Big) > 0,$$

using the inequality $(1+y)^{-1}\geqq\exp(-y)$ $(y\geqq 0)$. □

The following lemma, though very elementary, will be of tremendous use later on; arguments making use of it are often referred to as *partial summation* arguments.

2.3. Partial Summation Lemma. *Let G denote an arithmetical semigroup, and let $f\in\operatorname{Dir}(G)$. Let φ denote a complex-valued continuously differentiable function of real $y\geqq 1$, and let*

$$N(f, x) = \sum_{|a|\leqq x} f(a),$$

where $x\geqq 1$. Then

$$\sum_{|a|\leqq x} f(a)\varphi(|a|) = N(f, x)\,\varphi(x) - \int_1^x N(f, y)\,\varphi'(y)\,\mathrm{d}y.$$

If $N(f, x)\varphi(x)=o(1)$ as $x\to\infty$, and either the sum or the integral converges, then both converge and

$$\sum_{a\in G} f(a)\varphi(|a|) = -\int_1^\infty N(f, y)\,\varphi'(y)\,\mathrm{d}y.$$

This result is a consequence of the following more general lemma, stemming from Abel.

2.4. Lemma. *Let $a_1, a_2, \ldots$ denote a sequence of complex numbers, and let $q_1\leqq q_2\leqq\ldots$ be real numbers such that $q_n\to\infty$ as $n\to\infty$. Let φ denote a complex-valued continuously differentiable function of real $y\geqq q_1$, and let $A(x)=\sum_{q_n\leqq x} a_n$ for $x\geqq q_1$. Then*

$$\sum_{q_n\leqq x} a_n\varphi(q_n) = A(x)\,\varphi(x) - \int_{q_1}^x A(y)\,\varphi'(y)\,\mathrm{d}y.$$

If $A(x)\varphi(x)=o(1)$ as $x\to\infty$, and either the sum or the integral converges, then both converge and

$$\sum_{n=1}^{\infty} a_n\varphi(q_n) = -\int_{q_1}^{\infty} A(y)\varphi'(y)\,dy.$$

Proof. For the first statement, of course, φ need only be defined and differentiable within the range of summation. Although it may perhaps be most natural to interpret this lemma in terms of the partial integration formula for Stieltjes integrals, Riemann integrals will suffice for most of this book, and we may prove the lemma as follows.

Let $q_1=r_1<r_2<\ldots$ denote the sequence of *distinct* values taken by the numbers q_n. Let $r_k=\max\{r_i\leqq x\}$, and write $A(r_0)=0$. Then

$$\begin{aligned}\sum_{q_n\leqq x} a_n\varphi(q_n) &= \sum_{n=1}^{k}[A(r_n)-A(r_{n-1})]\varphi(r_n)\\ &= A(r_k)\varphi(r_k)-\sum_{n=1}^{k-1}A(r_n)[\varphi(r_{n+1})-\varphi(r_n)]\\ &= A(r_k)\varphi(r_k)-\sum_{n=1}^{k-1}A(r_n)\int_{r_n}^{r_{n+1}}\varphi'(y)\,dy\\ &= A(r_k)\varphi(r_k)-\sum_{n=1}^{k-1}\int_{r_n}^{r_{n+1}}A(r_n)\varphi'(y)\,dy\\ &= \Big\{A(x)\varphi(x)-\int_{r_k}^{x}A(r_k)\varphi'(y)\,dy\Big\}-\int_{r_1}^{r_k}A(y)\varphi'(y)\,dy.\end{aligned}$$

The first statement follows now, and the second one is immediate. □

2.5. Corollary. *Let G denote an arithmetical semigroup and let $f\in\mathrm{Dir}(G)$. Let φ denote a complex-valued continuously differentiable function of real $y\geqq x_1\geqq 0$. If $x_1<x_2$, then*

$$\sum_{x_1<|a|\leqq x_2} f(a)\varphi(|a|) = N(f,x_2)\varphi(x_2)-N(f,x_1)\varphi(x_1)-\int_{x_1}^{x_2}N(f,y)\varphi'(y)\,dy.$$

If $N(f,x)\varphi(x)=o(1)$ as $x\to\infty$, and either the sum or the integral converges, then both converge and

$$\sum_{|a|>x_1} f(a)\varphi(|a|) = -N(f,x_1)\varphi(x_1)-\int_{x_1}^{\infty}N(f,y)\varphi'(y)\,dy. \qquad \square$$

As a first application of partial summation, we now prove

2.6. Proposition. *Let G denote an arithmetical semigroup satisfying Axiom* A. *Then* $\sum_{a\in G}|a|^{-z}$ *is absolutely convergent for* $\operatorname{Re} z>\delta$, *and divergent for* $\operatorname{Re} z\leqq\delta$. *Hence* $\zeta_G(z)$ *is a non-zero analytic function satisfying the Euler product formula, for all complex z with* $\operatorname{Re} z>\delta$. *Further, when* $\operatorname{Re} z= =\sigma>\delta$,

$$\zeta_G(z) = z\int_1^{\infty} N_G(y)y^{-z-1}\,\mathrm{d}y,$$

$$\sum_{|a|\leqq x}|a|^{-z} = \zeta_G(z)+\mathrm{O}(x^{\delta-\sigma}) \quad \textit{as } x\to\infty.$$

In order to prove this proposition, consider:

2.7. Lemma. *Let G denote an arbitrary arithmetical semigroup, and let* $g\in\operatorname{Dir}(G)$ *be an arithmetical function such that*

$$N(g,x) = \sum_{|a|\leqq x} g(a) = \mathrm{O}(x^{\nu}) \quad \textit{as } x\to\infty.$$

Then the series

$$\tilde{g}(z) = \sum_{a\in G} g(a)\,|a|^{-z}$$

is uniformly convergent and defines an analytic function of z, for $\operatorname{Re} z\geqq\sigma_0>\nu$ *(*$\sigma_0>\nu$ *arbitrary). Further, for* $\operatorname{Re} z=\sigma>\nu$,

$$\tilde{g}(z) = z\int_1^{\infty} N(g,y)y^{-z-1}\,\mathrm{d}y,$$

$$\tilde{g}'(z) = -\sum_{a\in G}(\log|a|)g(a)\,|a|^{-z},$$

$$\sum_{|a|\leqq x} g(a)\,|a|^{-z} = \tilde{g}(z)+\mathrm{O}(x^{\nu-\sigma}) \quad \textit{as } x\to\infty.$$

Proof. For any fixed z with $\operatorname{Re} z=\sigma$, partial summation gives

$$\sum_{|a|\leqq x} g(a)\,|a|^{-z} = N(g,x)x^{-z}+z\int_1^x N(g,y)y^{-z-1}\,\mathrm{d}y$$
$$= \mathrm{O}(x^{\nu-\sigma})+\int_1^x \mathrm{O}(y^{\nu-\sigma-1})\,\mathrm{d}y,$$

by the assumption on $N(g,y)$. If $\sigma>\nu$, then it follows that the integral is absolutely convergent, and hence the series for $\tilde{g}(z)$ converges. Also, for $\operatorname{Re} z=\sigma\geqq\sigma_0>\nu$, Corollary 2.5 with $\varphi(x)=x^{-z}$ yields

$$\sum_{x_1<|a|\leqq x_2} g(a)\,|a|^{-z} = N(g,x_2)x_2^{-z}-N(g,x_1)x_1^{-z}+z\int_{x_1}^{x_2} N(g,y)y^{-z-1}\,\mathrm{d}y$$
$$= \mathrm{O}(x_2^{\nu-\sigma})+\mathrm{O}(x_1^{\nu-\sigma})+\mathrm{O}\Big(\int_{x_1}^{x_2} y^{\nu-\sigma-1}\,\mathrm{d}y\Big)$$
$$= \mathrm{O}(x_1^{\nu-\sigma}) = \mathrm{O}(x_1^{\nu-\sigma_0}).$$

Thus the series for $\tilde{g}(z)$ is uniformly convergent for $\sigma \geqq \sigma_0$, and since $\sigma_0 > \nu$ is arbitrary, this implies that $\tilde{g}(z)$ defines an analytic function of z for $\operatorname{Re} z > \nu$. This and the preceding discussion then implies the last three equations in the statement of the lemma. □

In order to deduce Proposition 2.6, now let $g = \zeta_G \in \operatorname{Dir}(G)$ and assume Axiom A, so that

$$N(g, x) = N_G(x) = O(x^\delta).$$

Then the only point still to be considered is the divergence of $\zeta_G(z)$ for $\operatorname{Re} z \leqq \delta$. This will follow from:

2.8. Proposition. *Let G denote an arithmetical semigroup satisfying Axiom* A. *Then:*

(i) $\sum_{|a| \leqq x} |a|^{-\delta} = \delta A \log x + \gamma_G + O(x^{\eta - \delta})$, *where*

$$\gamma_G = A + \delta \int_1^\infty [N_G(y) - Ay^\delta] y^{-\delta - 1} \, dy.$$

(ii) *For* $\operatorname{Re} z = \eta$,

$$\sum_{|a| \leqq x} |a|^{-z} = \frac{\delta A}{\delta - z} x^{\delta - z} + O(\log x).$$

(iii) *If* $\operatorname{Re} z \leqq \delta$, $\operatorname{Re} z = \sigma \neq \eta$ *and* $z \neq \delta$, *then*

$$\sum_{|a| \leqq x} |a|^{-z} = \frac{\delta A}{\delta - z} x^{\delta - z} + \alpha(z) + O(x^{\eta - \sigma}),$$

where $\alpha(z)$ *is constant as* $x \to \infty$.

Proof. It is convenient for later purposes to deal first with an arbitrary arithmetical semigroup G, and let $g \in \operatorname{Dir}(G)$ be a function such that

$$N(g, x) = \sum_{|a| \leqq x} g(a) = Bx^\delta + O(x^\eta) \quad \text{as } x \to \infty,$$

where B and δ are constants, and η is a constant with $0 \leqq \eta < \delta$. In each case, our starting point will be the equation

$$\sum_{|a| \leqq x} g(a) |a|^{-z} = N(g, x) x^{-z} + z \int_1^x N(g, y) y^{-z-1} \, dy, \tag{2.9}$$

which follows by partial summation as earlier.

Therefore, in the first place,

$$\begin{aligned} \sum_{|a| \leqq x} g(a) |a|^{-\delta} &= [Bx^\delta + O(x^\eta)] x^{-\delta} + \delta \int_1^x [By^\delta + O(y^\eta)] y^{-\delta - 1} \, dy \\ &= B + O(x^{\eta - \delta}) + \delta B \log x + I(x), \end{aligned}$$

where $I(x)$ is an absolutely convergent integral. In fact,

$$I(x) = \delta\int_1^\infty [N(g,y)-By^\delta]\,y^{-\delta-1}\,\mathrm{d}y - \delta\int_x^\infty [N(g,y)-By^\delta]\,y^{-\delta-1}\,\mathrm{d}y$$
$$= I - \mathrm{O}\Big(\int_x^\infty y^{\eta-\delta-1}\,\mathrm{d}y\Big),$$

say, and the last term is $\mathrm{O}(x^{\eta-\delta})$. Therefore

$$\sum_{|a|\leqq x} g(a)\,|a|^{-\delta} = \delta B\log x + \gamma_g + \mathrm{O}(x^{\eta-\delta}) \quad \text{as } x\to\infty,$$

where $\gamma_g = B+I$. Part (i) of Proposition 2.8 follows as a special case.
Next, for $\operatorname{Re} z=\eta$, eq. (2.9) gives

$$\sum_{|a|\leqq x} g(a)\,|a|^{-z} = [Bx^\delta + \mathrm{O}(x^\eta)]x^{-z} + \frac{zB}{\delta - z}[x^{\delta-z}-1] + \mathrm{O}\left(\int_1^x y^{-1}\,\mathrm{d}y\right)$$
$$= \frac{\delta B}{\delta - z}x^{\delta-z} + \mathrm{O}(1) + \mathrm{O}(\log x),$$

which implies part (ii) of the proposition.
Lastly, for $\operatorname{Re} z=\sigma\leqq\delta$, $\sigma\neq\eta$ and $z\neq\delta$, eq. (2.9) gives

$$\sum_{|a|\leqq x} g(a)\,|a|^{-z} = [Bx^\delta + \mathrm{O}(x^\eta)]x^{-z} + \frac{zB}{\delta - z}[x^{\delta-z}-1] + J(x),$$

where $J(x)$ is an integral which, for $\eta>\sigma$, is $\mathrm{O}(x^{\eta-\sigma})$. On the other hand, for $\eta<\sigma$, $J(x)$ is an absolutely convergent integral and

$$J(x) = z\Big(\int_1^\infty - \int_x^\infty\Big)[N(g,y)-By^\delta]y^{-z-1}\,\mathrm{d}y$$
$$= J + \mathrm{O}\Big(\int_x^\infty y^{\eta-\sigma-1}\,\mathrm{d}y\Big),$$

say. Part (iii) follows, and in fact we have:

2.10. Corollary. *Let G denote an arbitrary arithmetical semigroup, and let $g\in\operatorname{Dir}(G)$ be a function such that*

$$N(g,x) = Bx^\delta + \mathrm{O}(x^\eta) \quad \textit{as } x\to\infty,$$

where B and δ are constants, and η is a constant with $0\leqq\eta<\delta$. Then:

(i) $\sum_{|a|\leqq x} g(a)\,|a|^{-\delta} = \delta B\log x + \gamma_g + \mathrm{O}(x^{\eta-\delta})$, *where*

$$\gamma_g = B + \delta\int_1^\infty [N(g,y)-By^\delta]y^{-\delta-1}\,\mathrm{d}y.$$

(ii) *For* $\operatorname{Re} z=\eta$,

$$\sum_{|a|\leqq x} g(a)\,|a|^{-z} = \frac{\delta B}{\delta - z}x^{\delta-z} + \mathrm{O}(\log x).$$

(iii) *If* $\mathrm{Re}\, z \leqq \delta$, $\mathrm{Re}\, z = \sigma \neq \eta$ *and* $z \neq \delta$, *then*

$$\sum_{|a| \leqq x} g(a)\, |a|^{-z} = \frac{\delta B}{\delta - z}\, x^{\delta - z} - \alpha_g(z) + \mathrm{O}(x^{\eta - \sigma}),$$

where $\alpha_g(z)$ *is constant as* $x \to \infty$. □

Before considering one further analytic property of the zeta function, we note an interesting arithmetical consequence of Proposition 2.6.

2.11. Proposition. *Let G denote an arithmetical semigroup satisfying Axiom* A. *Then the set P of primes in G is infinite and has 'asymptotic density'* 0, *i.e.*

$$\pi_G(x) = \mathrm{o}\big(N_G(x)\big) \quad \text{as } x \to \infty.$$

Proof. By Propositions 2.6 and 2.1, the series $\sum_{p \in P} |p|^{-\delta}$ and the product $\prod_{p \in P} (1 - |p|^{-\delta})^{-1}$ are both divergent. Therefore P must be infinite.

Now arrange P in a sequence $p_1, p_2, \ldots$, and let G_n denote the arithmetical semigroup of all elements $a \in G$ that are not divisible by any p_i with $i \leqq n$. Then Proposition 1.3 implies that

$$N_{G_n}(x) = A \prod_{i=1}^{n} (1 - |p_i|^{-\delta})\, x^{\delta} + \mathrm{O}(x^{\eta}).$$

Therefore, since

$$\pi_G(x) \leqq n + \pi_{G_n}(x) \leqq n + N_{G_n}(x),$$

we have

$$\frac{\pi_G(x)}{N_G(x)} \leqq \frac{\prod_{i=1}^{n} (1 - |p_i|^{-\delta}) + \mathrm{O}(x^{\eta - \delta})}{1 + \mathrm{O}(x^{\eta - \delta})}$$

$$\sim \prod_{i=1}^{n} (1 - |p_i|^{-\delta})$$

as $x \to \infty$. Since the last product is arbitrarily small for n sufficiently large, the 'density' statement follows. □

2.12. Corollary. *Let G denote an arithmetical semigroup satisfying. Axiom A. Then the series* $\sum_{p \in P} |p|^{-\sigma}$ *and the product* $\prod_{p \in P} (1 - |p|^{-\sigma})^{-1}$ *diverge for all real* $\sigma \leqq \delta$. □

The above facts about $\zeta_G(z)$ will suffice for most later purposes of this book, but the following result will also be used at one stage.

2.13. Proposition. *Let G denote an arithmetical semigroup satisfying Axiom* A. *Then* $\zeta_G(z)$ *may be continued analytically into the entire region* $\mathrm{Re}\, z > \eta$, $z \neq \delta$, *and has a simple pole with residue* δA *at* $z = \delta$.

Proof. For $x \geqq 1$, let

$$R(x) = N_G(x) - Ax^\delta.$$

Then, for $\operatorname{Re} z > \delta$, Proposition 2.6 implies that

$$\begin{aligned} \zeta_G(z) &= z\int_1^\infty N_G(y) y^{-z-1}\,\mathrm{d}y \\ &= z\int_1^\infty [Ay^\delta + R(y)] y^{-z-1}\,\mathrm{d}y \\ &= \frac{\delta A}{z-\delta} + A + z\int_1^\infty R(y) y^{-z-1}\,\mathrm{d}y. \end{aligned}$$

Since $R(y) = O(y^\eta)$ by Axiom A, the last integral is absolutely convergent for $\operatorname{Re} z = \sigma > \eta$. Also, for such z,

$$\int_1^\infty R(y) y^{-z-1}\,\mathrm{d}y = \sum_{n=1}^\infty \int_n^{n+1} R(y) y^{-z-1}\,\mathrm{d}y = \sum_{n=1}^\infty R_n(z),$$

say.

Now, the order estimate for $R(y)$ above implies that there is some constant $C > 0$ such that $|R(y)| \leqq Cy^\eta$ $(y \geqq 1)$, and therefore

$$|R_n(z)| \leqq C\int_n^{n+1} y^{\eta-\sigma-1}\,\mathrm{d}y = \frac{C}{\sigma-\eta}[n^{\eta-\sigma} - (n+1)^{\eta-\sigma}].$$

Therefore

$$\begin{aligned} \left|\sum_{n=M}^N R_n(z)\right| &\leqq \frac{C}{\sigma-\eta}\sum_{n=M}^N [n^{\eta-\sigma} - (n+1)^{\eta-\sigma}] = \frac{C}{\sigma-\eta}\left[M^{\eta-\sigma} - (N+1)^{\eta-\sigma}\right] \\ &\leqq \frac{C}{\sigma-\eta} M^{\eta-\sigma} \leqq \frac{C}{r_0} M^{-r_0} \end{aligned}$$

whenever $\sigma - \eta \geqq r_0 > 0$. This shows that the series $\sum_{n=1}^\infty R_n(z)$ converges uniformly in the region $\operatorname{Re} z \geqq \eta + r_0$, for each fixed $r_0 > 0$. Therefore it will follow that this series defines an analytic function of z for $\operatorname{Re} z > \eta$, once it has been shown that each $R_n(z)$ is analytic in that region.

For the latter purpose, now note that

$$\begin{aligned} R_n(z) &= \int_n^{n+1} [N_G(y) - Ay^\delta] y^{-z-1}\,\mathrm{d}y \\ &= \int_n^{n+1} N_G(y) y^{-z-1}\,\mathrm{d}y + F(z), \end{aligned}$$

say. Now, since $N_G(x)$ is a step function, the first integral can be written as a finite sum of terms of the form

$$c_i z^{-1}[n_i^{-z} - n_{i+1}^{-z}] \qquad [z \neq 0,\ c_i \text{ constant}].$$

Therefore it is certainly an analytic function of z for $z \neq 0$. Also, $F(z)$ is the integral of the continuous function $f(z, y) = -Ay^{\delta-z-1}$ over the real contour $n \leqq y \leqq n+1$. Since $f(z, y)$ is an analytic function of z for every such y, a well known theorem on integrals of analytic functions (see Titchmarsh [2] p. 99, say) implies that $F(z)$ is also analytic. Thus $R_n(z)$ is certainly analytic for $\operatorname{Re} z > \eta$ (in fact, this can be proved for all z).

The equation

$$\zeta_G(z) = \frac{\delta A}{z-\delta} + A + z\int_1^\infty R(y)y^{-z-1}\mathrm{d}y$$

now shows that $\zeta_G(z)$ can be continued analytically into the entire region $\operatorname{Re} z > \eta$, $z \neq \delta$, and that it has a simple pole with residue δA at $z = \delta$. In addition it implies:

2.14. Corollary. *If G satisfies Axiom* A, *then*

$$\lim_{z\to\delta}\left\{\zeta_G(z) - \frac{\delta A}{z-\delta}\right\} = \gamma_G.$$

the constant defined in Proposition 2.8 (i). □

As a general comment, we note that Proposition 2.8 (i) implies that

$$\gamma_G = \lim_{x\to\infty}\left\{\sum_{|a|\leqq x}|a|^{-\delta} - \delta A \log x\right\},$$

and that for the semigroup $G_{\mathbf{Z}}$ of positive integers the right-hand limit is the well-known Euler constant $\gamma = 0.57721\ldots$. For this reason, in general, γ_G will be called the **Euler constant** of the arithmetical semigroup G (or the category to which G is associated, if appropriate); it will appear again in later contexts.

For later purposes, we note that Lemma 2.7 together with a very minor modification of the proof of Proposition 2.13 yields

2.15. Proposition. *Let G denote an arbitrary arithmetical semigroup, and let $g \in \operatorname{Dir}(G)$ be a function such that*

$$N(g, x) = Bx^\delta + O(x^\eta) \quad \text{as } x \to \infty,$$

where B, δ are constants and η is a constant with $0 \leqq \eta < \delta$. Then the series

$$\tilde{g}(z) = \sum_{a\in G} g(a)|a|^{-z}$$

defines an analytic function of z for Re $z>\delta$ *that may be continued analytically into the entire region* Re $z>\eta$, $z\neq\delta$, *and has a simple pole with residue* δB *at* $z=\delta$.

As an exercise, the reader may like to determine 'Stirling's formula' for $\sum_{|a|\leq x} \log |a|$ as $x\to\infty$, when G is an arithmetical semigroup satisfying Axiom A.

§ 3. Average values of arithmetical functions

Throughout §§ 3—5, G will denote a given arithmetical semigroup satisfying Axiom A.

Given some subset H of G, and $x\geqq 1$, let

$$H[x] = \{a\in H\colon\ |a|\leqq x\}, \qquad N_H(x) = \text{card}\, H[x].$$

If $f\in \text{Dir}\,(G)$, let

$$N_H(f, x) = \sum_{a\in H[x]} f(a), \qquad m_H(f, x) = \frac{N_H(f, x)}{N_H(x)}.$$

Then $m_H(f, x)$ may be called the **average value** of f on $H[x]$. If $m_H(f, x)\sim \sim F(x)$ as $x\to\infty$, we shall say that f has the **approximate** average value $F(x)$ on $H[x]$; if $m_H(f, x)$ tends to some limit $m_H(f)$ as $x\to\infty$, we say that f has the **asymptotic mean-value** $m_H(f)$ on H.

In the case when f is the *characteristic* function of some subset of G, so that $f(a)=1$ if $a\in E$ and $f(a)=0$ otherwise, we shall call

$$d_H(E, x) = m_H(f, x)$$

the **relative density** of E in $H[x]$. If

$$d_H(E, x) \sim F(x) \quad \text{as } x\to\infty,$$

we say that E has **approximate** relative density $F(x)$ in $H[x]$; lastly, if $d_H(E, x)$ tends to some limit $d_H(E)$ as $x\to\infty$, we call $d_H(E)$ the **asymptotic** relative density of E in H.

In Part II, we shall only be concerned with the "absolute" case in which $H=G$, and here the suffix or references to H and use of the word 'relative' will usually be omitted. (Of course, these definitions do not require Axiom A, but in other circumstances slightly different definitions might sometimes possibly be preferable; compare Chapter 8, § 5.)

As an example, it may be noted that, while Proposition 2.11 has already shown that the set P of primes in G has zero asymptotic density, the abstract prime number theorem to be proved in Chapter 6 states that

$$\pi_G(x) \sim x^\delta/\delta \log x \quad \text{as } x \to \infty.$$

Hence P has the approximate density $1/\delta A \log x$ in $G[x]$.

As another example, consider the set $G^{(k)}$ of all k^{th} powers a^k of elements $a \in G$, where k is a positive integer. Then

$$\begin{aligned} d(G^{(k)}, x) &= \frac{1}{N_G(x)} \sum_{\substack{a \in G^{(k)} \\ |a| \leq x}} 1 = \frac{1}{N_G(x)} \sum_{|a^k| \leq x} 1 \\ &= \frac{1}{N_G(x)} \sum_{|a| \leq x^{1/k}} 1 = \frac{N_G(x^{1/k})}{N_G(x)} \\ &\sim x^{\delta/k - \delta} \quad \text{as } x \to \infty. \end{aligned}$$

Hence $G^{(k)}$ has approximate density $x^{\delta/k-\delta}$ in $G[x]$, and, for $k>1$, this implies that $G^{(k)}$ also has asymptotic density zero in G.

If the asymptotic behaviour of $N_H(x)$ as $x \to \infty$ is already known, it is clear that approximate average values or relative densities can all be deduced from a study of the functions $N_H(f, x)$ as $x \to \infty$. In particular, in the absolute case of $H=G$, we shall be concerned mainly with $N(f, x)$ as $x \to \infty$; by using Axiom A, it is then usually trivial to find the approximate average value in question. The following theorem concerning $N(f, x)$ has applications to many natural arithmetical functions.

3.1. Theorem. *Let $f, g \in \text{Dir}(G)$ be functions such that*

$$f(z) = [\zeta_G(mz)]^k g(z),$$

where m and k are positive integers. Suppose that $\tilde{g}(z)$ is absolutely convergent whenever $\text{Re}\, z > \nu$, for some $\nu < \delta/m$. Then as $x \to \infty$

$$N(f, x) = \big(A/(k-1)!\big)(\delta A/m)^{k-1}[\tilde{g}(\delta/m) + o(1)]x^{\delta/m}(\log x)^{k-1}.$$

In particular, if $m=k=1$, then f has the asymptotic mean-value $\tilde{g}(\delta)$ on G.

In order to prove this theorem, first consider the case when $f(z)=d_k(z)= =[\zeta_G(z)]^k$; for later purposes, we shall treat this in greater detail than strictly necessary at present.

3.2. Proposition. *The approximate average value of d_k on $G[x]$ is*

$$\frac{(\delta A)^{k-1}}{(k-1)!}(\log x)^{k-1}.$$

More precisely, there exist constants c_i *with* $c_1=(\delta A)^{k-1}/(k-1)!$, *such that, for* $k>1$,

$$N(d_k, x) = Ax^\delta \sum_{i=1}^{k} c_i(\log x)^{k-i} + O\left(x^{\delta-(\delta-\eta)/k}(\log x)^{k-2}\right).$$

In particular, for the divisor function $d=d_2$, this proposition yields **"Dirichlet's formula"**:

$$N(d, x) = Ax^\delta(\delta A \log x + 2\gamma_G - A) + O(x^{(\delta+\eta)/2}).$$

First consider the following lemma, which does not require Axiom A.

3.3. Lemma. *Let* $f,g \in \mathrm{Dir}\,(G)$. *Then*

$$N(f*g, x) = \sum_{|a|\leqq x} f(a)N(g, x/|a|) = \sum_{|a|\leqq x} g(a)N(f, x/|a|).$$

Also, for positive u, v,

$$N(f*g, uv) = \sum_{|a|\leqq u} f(a)N(g, uv/|a|) + \sum_{|a|\leqq v} g(a)N(f, uv/|a|)$$
$$-N(f, u)N(g, v).$$

Proof. Firstly, by definition,

$$N(f*g, x) = \sum_{|a|\leqq x}\sum_{bc=a} f(b)g(c) = \sum_{|bc|\leqq x} f(b)g(c)$$
$$= \sum_{|b|\leqq x} f(b) \sum_{|c|\leqq x/|b|} g(c) = \sum_{|c|\leqq x} g(c) \sum_{|b|\leqq x/|c|} f(b).$$

This proves the first two equations. Also, for $x=uv$, note that the set of all ordered pairs (b, c) of elements $b,c \in G$ such that $|bc|\leqq x$ can be expressed as $B\cup C$, where B is the subset of those pairs such that $|b|\leqq u$, and C is the subset of those pairs satisfying $|c|\leqq v$. Therefore

$$N(f*g, uv) = \sum_{|bc|\leqq uv} f(b)g(c)$$
$$= \left(\sum_{B} + \sum_{C} - \sum_{B\cap C}\right) f(b)g(c)$$
$$= \sum_{|b|\leqq u} f(b) \sum_{|bc|\leqq uv} g(c) + \sum_{|c|\leqq v} g(c) \sum_{|bc|\leqq uv} f(b) - \sum_{\substack{|b|\leqq u\\ |c|\leqq v}} f(b)g(c)$$
$$= \sum_{|b|\leqq u} f(b)N(g, uv/|b|) + \sum_{|c|\leqq v} g(c)N(f, uv/|c|) - N(f, u)N(g, v).$$

□

Now note that Theorem 2.6.1 shows that $d_k = \zeta_G * d_{k-1}$ for $k > 1$. For $x \geqq 1$, let

$$u = x^{1/k}, \qquad v = x^{1-(1/k)}.$$

Then Lemma 3.3 gives

$$\begin{aligned} N(d_k, x) &= N(\zeta_G * d_{k-1}, uv) \\ &= \sum_{|a| \leqq u} N(d_{k-1}, x/|a|) + \sum_{|a| \leqq v} d_{k-1}(a) N_G(x/|a|) - N_G(u) N(d_{k-1}, v) \\ &= \Sigma_1 + \Sigma_2 - \Delta, \end{aligned}$$

say.

In the special case when $k=2$, this becomes

$$\begin{aligned} N(d, x) &= 2 \sum_{|a| \leqq \sqrt{x}} N_G(x/|a|) - [N_G(\sqrt{x})]^2 \\ &= 2 \sum_{|a| \leqq \sqrt{x}} \left(A(x/|a|)^\delta + O(x/|a|)^\eta\right) - [Ax^{\delta/2} + O(x^{\eta/2})]^2 \\ &= 2Ax^\delta \sum_{|a| \leqq \sqrt{x}} |a|^{-\delta} + O\Big(x^\eta \sum_{|a| \leqq \sqrt{x}} |a|^{-\eta}\Big) - A^2 x^\delta + O(x^{(\delta+\eta)/2}). \end{aligned}$$

By Proposition 2.8 it follows that

$$\begin{aligned} N(d, x) = 2Ax^\delta(\delta A \log \sqrt{x} + \gamma_G + O(x^{(\eta-\delta)/2})) + O(x^\eta x^{(\delta-\eta)/2}) \\ - A^2 x^\delta + O(x^{(\delta+\eta)/2}). \end{aligned}$$

This proves the stated asymptotic formula for $N(d, x)$.

Now suppose that $k > 2$, and, as an inductive hypothesis, assume that the required asymptotic formula has already been established for $N(d_{k-1}, x)$. Then this assumption, Proposition 2.8 and partial summation show that there exist constants a_i', b_i', a_{1i}, b_{1i} such that:

$$(1) \quad \sum_{|a| \leqq x} d_{k-1}(a) |a|^{-\delta} = \sum_{i=1}^{k-1} a_i' (\log x)^{k-i} + O\big(x^{(\eta-\delta)/(k-1)} (\log x)^{k-3}\big);$$

$$\begin{aligned} (2) \quad \sum_{|a| \leqq x} d_{k-1}(a) |a|^{-\eta} = x^{\delta-\eta} \sum_{i=1}^{k-1} b_i' (\log x)^{k-1-i} \\ + O\big(x^{\delta-\eta-(\delta-\eta)/(k-1)} (\log x)^{k-3}\big); \end{aligned}$$

$$(3) \quad \sum_{|a| \leqq x} (\log |a|)^i |a|^{-\delta} = a_{1i} (\log x)^{i+1} + b_{1i} (\log x)^i + O\big(x^{\eta-\delta} (\log x)^{i-1}\big);$$

$$(4) \quad \sum_{|a| \leqq x} (\log |a|)^i |a|^{-\delta+(\delta-\eta)/(k-1)} = O\big(x^{(\delta-\eta)/(k-1)} (\log x)^i\big).$$

Now consider the sum Σ_1 above, and, in applying the inductive hypothesis to $N(d_{k-1}, x/|a|)$, expand every term of the form

$$c_i'(\log(x/|a|))^i = c_i'(\log x - \log|a|)^i$$

by means of the binomial theorem. Then, with the aid of the estimates (1)—(4), one may verify that there exist constants A_i, B_i and C_i such that:

$$(5)\quad \Sigma_1 = x^\delta \sum_{i=1}^{k} A_i(\log x)^{k-i} + O\big(x^{\delta-(\delta-\eta)/k}(\log x)^{k-3}\big);$$

$$(6)\quad \Sigma_2 = x^\delta \sum_{i=1}^{k} B_i(\log x)^{k-i} + O\big(x^{\delta-(\delta-\eta)/k}(\log x)^{k-2}\big);$$

$$(7)\quad \Delta = x^\delta \sum_{i=1}^{k-1} C_i(\log x)^{k-1-i} + O\big(x^{\delta-(\delta-\eta)/k}(\log x)^{k-2}\big).$$

It follows now that there are constants c_i such that Proposition 3.2 is true for $N(d_k, x)$, and it remains to evaluate c_1. (Note that it was not necessary to assume the required value for c_1' in the inductive hypothesis about $N(d_{k-1}, x)$.) The following lemma shows that the constants c_i may be determined from the coefficients of $[\zeta_G(z)]^k$ in its Laurent expansion about the pole at $z=\delta$. In particular, the required evaluation of c_1 is obtained by noting that Proposition 2.13 implies that the coefficient of $(z-\delta)^{-k}$ in this Laurent expansion is $(\delta A)^k$. □

3.4. Lemma. *If*

$$N(d_k, x) = Ax^\delta \sum_{i=1}^{k} c_i(\log x)^{k-i} + O\big(x^{\delta-(\delta-\eta)/k}(\log x)^{k-2}\big),$$

then, for $|z-\delta|$ sufficiently small and non-zero,

$$[\zeta_G(z)]^k = \frac{(k-1)!\,\delta A c_1}{(z-\delta)^k} + \sum_{i=1}^{k-1} \frac{(k-i-1)!\,A[\delta c_{i+1} + (k-i)c_i]}{(z-\delta)^{k-i}} + \sum_{j=0}^{\infty} b_j(z-\delta)^j,$$

where the b_j are constants.

Proof. The proof follows the same general lines as that of Proposition 2.13. First let

$$R_k(x) = N(d_k, x) - Ax^\delta \sum_{i=1}^{k} c_i(\log x)^{k-i},$$

so that

$$R_k(x) = O\left(x^{\delta-(\delta-\eta)/k}(\log x)^{k-2}\right).$$

Then note that, for $\operatorname{Re} z=\sigma$,

$$N(d_k, x)x^{-z} = O\left(x^{\delta-\sigma}(\log x)^{k-1}\right) = O(x^{\delta-\sigma+\varepsilon})$$

for every $\varepsilon>0$. Therefore the following integral converges and the Partial Summation Lemma 2.3 implies that, for $\operatorname{Re} z>\delta$,

$$\tilde{d}_k(z) = \sum_{a\in G} d_k(a)\,|a|^{-z} = z\int_1^\infty N(d_k, y)\,y^{-z-1}\,\mathrm{d}y.$$

Now, if r is a positive integer and $\operatorname{Re} z>\delta$, integration by parts shows that

$$\int_1^\infty (\log y)^r\, y^{\delta-z-1}\,\mathrm{d}y = \frac{r}{z-\delta}\int_1^\infty (\log y)^{r-1}\, y^{\delta-z-1}\,\mathrm{d}y = \ldots$$

$$= \frac{r!}{(z-\delta)^{r+1}}.$$

Therefore, for $\operatorname{Re} z>\delta$,

$$\tilde{d}_k(z) = Az\sum_{i=1}^{k}\frac{(k-i)!\,c_i}{(z-\delta)^{k-i+1}} + z\int_1^\infty R_k(y)\,y^{-z-1}\,\mathrm{d}y$$

$$= \frac{(k-1)!\,\delta A c_1}{(z-\delta)^k} + \sum_{i=1}^{k-1}\frac{(k-i-1)!\,A[\delta c_{i+1}+(k-i)c_i]}{(z-\delta)^{k-i}}$$

$$+ Ac_k + z\int_1^\infty R_k(y)\,y^{-z-1}\,\mathrm{d}y.$$

Next, as in the proof of Proposition 2.13, the order-of-magnitude estimate for $R_k(x)$ implies that the last integral is absolutely convergent for $\operatorname{Re} z \geqq \delta-(\delta-\eta)/k+\varepsilon$, where $\varepsilon>0$ is arbitrary, and that in every such region it is the sum of the uniformly convergent series

$$\sum_{n=1}^\infty \int_n^{n+1} R_k(y)\,y^{-z-1}\,\mathrm{d}y = \sum_{n=1}^\infty R_{kn}(z),$$

say. Then, just as in the earlier case, one sees that $R_{kn}(z)$ is an analytic function of z for $\operatorname{Re} z>0$ (in fact, this is so for all z).

It follows now that $\tilde{d}_k(z)$ may be continued analytically into the entire region $\operatorname{Re} z>\delta-(\delta-\eta)/k$, $z\neq\delta$, that it has a pole of order k at $z=\delta$, and that it has the Laurent expansion about $z=\delta$ indicated in the statement of Lemma 3.4. Finally, it must be verified that $\tilde{d}_k(z)=[\zeta_G(z)]^k$ in the region in question. For this purpose, observe that $\tilde{d}_k(z)$ is the formal k^{th} power of

$\zeta_G(z)$ for $\operatorname{Re} z > \delta$, and that $\zeta_G(z)$ is absolutely convergent when $\operatorname{Re} z > \delta$. The conclusion therefore follows from:

3.5. Lemma. *Let $f, g \in \operatorname{Dir}(G)$, and suppose that the series*

$$\tilde{f}(z) = \sum_{a \in G} f(a)|a|^{-z}, \qquad \tilde{g}(z) = \sum_{a \in G} g(a)|a|^{-z}$$

are absolutely convergent for all complex z with $\operatorname{Re} z > \alpha$. Then the series $\sum_{a \in G} (f * g)(a)|a|^{-z}$ is also absolutely convergent for $\operatorname{Re} z > \alpha$, and then its sum is the ordinary product $\tilde{f}(z)\tilde{g}(z)$.

Proof. Let $\sum a_n = A$ and $\sum b_n = B$ be absolutely convergent series, and let $\sum c_n$ be the series obtained by bracketing and arranging the products $a_i b_j$ $(i,j = 1,2,\ldots)$ in any convenient way. The lemma follows from the well-known fact that $\sum c_n$ must then also be absolutely convergent and have the sum $C = AB$. □

In order to prove Theorem 3.1, consider one more lemma.

3.6. Lemma. *Let $f, F, g \in \operatorname{Dir}(G)$ be arbitrary functions such that $f(z) = F(z)g(z)$. Suppose that*

$$N(F, x) = Bx^{\alpha}(\log x)^{r} + O\left(x^{\beta}(\log x)^{s}\right),$$

where $\alpha > 0$, $0 \leqq \beta \leqq \alpha$, and r and s are non-negative integers with the property that $\beta < \alpha$ if $r = 0$, while $s < r$ if $\beta = \alpha$. Suppose also that the series $\tilde{g}(z)$ is absolutely convergent for all complex z with $\operatorname{Re} z > v$, where $v < \alpha$. Then as $x \to \infty$,

$$N(f, x) = [B\tilde{g}(\alpha) + o(1)]x^{\alpha}(\log x)^{r}.$$

Proof. Since $f(z) = F(z)g(z)$,

$$\begin{aligned} N(f, x) &= \sum_{|bc| \leqq x} F(b)g(c) \\ &= \sum_{|c| \leqq x} g(c) \sum_{|b| \leqq x/|c|} F(b) \\ &= \sum_{|c| \leqq x} g(c)\left\{B(x/|c|)^{\alpha}(\log(x/|c|))^{r} + O\left((x/|c|)^{\beta}(\log(x/|c|)^{s})\right)\right\}. \end{aligned}$$

Now suppose that $r = 0$. Then $\beta < \alpha$ and $x^{\beta}(\log x)^{s} = O(x^{\beta'})$ for a suitable β' with $v < \beta' < \alpha$. Hence

$$\begin{aligned} N(f, x) &= Bx^{\alpha} \sum_{|c| \leqq x} g(c)|c|^{-\alpha} + O\left(x^{\beta'} \sum_{|c| \leqq x} |g(c)|\,|c|^{-\beta'}\right) \\ &= Bx^{\alpha}[\tilde{g}(\alpha) + o(1)] + O(x^{\beta'}) = x^{\alpha}[B\tilde{g}(\alpha) + o(1)]. \end{aligned}$$

Alternatively, if $r>0$, then

$$x^{\beta}(\log x)^{s} = O\big(x^{\alpha}(\log x)^{s'}\big)$$

for some non-negative $s'<r$ in the case when $\beta<\alpha$, while otherwise this would be true with $s'=s$ by an assumption of the lemma. Hence now

$$\begin{aligned} N(f,x) &= Bx^{\alpha} \sum_{|c|\leq x}' g(c)|c|^{-\alpha}(\log x - \log|c|)^{r} \\ &\quad + O\big(x^{\alpha} \sum_{|c|\leq x} |g(c)|\,|c|^{-\alpha}(\log(x/|c|))^{s'}\big) \\ &= Bx^{\alpha}\left\{(\log x)^{r}[\tilde{g}(\alpha)+o(1)] + O\left(\sum_{t=0}^{r-1}(\log x)^{t} \sum_{|c|\leq x}|g(c)|\,|c|^{\varepsilon-\alpha}\right)\right\} \\ &\quad + O\big(x^{\alpha}(\log x)^{s'}O(1)\big), \end{aligned}$$

where $\varepsilon>0$ is arbitrary and may be chosen so that $\alpha-\varepsilon>\nu$. Therefore

$$\begin{aligned} N(f,x) &= Bx^{\alpha}\{(\log x)^{r}[\tilde{g}(\alpha)+o(1)] + O((\log x)^{r-1}O(1))\} + O(x^{\alpha}(\log x)^{s'}) \\ &= x^{\alpha}(\log x)^{r}[B\tilde{g}(\alpha)+o(1)] \quad \text{as } x\to\infty. \end{aligned}$$

□

In order to deduce Theorem 3.1 from the above lemma, it remains only to verify that, if

$$F(z) = [\zeta_G(mz)]^{k} = d_k(mz),$$

then

$$\begin{aligned} N(F,x) &= \sum_{|a^m|\leq x} d_k(a) = \sum_{|a|\leq x^{1/m}}' d_k(a) \\ &= Ax^{\delta/m}\sum_{i=1}^{k} c_i\big((1/m)\log x\big)^{k-i} + O\big(x^{(\delta-(\delta-\eta)/k)/m}(\log x)^{k-2}\big). \end{aligned}$$

□

Given an arithmetical function f, Theorem 3.1 shows that, if $f(z)$ can be factorized in a suitable way, then in principle it is possible to calculate the approximate average value of f on $G[x]$. It is therefore of interest to obtain sufficient conditions for the existence of such a factorization. The next proposition provides one such set of conditions.

3.7. Proposition. *Let $f\in \mathrm{Dir}\,(G)$ be a* PIM-*function such that*

$$f(p^{r}) = O(c^{r}) \quad \text{as } r\to\infty \qquad (p\in P),$$

for some constant c satisfying $1<c<q_0^{\delta/m}$, $q_0=\min\{|p|: p\in P\}$. If

$$\hat{f} = 1 + kt^{m} + \sum_{r>m} c_r t^{r},$$

then

$$f(z) = [\zeta_G(mz)]^k \tilde{g}(z),$$

where $\tilde{g}(z)$ *is absolutely convergent whenever* $\operatorname{Re} z > \nu$ *for some* $\nu < \delta/m$.

Proof. If $\hat{f}$ is of the form indicated, the proof of Theorem 2.7.1 shows that

$$f(z) = [\zeta_G(mz)]^k g(z),$$

where g is a PIM-function such that, if $c_r = f(p^r)$ ($r \geqq 0$, $p \in P$),

$$\hat{g} = \left(\sum_{r=0}^{\infty} (-1)^r \binom{k}{r} t^{rm}\right)\left(\sum_{r=0}^{\infty} c_r t^r\right)$$

$$= \sum_{r=0}^{\infty} c_r' t^r,$$

say. Then $c_0' = 1$, $c_r' = 0$ ($0 < r \leqq m$), and for r sufficiently large the order-of-magnitude assumption about c_r implies that

$$|c_r'| = \left|\sum_{j \leqq r/m} (-1)^j \binom{k}{j} c_{r-mj}\right| = O\left(\sum_{j \leqq r/m} \left|\binom{k}{j}\right| c^{r-mj}\right)$$

$$= O\left(c^r \sum_{j \leqq r/m} \binom{-|k|}{j} (-c^{-m})^j\right) = O(c^r (1 - c^{-m})^{-|k|})$$

$$= O(c^r) \quad \text{as } r \to \infty,$$

since $0 < c^{-m} < 1$.

Now choose $\nu \geqq \delta/(m+1)$ such that $c < q_0^{\nu}$ and, given $\sigma > \nu$, let $\tau = (m+1)\sigma > \delta$. Then for $p \in P$ and $\operatorname{Re} z = \sigma > \nu$,

$$\left|\sum_{r=1}^{\infty} |g(p^r)|\, |p|^{-rz}\right| = \left|\sum_{r=m+1}^{\infty} |c_r'|\, |p|^{-rz}\right|$$

$$\leqq \sum_{r=m+1}^{\infty} |c_r'|\, |p|^{-r\sigma}$$

$$= O\left(\sum_{r=m+1}^{\infty} c^r |p|^{-r\sigma}\right) = O\left((c|p|^{-\sigma})^{m+1} \sum_{r=0}^{\infty} (c|p|^{-\sigma})^r\right)$$

$$= O(|p|^{-\sigma(m+1)}) = O(|p|^{-\tau}),$$

since $c < q_0^{\nu} < q_0^{\sigma} \leqq |p|^{\sigma}$.

By Propositions 2.1 and 2.6, $\sum_{p \in P} |p|^{-\tau}$ converges, and therefore the preceding estimate implies that the product

$$\prod_{p \in P} \left(1 + \sum_{r=1}^{\infty} |g(p^r)|\, |p|^{-rz}\right)$$

is absolutely convergent whenever $\operatorname{Re} z > \nu$. In order to deduce that the series $\tilde{g}(z)$ is absolutely convergent for $\operatorname{Re} z > \nu$, now arrange the elements of P in a sequence as in the proof of Proposition 2.1. Then, for $\sigma > \nu$,

$$\begin{aligned}\sum_{|a|<|p_N|} |g(a)|\,|a|^{-\sigma} &\leqq \sum_{\alpha_i \geqq 0} |g(p_1^{\alpha_1} p_2^{\alpha_2} \ldots p_N^{\alpha_N})|\,|p_1^{\alpha_1} p_2^{\alpha_2} \ldots p_N^{\alpha_N}|^{-\sigma} \\ &= \sum_{\alpha_i \geqq 0} |c_{\alpha_1}'|\,|c_{\alpha_2}'| \ldots |c_{\alpha_N}'|\,|p_1|^{-\alpha_1\sigma}\,|p_2|^{-\alpha_2\sigma} \ldots |p_N|^{-\alpha_N\sigma} \\ &= \prod_{i=1}^{N} \left(\sum_{r=0}^{\infty} |c_r'|\,|p_i|^{-r\sigma} \right) \\ &= \mathrm{O}(1) \quad \text{as } N \to \infty.\end{aligned}$$

This establishes the absolute convergence of $\tilde{g}(z)$, but for later purposes we also include:

3.8. Lemma. *Let g be an arbitrary* PIM-*function such that, for $p \in P$, either*

(i) $g(p^r) = 0$ $(0 < r \leqq m)$ *and* $g(p^r) = \mathrm{O}(c^r)$ *for some constant* $c < q_0^{\nu}$, *where* $\nu \geqq \delta/(m+1)$, *or*

(ii) $g(p) \neq 0$ *and* $g(p^r) = \mathrm{O}(c^r)$ *for a constant* $c < q_0^{\nu}$, *where* $\nu \geqq \delta$. *Then*

$$\tilde{g}(z) = \prod_{p \in P} \left(\sum_{r=0}^{\infty} g(p^r)\,|p|^{-rz} \right) \qquad [\operatorname{Re} z > \nu],$$

and the series and product are both absolutely convergent in this range.

Proof. The above discussion shows that the product is absolutely convergent when $\operatorname{Re} z > \nu$. Then, as in the proof of Proposition 2.1, note that for such z

$$\begin{aligned}\prod_{n=1}^{N} \left(\sum_{r=0}^{\infty} c_r'\,|p_n|^{-rz} \right) &= \sum_{\alpha_i \geqq 0} c_{\alpha_1}' c_{\alpha_2}' \ldots c_{\alpha_N}'\,|p_1|^{-\alpha_1 z}\,|p_2|^{-\alpha_2 z} \ldots |p_N|^{-\alpha_N z} \\ &= \sum_{|a|<|p_N|} g(a)\,|a|^{-z} + \sum_{|a| \geqq |p_N|}{}' g(a)\,|a|^{-z},\end{aligned}$$

where $\sum'$ indicates a sum running only over elements a divisible by primes p_i with $i \leqq N$. Therefore, for $\operatorname{Re} z = \sigma > \nu$,

$$\begin{aligned}\left| \prod_{n=1}^{N} \left(\sum_{r=0}^{\infty} c_r'\,|p_n|^{-rz} \right) - \sum_{|a|<|p_N|} g(a)\,|a|^{-z} \right| &\leqq \sum_{|a| \geqq |p_N|} |g(a)|\,|a|^{-\sigma} \\ &< \tfrac{1}{2}\varepsilon\end{aligned}$$

for N sufficiently large. The final conclusion follows now in the same general way as for Proposition 2.1. □

As an exercise, the reader may perhaps care to extend Proposition 3.7 to the case when $f \in M(G)$ is not necessarily prime-independent, but $\mathrm{O}(c^r)$ indicates a quantity independent of p, and

$$f_p(z) = 1 + kp^{-mz} + \sum_{r>m} c_{pr}\, p^{-rz}$$

(k, m independent of p). Lemma 3.8 may be extended in a similar way.

§ 4. Approximate average values of special arithmetical functions

Given an arithmetical function $f \in \mathrm{Dir}\,(G)$, let f^k denote the point-wise k^{th} power of f: $f^k(a) = [f(a)]^k$ for $a \in G$. If f^k possesses an asymptotic mean-value $m_H(f^k)$ on some subset H of G, we shall say that f has the **asymptotic k^{th} moment** $m_H(f^k)$ on H; more generally, if f^k has the approximate average value $F(x)$ on $H[x]$, we shall call $F(x)$ the **approximate** k^{th} moment of f on $H[x]$.

4.1. Proposition. *For $k = 1, 2, \dots$, there is a constant $B_k > 0$ such that the divisor function d has the approximate k^{th} moment $B_k(\log x)^{2^k-1}$ on $G[x]$. In particular, $B_1 = \delta A$ and $B_2 = (\delta A)^3/6\zeta_G(2\delta)$.*

Proof. The case $k = 1$ of this proposition has already been covered by Proposition 3.2. Also, for $k = 2$, Theorem 2.6.1 shows that

$$d^2(z) = [\zeta_G(z)]^4 \mu(2z).$$

Since $\tilde{\zeta}_G(z)$ is absolutely convergent for all complex z with $\mathrm{Re}\, z > \delta$, it is easy to see that $\tilde{\mu}(2z)$ is absolutely convergent when $\mathrm{Re}\, z > \frac{1}{2}\delta$. Therefore Theorem 3.1 implies that as $x \to \infty$

$$N(d^2, x) = \tfrac{1}{6} A(\delta A)^3 [\tilde{\mu}(2\delta) + \mathrm{o}(1)] x^\delta (\log x)^3.$$

This proves the statement for the case $k = 2$, since Lemma 3.8 and Propositions 2.1 and 2.6 show that $\tilde{\mu}(2\delta) = 1/\zeta_G(2\delta)$.

For the general case, now note that d^k is a PIM-function such that

$$\hat{d^k} = 1 + 2^k t + 3^k t^2 + \dots + (r+1)^k t^r + \dots .$$

Therefore the coefficient $(r+1)^k$ of t^r is $\mathrm{O}(c^r)$ as $r \to \infty$, for every $c > 1$, and so Proposition 3.7 implies that

$$d^k(z) = [\zeta_G(z)]^{2^k} g_k(z),$$

for some function g_k such that $\tilde{g}_k(z)$ is absolutely convergent whenever $\operatorname{Re} z > v$ for suitable $v < \delta$. By Theorem 3.1, it follows that

$$N(d^k, x) = \left(A/(2^k-1)!\right)(\delta A)^{2^k-1}[\tilde{g}_k(\delta)+o(1)]x^{\delta}(\log x)^{2^k-1}$$

as $x \to \infty$. This proves the proposition. □

If G is some specific arithmetical semigroup satisfying Axiom A for which the constants A, δ and η have been determined, it is usually a trivial matter to substitute the values of these constants in the various asymptotic formulae that occur in Part II of this book, and for this reason we shall not usually make such substitutions explicitly. (Note that the constant η is not unique, and in some specific cases it is at least conceivable that a smaller value than the one found may also be admissible; the problem of sharpening such estimates is likely to be very delicate in general.) For the sake of illustration, it may, however, at least be interesting to note occasionally what certain formulae look like for the fundamental semigroup $G_{\mathbf{Z}}$. (In some of these cases, the reader may perhaps also like to refer to a classical book like that of Hardy and Wright [1], say.)

Thus, for example, part of Proposition 3.2 yields the original 'Dirichlet divisor formula' for $G_{\mathbf{Z}}$:

$$\sum_{n \leqq x} d(n) = x(\log x + 2\gamma - 1) + O(\sqrt{x}),$$

where γ is Euler's constant 0.57721… . Also, it is well known that

$$\zeta(2) = \sum_{n=1}^{\infty} n^{-2} = \tfrac{1}{6}\pi^2.$$

Hence, Proposition 4.1 implies that the approximate second moment of d on $\{1, 2, \ldots, [x]\}$ is $\pi^{-2}(\log x)^3$, a conclusion which goes back to Ramanujan. This result, and Proposition 4.1 in general, provides an illustration of the obvious conclusion that the average value indicates only a special aspect of the "statistical" behaviour of a given arithmetical function. The study of higher moments and other statistical properties can significantly increase one's understanding of such functions; see also Chapter 5 below.

The next result shows that the unitary-divisor function d_* behaves quite similarly to d as regards moments.

4.2. Proposition. *For each integer $k \geqq 1$, there is a constant $B_k^* > 0$ such that the unitary-divisor function d_* has the approximate k^{th} moment $B_k^*(\log x)^{2^k-1}$ on $G[x]$. In particular,*

$$B_1^* = \delta A/\zeta_G(2\delta).$$

Proof. By Theorem 2.6.1,

$$d_*(z) = [\zeta_G(z)]^2 \mu(2z).$$

Therefore the case $k=1$ follows, since, by the same type of argument as for the case $k=2$ of Proposition 4.1,

$$N(d_*, x) = A(\delta A)\,[\tilde{\mu}(2\delta) + o(1)]\,x^\delta \log x.$$

In general, $d_*{}^k$ is a PIM-function with

$$\hat{d}_*{}^k = 1 + 2^k(t + t^2 + \ldots).$$

The coefficients here are constant after the first, and so Proposition 3.7 implies that

$$d_*{}^k(z) = [\zeta_G(z)]^{2^k} g_k{}^*(z),$$

where $\tilde{g}_k{}^*(z)$ is absolutely convergent whenever $\operatorname{Re} z > \nu$, for some $\nu < \delta$. The conclusion follows now in the same way as for Proposition 4.1. □

If $f, g \in \operatorname{Dir}(G)$ and $g(a) \neq 0$ for all $a \in G$, we define the point-wise quotient f/g by letting

$$(f/g)(a) = f(a)/g(a) \quad \text{for } a \in G.$$

Then, comparison of Propositions 4.1 and 4.2 suggests that the particular quotients d/d_* and d_*/d should both possess asymptotic k^{th} moments for each $k=1, 2, \ldots$. The next proposition shows that this is indeed the case, but also illustrates the common observation that plausibility in number theory can nevertheless be misleading on occasion. For, by way of example, one might perhaps conjecture on the basis of Propositions 4.1 and 4.2 that d/d_* should have the asymptotic mean value

$$\begin{aligned} B_1/B_1{}^* &= \zeta_G(2\delta) \\ &= \prod_{p \in P} (1 - |p|^{-2\delta})^{-1} \\ &= \prod_{p \in P} \{1 + (1 - |p|^{-2\delta})^{-1} |p|^{-2\delta}\}. \end{aligned}$$

However, Proposition 4.3 below shows that in fact

$$\begin{aligned} m(d/d_*) &= \prod_{p \in P} \{1 + \tfrac{1}{2}(1 - |p|^{-\delta})^{-1} |p|^{-2\delta}\} \\ &< B_1/B_1{}^*. \end{aligned}$$

4.3. Proposition. *For $k=1, 2, \ldots$, the point-wise quotients d/d_* and d_*/d possess the asymptotic k^{th} moments*

$$m((d/d_*)^k) = \prod_{p \in P} \left\{1+2^{-k} \sum_{r=2}^{\infty} [(r+1)^k - r^k] |p|^{-r\delta}\right\},$$

$$m((d_*/d)^k) = \prod_{p \in P} \left\{1+2^{k} \sum_{r=2}^{\infty} [(r+1)^{-k} - r^{-k}] |p|^{-r\delta}\right\}.$$

Proof. Let $f_1=(d/d_*)^k$ and $f_2=(d_*/d)^k$. Then the preceding discussion shows that f_1 and f_2 are PIM-functions such that

$$\hat{f}_1 = 1+\sum_{r=1}^{\infty} 2^{-k}(r+1)^k t^r, \qquad \hat{f}_2 = 1+\sum_{r=1}^{\infty} 2^{k}(r+1)^{-k} t^r.$$

The coefficients of these power series satisfy the requirements of Proposition 3.7, and therefore

$$f_1(z) = \zeta_G(z) g_1(z), \qquad f_2(z) = \zeta_G(z) g_2(z),$$

where both $\tilde{g}_1(z)$ and $\tilde{g}_2(z)$ are absolutely convergent whenever $\operatorname{Re} z > \nu$, for some $\nu < \delta$. Therefore Theorem 3.1 and Lemma 3.8 imply that f_i has the asymptotic mean-value

$$\tilde{g}_i(\delta) = \prod_{p \in P} \left(\sum_{r=0}^{\infty} g_i(p^r)|p|^{-r\delta}\right) \qquad [i=1, 2].$$

The stated formulae for $m((d/d_*)^k)$ and $m((d_*/d)^k)$ then follow from the fact that $\hat{g}_i=(1-t)\hat{f}_i$. □

Now consider the function β such that $\beta(1)=1$ and

$$\beta(p_1^{\alpha_1} p_2^{\alpha_2} \ldots p_r^{\alpha_r}) = \alpha_1 \alpha_2 \ldots \alpha_r$$

for distinct $p_i \in P$, $\alpha_i > 0$. By using an argument similar to the above, and the formula

$$\beta(z) = \zeta_G(z)\zeta_G(2z)\zeta_G(3z)/\zeta_G(6z)$$

given in Theorem 2.6.1, one obtains:

4.4. Proposition. *For $k=1, 2, \ldots$, the function β has the asymptotic k^{th} moment*

$$m(\beta^k) = \prod_{p \in P} \left\{1+\sum_{r=2}^{\infty} [r^k - (r-1)^k] |p|^{-r\delta}\right\}.$$

In particular, β has the asymptotic mean-value

$$m(\beta) = \zeta_G(2\delta)\zeta_G(3\delta)/\zeta_G(6\delta).$$

For the semigroup G_Z of positive integers, this proposition implies that β has the surprisingly small asymptotic mean-value $\zeta(2)\zeta(3)/\zeta(6)=1.94359\ldots$.

4.5. Proposition. *Let G_k denote the set of all k-free elements in G. Then G has* the *asymptotic density*

$$d(G_k) = 1/\zeta_G(k\delta)$$

in G.

Proof. By Theorem 2.6.1, the characteristic function q_k of G_k satisfies the formula $q_k(z)=\zeta_G(z)\mu(kz)$. In the same way as in the proof of Proposition 4.1, one sees that $\tilde{\mu}(kz)$ is absolutely convergent for $\operatorname{Re} z>\delta/k$ and hence that q_k has the asymptotic mean-value

$$\tilde{\mu}(k\delta) = 1/\zeta_G(k\delta). \qquad \square$$

For the special case of the semigroup G_Z, this proposition implies that the set of all square-free positive integers possesses the asymptotic density $1/\zeta(2)=6/\pi^2$.

§ 5. Asymptotic formulae with error estimates

If $f\in\operatorname{Dir}(G)$ is a given arithmetical function, it is sometimes interesting or useful for other purposes to study its approximate average value on $G[x]$ more carefully than in the preceding sections. Thus one might seek to determine the order of magnitude or even a more precise asymptotic formula for the 'error' $N(f, x)-V(x)$ as $x\to\infty$, given that $N(f, x)\sim V(x)$ as $x\to\infty$. In general, this problem increases in difficulty with the degree of precision required. One purpose of this section is to prove a theorem that provides a reasonable amount of information in various specific cases. However, there is no attempt at completeness in this direction, and the question of obtaining "best possible" estimates or asymptotic formulae is not considered.

The basic theorem to be considered here arises by slightly sharpening the assumption made about the arithmetical function g in Theorem 3.1:

5.1. Theorem. *Let $f,g\in\operatorname{Dir}(G)$ be functions such that*

$$f(z) = [\zeta_G(mz)]^k g(z),$$

where m and k are positive integers. Suppose that

$$N(|g|, x) = \sum_{|a|\leq x} |g(a)| = O(x^\nu),$$

where $\nu<\delta/m$. *Then there exist constants* b_i *such that*

$$N(f,x) = Ax^{\delta/m}\sum_{i=1}^{k} b_i(\log x)^{k-i}$$

$$+\begin{cases} O(x^{\eta/m}) & \text{if } k=1 \text{ and } \nu<\eta/m,\\ O\big(x^{(\delta-(\delta-\eta)/k)/m}(\log x)^{k-2}\big) & \text{if } k>1 \text{ and } \nu<(\delta-(\delta-\eta)/k)/m,\\ O(x^{\nu+\varepsilon}) & \text{if } \nu\geqq(\delta-(\delta-\eta)/k)/m,\end{cases}$$

where $\varepsilon>0$ *is arbitrary. In particular*

$$b_1 = \frac{1}{(k-1)!}\left(\frac{\delta A}{m}\right)^{k-1}\tilde{g}(\delta/m),$$

$$b_2 = \frac{1}{(k-2)!}\left(\frac{\delta A}{m}\right)^{k-1}\tilde{g}'(\delta/m)+\frac{c_2\tilde{g}(\delta/m)}{m^{k-2}}$$

where c_2 *is the constant occurring in Proposition* 3.2.

Proof. In view of Proposition 3.2, if

$$F(z) = [\zeta_G(mz)]^k = d_k(mz)$$

then, as in the proof of Theorem 3.1,

$$N(F,x) = Ax^{\delta/m}\sum_{i=1}^{k} c_i(m^{-1}\log x)^{k-i}+O\big(x^{(\delta-(\delta-\eta)/k)/m}(\log x)^{k-2}\big).$$

Therefore the present theorem will follow from:

5.2. Lemma. *Let* f, F *and* $g\in\text{Dir}(G)$ *be arbitrary functions such that* $f(z)=$ $=F(z)g(z)$, *Suppose that*

$$N(F,x) = Bx^{\alpha}\sum_{i=0}^{r} a_i(\log x)^{r-i}+O\big(x^{\beta}(\log x)^{s}\big)$$

for certain constants B, a_i, α, β *with* $0\leqq\beta<\alpha$, *and* r, s, *where* r *and* s *are non-negative integers. Suppose also that* $N(|g|,x)=O(x^{\nu})$, *where* $\nu<\alpha$. *Then there exist constants* a_i' *such that*

$$N(f,x) = Bx^{\alpha}\sum_{i=0}^{r} a_i'(\log x)^{r-i}$$

$$+\begin{cases} O\big(x^{\beta}(\log x)^{s}\big) & \text{if } \beta>\nu,\\ O(x^{\nu+\varepsilon}) & \text{if } \beta\leqq\nu,\end{cases}$$

where $\varepsilon>0$ *is arbitrary. In particular,*

$$a_0' = a_0\tilde{g}(\alpha),\qquad a_1' = ra_0\tilde{g}'(\alpha)+a_1\tilde{g}(\alpha).$$

Proof. Since $f(z)=F(z)g(z)$

$$N(f,x) = \sum_{|bc|\leqq x} F(b)g(c) = \sum_{|c|\leqq x} g(c) \sum_{|b|\leqq x/|c|} F(b)$$

$$= \sum_{|c|\leqq x} g(c)B(x/|c|)^{\alpha} \sum_{i=0}^{r} a_i(\log(x/|c|))^{r-i}$$

$$+ \sum_{|c|\leqq x} g(c)\mathrm{O}((x/|c|)^{\beta}(\log(x/|c|))^{s}).$$

If one expands terms of the form $(\log x-\log|c|)^{r-i}$ by means of the Binomial theorem, one is lead to consider sums of the form

$$\sum_{|c|\leqq x} g(c)\,|c|^{-\alpha}(-\log|c|)^m = \tilde{g}^{(m)}(\alpha) - \sum_{|c|>x} g(c)\,|c|^{-\alpha}(-\log|c|)^m,$$

by Lemma 2.7. Therefore

$$\sum_{|c|\leqq x} g(c)\,|c|^{-\alpha}(-\log|c|)^m = \tilde{g}^{(m)}(\alpha) + \mathrm{O}\Big(\sum_{|c|>x} |g(c)|\,|c|^{-\alpha}|c|^{\varepsilon}\Big)$$

$$= \tilde{g}^{(m)}(\alpha) + \mathrm{O}(x^{\nu-(\alpha-\varepsilon)}),$$

by Lemma 2.7 applied to $|g|$, where ε is any positive number such that $\alpha-\varepsilon>\nu$. (If $m=0$, one may take $\varepsilon=0$, by Lemma 2.7.)

Hence, after expanding and collecting terms,

$$N(f,x) = a_0Bx^{\alpha}(\log x)^r\,[\tilde{g}(\alpha)+\mathrm{O}(x^{\nu-\alpha})]$$

$$+Bx^{\alpha}\sum_{i=1}^{r}(\log x)^{r-i}[a_i'+\mathrm{O}(x^{\nu+\varepsilon-\alpha})]$$

$$+\mathrm{O}\Big(x^{\beta}(\log x)^s \sum_{|c|\leqq x} |g(c)|\,|c|^{-\beta}\Big),$$

where

$$a_1' = ra_0\tilde{g}'(\alpha)+a_1\tilde{g}(\alpha).$$

(Similar formulae could be given for the other coefficients, if desired.) In order to estimate the last term, one may note that partial summation gives

$$\sum_{|c|\leqq x} |g(c)|\,|c|^{-\beta} = \mathrm{O}(x^{\nu-\beta}) + \begin{cases} \mathrm{O}(x^{\nu-\beta}) & \text{if } \nu>\beta, \\ \mathrm{O}(\log x) & \text{if } \nu=\beta, \\ \mathrm{O}(1) & \text{if } \nu<\beta. \end{cases}$$

The lemma follows now. □

For later purposes, we note particularly that the above proof also yields:

5.3. Corollary. *Let $f, g \in \mathrm{Dir}\,(G)$ be functions such that $f(z) = \zeta_G(z) g(z)$, and suppose that*

$$N(|g|, x) = O(x^{\nu}),$$

where $\nu < \delta$. Then

$$N(f, x) = A\tilde{g}(\delta)x^{\delta} + \begin{cases} O(x^{\eta}) & \text{if } \nu < \eta, \\ O(x^{\eta} \log x) & \text{if } \nu = \eta, \\ O(x^{\nu}) & \text{if } \nu > \eta. \end{cases}$$

Now consider some applications of the above results to special arithmetical functions:

5.4. Proposition. *The unitary-divisor function satisfies*

$$N(d_*, x) = \frac{Ax^{\delta}}{\zeta_G(2\delta)} \left[\delta A \log x - \frac{2\delta A \zeta_G{}'(2\delta)}{\zeta_G(2\delta)} + 2\gamma_G - A\right]$$

$$+ \begin{cases} O(x^{(\delta+\eta)/2}) & \text{if } \eta > 0, \\ O(x^{\delta/2+\varepsilon}) & \text{if } \eta = 0,\ \varepsilon > 0 \text{ arbitrary.} \end{cases}$$

Proof. As noted in the proof of Proposition 4.2 (which gave the dominant term of the present statement),

$$d_*(z) = [\zeta_G(z)]^2 \mu(2z),$$

where, by Lemma 3.8 and Propositions 2.1 and 2.6, $\tilde{\mu}(2z)$ is absolutely convergent with sum equal to $1/\zeta_G(2z)$ when $\operatorname{Re} z > \frac{1}{2}\delta$. Further, if $g(z) = \mu(2z)$, then

$$N(|g|, x) \leqq \sum_{|a^2| \leqq x} 1 = N_G(x^{1/2}) = O(x^{\delta/2}).$$

Therefore Theorem 5.1 applies, with $m = 1$, $k = 2$, $\nu = \frac{1}{2}\delta$. In particular, over here,

$$b_2 = \delta A \tilde{\mu}'(2\delta) + (2\gamma_G - A)\tilde{\mu}(2\delta) = -\frac{\delta A \zeta_G{}'(2\delta)}{[\zeta_G(2\delta)]^2} + \frac{2\gamma_G - A}{\zeta_G(2\delta)}. \qquad \square$$

In a similar way, one may refine part of Proposition 4.1 so as to obtain an asymptotic expansion of $N(d^2, x)$, where d is the divisor function. This expansion involves a polynomial of degree 3 in $\log x$, and Theorem 5.1 allows one to determine the two highest coefficients. (With a little extra effort, the above results and methods would provide all the coefficients in terms of constants depending on $\zeta_G(z)$.)

By bearing in mind Corollary 5.3 and the equations

$$q_k(z) = \zeta_G(z)/\zeta_G(kz), \qquad \beta(z) = \zeta_G(z)\zeta_G(2z)\zeta_G(3z)/\zeta_G(6z),$$

similar arguments now lead to the following refinements of Propositions 4.4 and 4.5:

5.5. Proposition. *Let $Q_k(x)$ denote the total number of k-free elements of norm at most x in G. Then*

$$Q_k(x) = \frac{Ax^\delta}{\zeta_G(k\delta)} + \begin{cases} O(x^{\delta/k}) & \text{if } \eta < \delta/k, \\ O(x^{\delta/k}\log x) & \text{if } \eta = \delta/k, \\ O(x^\eta) & \text{if } \eta > \delta/k. \end{cases}$$ □

5.6. Proposition.

$$N(\beta, x) = \frac{A\zeta_G(2\delta)\zeta_G(3\delta)}{\zeta_G(6\delta)} x^\delta + \begin{cases} O(x^{\delta/2}) & \text{if } \eta < \frac{1}{2}\delta, \\ O(x^{\delta/2}\log x) & \text{if } \eta = \frac{1}{2}\delta, \\ O(x^\eta) & \text{if } \eta > \frac{1}{2}\delta. \end{cases}$$

In considering the function β, it is useful to note that, if

$$g_1(z) = \zeta_G(2z)\zeta_G(3z)\zeta_G(6z),$$

then

$$g_1(z) = \zeta_G(2z)h(z)$$

say, where the previous type of discussion together with Lemma 3.5 shows that $\tilde{h}(z)$ is absolutely convergent for Re $z > \frac{1}{3}\delta$. Then Theorem 3.1 shows that $N(g_1, x) = O(x^{\delta/2})$, and so, if

$$g(z) = \zeta_G(2z)\zeta_G(3z)/\zeta_G(6z),$$

then

$$|g(a)| \leqq g_1(a)$$

for $a \in G$; thus $N(|g|, x) = O(x^{\delta/2})$ also. □

Now consider the *Euler-type* function φ_w such that

$$\varphi_w(a) = \sum_{d|a} \mu(d)\,|a/d|^w \qquad (a \in G),$$

which was discussed in § 1 above. Also, for any real w, consider the **divisor-sum** function σ_w defined by

$$\sigma_w(a) = \sum_{d|a} |d|^w \qquad (a \in G).$$

Given any function $f \in \text{Dir}(G)$ and a complex number w, now formally *define*

$$f(z+w) = \sum_{a \in G} f(a)\,|a|^{-w} a^{-z}.$$

Then it follows from the definitions that

$$\varphi_w(z) = \mu(z)\zeta_G(z-w) = \zeta_G(z-w)/\zeta_G(z),$$
$$\sigma_w(z) = \zeta_G(z)\zeta_G(z-w).$$

In considering σ_w, special interest attaches to the case when w is positive, since then $\sigma_w(a)$ provides some kind of measure of the "average size" of a divisor of $a \in G$. For example, if $\sigma=\sigma_1$ then $\sigma_*(a)=\sigma(a)/d(a)$ is the average norm of a divisor of $a \in G$. Since the function φ_δ was of particular interest in § 1, the following discussion of φ_w and σ_w confines attention to the case when $w>0$.

5.7. Proposition. *For* $w>0$, *let*

$$\bar{\varphi}_w(a) = |a|^{-w}\varphi_w(a), \qquad \bar{\sigma}_w(a) = |a|^{-w}\sigma_w(a).$$

Then $\bar{\varphi}_w$ *and* $\bar{\sigma}_w$ *have the asymptotic mean-values* $1/\zeta_G(\delta+w)$ *and* $\zeta_G(\delta+w)$, *respectively.*

Proof. It follows from the definitions that

$$\bar{\varphi}_w(z) = \varphi_w(z+w) = \zeta_G(z)\mu(z+w),$$
$$\bar{\sigma}_w(z) = \sigma_w(z+w) = \zeta_G(z)\zeta_G(z+w).$$

Since $\tilde{\mu}(z+w)$ and $\zeta_G(z+w)$ are both absolutely convergent for Re $z>\delta-w$, the proposition follows from Theorem 3.1. □

This proposition shows in particular that, although $\sigma_w(a) \geqq |a|^w$, 'on average' $\sigma_w(a)$ is not much larger than $|a|^w$. For the semigroup $G_{\mathbf{Z}}$ of positive integers, it shows in particular that $\bar{\varphi}(n)=n^{-1}\varphi(n)$ and $\bar{\sigma}(n)= =n^{-1}\sigma(n)$ $(n=1, 2, \ldots)$ have the asymptotic mean-values $6\pi^{-2}$ and $\pi^2/6$, respectively.

Although it is possible to use Corollary 5.3 in order to sharpen Proposition 5.7, in the present case a direct discussion leads to even better estimates for $N(\varphi_w, x)$ and $N(\sigma_w, x)$:

5.8. Proposition. *For* $w>0$,

$$N(\varphi_w, x) = \frac{\delta A x^{\delta+w}}{(\delta+w)\zeta_G(\delta+w)} + R_1(x),$$

$$N(\sigma_w, x) = \frac{\delta A}{\delta+w}\zeta_G(\delta+w)x^{\delta+w} + R_2(x),$$

where

$$R_i(x) = \begin{cases} O(x^{\delta}) & \text{if } w < \delta - \eta, \\ O(x^{\delta}\log x) & \text{if } w = \delta - \eta, \\ O(x^{\eta+w}) & \text{if } w > \delta - \eta. \end{cases}$$

Proof. By definition of φ_w, Proposition 2.8 leads to

$$\begin{aligned} N(\varphi_w, x) &= \sum_{|bc| \leqq x} \mu(b)\,|c|^w = \sum_{|b| \leqq x} \mu(b) \sum_{|c| \leqq x/|b|} |c|^w \\ &= \sum_{|b| \leqq x} \mu(b) \left\{ \frac{\delta A}{\delta + w} (x/|b|)^{\delta + w} + O\big((x/|b|)^{\eta + w}\big) \right\} \\ &= \frac{\delta A}{\delta + w} x^{\delta + w} \Big\{ \tilde{\mu}(\delta + w) - \sum_{|b| > x} \mu(b)\,|b|^{-\delta - w} \Big\} \\ &\quad + O\Big(x^{\eta + w} \sum_{|b| \leqq x} |b|^{-\eta - w}\Big) \\ &= \frac{\delta A}{\delta + w} \tilde{\mu}(\delta + w)\, x^{\delta + w} + O\Big(x^{\delta + w} \sum_{|b| > x} |b|^{-\delta - w}\Big) \\ &\quad + O\Big(x^{\eta + w} \sum_{|b| \leqq x} |b|^{-\eta - w}\Big). \end{aligned}$$

The conclusion about φ_w then follows from Propositions 2.6 and 2.8. One may deal with σ_w similarly. □

It may be interesting to contrast the above conclusion in the case of the particular function φ_δ with the following one concerning the Euler function φ on G defined in Chapter 2, § 6. (Of course, for $G_{\mathbf{Z}}$, the two results coincide.)

5.9. Proposition. *The Euler function* φ *satisfies*

$$N(\varphi, x) = \frac{A^2 x^{2\delta}}{2\zeta_G(2\delta)} + \begin{cases} O(x^{\delta+\eta}) & \text{if } \eta > 0, \\ O(x^{\delta}\log x) & \text{if } \eta = 0. \end{cases}$$

Proof. By Lemma 2.6.2, $\varphi(a) = \sum_{d|a} \mu(d) N_G(|a/d|)$ for $a \in G$. Therefore the proposition is a consequence of the next lemma.

5.10. Lemma. *Let* $g \in \mathrm{Dir}\,(G)$ *denote a bounded function on* G, *and let*

$$\varphi^{(g)}(a) = \sum_{d|a} g(d) N_G(|a/d|) \quad \text{for } a \in G.$$

Then

$$N(\varphi^{(g)}, x) = \tfrac{1}{2}A^2\tilde{g}(2\delta)x^{2\delta} + \begin{cases} O(x^{\delta+\eta}) & \textit{if } \eta > 0, \\ O(x^{\delta}\log x) & \textit{if } \eta = 0. \end{cases}$$

Proof. Since g is bounded on G, $\tilde{g}(2\delta)$ is well-defined and the lemma may be proved by essentially the same type of argument as for Proposition 5.8. □

We note that, when $g=\zeta_G$, then

$$\varphi^{(g)}(a) = \sum_{d|a} N(|d|),$$

and on $G_{\mathbf{Z}}$ this coincides with the function σ. Hence it is of some interest to record:

5.11. Corollary. *Let σ' denote the above function $\varphi^{(g)}$ in the case when $g=\zeta_G$. Then*

$$N(\sigma', x) = \tfrac{1}{2}A^2\zeta_G(2\delta)x^{2\delta} + \begin{cases} O(x^{\delta+\eta}) & \textit{if } \eta > 0, \\ O(x^{\delta}\log x) & \textit{if } \eta = 0. \end{cases}$$

This corollary, and also Proposition 5.8, implies that the approximate average value of $\sigma(n)$ for a positive integer $n \leqq x$ is $\frac{1}{12}\pi^2 x$. Since $\sigma(n) \geqq n$, this conclusion appears slightly paradoxical and hence suggests that in some cases it may be desirable to modify the above concept of 'approximate average value'. One way of doing this over a general arithmetical semigroup G, or over a subset H of G, would be to say that an arithmetical function f has the **same average order** as the arithmetical function g over H whenever

$$N_H(f, x) \sim N_H(g, x) \quad \text{as } x \to \infty.$$

Then, for example, Proposition 5.7 implies that $\sigma(n)$ has the 'average order' $\frac{1}{6}\pi^2 n$ over $G_{\mathbf{Z}}$ — a form of words which may look more attractive than the previous one. Since, however, the previous terminology is often quite useful or natural nevertheless, and the present question is only terminological, the concept of 'average order' will not be stressed in the later discussions.

For certain semigroups G, like the familiar $G_{\mathbf{Z}}$, the above result about φ may be used to derive a conclusion about the asymptotic density of the coprime pairs of elements of G. However, in the general case, it appears to be easier to do this and prove a more general theorem, without direct reference to φ (or to a certain natural generalization of φ suggested by the problem).

In order to explain the question and concepts involved, first consider some subset H of G, and let $H^k[x]$ denote the set of all (ordered) k-tuples $(a_1, \ldots, a_k)$ of elements $a_i \in H$ such that $|a_i| \leqq x$, where $x > 0$. If W is any set of k-tuples of elements of G, call

$$\frac{\operatorname{card}(W \cap H^k[x])}{\operatorname{card}(H^k[x])}$$

the **relative density** of W in $H^k[x]$; if this ratio tends to a limit L as $x \to \infty$, L will be called the **asymptotic** relative density of W in H^k. In the "absolute" case $H = G$ to be considered below,

$$\operatorname{card} H^k[x] = [N_G(x)]^k,$$

and the word 'relative' will be omitted. (If W has asymptotic relative density L in H^k, it is sometimes suggestive to express this fact by saying that "the probability that a random element of H^k should belong to W is equal to L". However, it should be emphasized that, in general, asymptotic density does not give rise to a probability measure in the technical sense. For example, if H is infinite, one verifies that asymptotic density is not countably additive, by noting that H^k itself has asymptotic density 1 while every finite subset of H^k has asymptotic density 0.)

Now, just as for two elements, there is an obvious definition of the *greatest common divisor* (or *g.c.d.*) of any finite number of elements $a_1, \ldots, a_k \in G$. If confusion with the corresponding k-tuple seems unlikely, we shall follow ordinary arithmetical practice and denote this g.c.d. by $(a_1, \ldots, a_k)$ also. Then $a_1, \ldots, a_k$ are *coprime* (or *relatively prime*) if and only if $(a_1, \ldots, a_k) = 1$. It is now clear how to define the 'density' $\Delta_k(x)$ of the relatively prime k-tuples of elements of G in $G^k[x]$, and (assuming it exists) the 'asymptotic density' Δ_k of the relatively prime k-tuples of elements of G. (Of course, none of the above definitions depend on Axiom A.)

5.12. Theorem. *The asymptotic density Δ_k of the relatively prime k-tuples of elements of G exists and is equal to $1/\zeta_G(k\delta)$ $(k>1)$. More precisely,*

$$\Delta_k(x) = \frac{1}{\zeta_G(k\delta)} + \begin{cases} O(x^{\eta-\delta}) & \text{if } k > 2, \text{ or } k = 2 \text{ and } \eta > 0, \\ O(x^{-\delta}\log x) & \text{if } k = 2,\ \eta = 0. \end{cases}$$

Proof. First consider:

5.13. Lemma. *Let $F_k(x)$ denote the total number of relatively prime k-tuples $(a_1, \ldots, a_k)$ in $G^k[x]$. Then*

$$F_k(x) = \sum_{\substack{a \in G \\ |a| \leq x}} \mu(a)\,[N_G(x/|a|)]^k,$$

where μ is the Möbius function on G.

Proof. (This lemma does not require Axiom A.) Firstly, note that

$$[N_G(x)]^k = \text{card}\, G^k[x] = \sum_{|b| \leq x} \sum_{\substack{(a_1, \ldots, a_k) = b \\ |a_i| \leq x}} 1,$$

and that $(a_1, \ldots, a_k) = b$ if and only if b divides each a_i and $(a_1/b, \ldots, a_k/b) = 1$. Hence

$$\begin{aligned}[N_G(x)]^k &= \sum_{|b| \leq x} \sum_{\substack{(c_1 b, \ldots, c_k b) = b \\ |c_i| \leq x/|b|}} 1 \\ &= \sum_{|b| \leq x} \sum_{\substack{(c_1, \ldots, c_k) = 1 \\ |c_i| \leq x/|b|}} 1 \\ &= \sum_{|b| \leq x} F_k(x/|b|).\end{aligned}$$

It therefore follows from the inversion principle of Proposition 2.5.4 that

$$F_k(x) = \sum_{\substack{a \in G \\ |a| \leq x}} \mu(a)\,[N_G(x/|a|)]^k.$$

□

In order to prove the theorem, we may now deduce that

$$\begin{aligned}F_k(x) &= \sum_{|a| \leq x} \mu(a)\left[A(x/|a|)^\delta + O\left((x/|a|)^\eta\right)\right]^k \\ &= A^k x^{k\delta} \sum_{|a| \leq x} \mu(a)|a|^{-k\delta} + \sum_{r=2}^{k-1} O\Big(x^{r\delta + (k-r)\eta} \sum_{|a| \leq x} |a|^{-[r\delta + (k-r)\eta]}\Big) \\ &\quad + O\Big(x^{\delta + (k-1)\eta} \sum_{|a| \leq x} |a|^{-[\delta + (k-1)\eta]}\Big) + O\Big(x^{k\eta} \sum_{|a| \leq x} |a|^{-k\eta}\Big) \\ &= A^k x^{k\delta}[\tilde{\mu}(k\delta) + O(x^{\delta - k\delta})] + \sum_{r=2}^{k-1} O(x^{r\delta + (k-r)\eta}) \\ &\quad + \begin{cases} O(x^\delta \log x) & \text{if } \eta = 0, \\ O(x^{\delta + (k-1)\eta}) & \text{if } \eta > 0, \end{cases} \\ &\quad + \begin{cases} O(x^{k\eta}) & \text{if } \eta > \delta/k, \\ O(x^\delta \log x) & \text{if } \eta = \delta/k, \\ O(x^\delta) & \text{if } \eta < \delta/k, \end{cases}\end{aligned}$$

by Lemma 2.7 and Proposition 2.8. (If $k=2$, the sum $\sum_{r=2}^{k-1}$ does not occur, of course.) Now, for $k>2$, this estimate reduces to

$$F_k(x) = \frac{A^k x^{k\delta}}{\zeta_G(k\delta)} + O(x^{(k-1)\delta+\eta})$$

$$= \frac{A^k x^{k\delta}}{\zeta_G(k\delta)}[1+O(x^{\eta-\delta})].$$

On the other hand, the estimate for $F_2(x)$ reduces to

$$F_2(x) = \frac{A^2 x^{2\delta}}{\zeta_G(2\delta)} + \begin{cases} O(x^\delta \log x) & \text{if } \eta = 0, \\ O(x^{\delta+\eta}) & \text{if } \eta > 0. \end{cases}$$

Also,

$$[N_G(x)]^k = [Ax^\delta + O(x^\eta)]^{-k} = [A^k x^{k\delta} + O(x^{(k-1)\delta+\eta})]^{-1}$$

$$= A^{-k}x^{-k\delta}[1+O(x^{\eta-\delta})]^{-1} = A^{-k}x^{-k\delta}[1+O(x^{\eta-\delta})].$$

Since

$$\Delta_k(x) = F_k(x)\,[N_G(x)]^{-k},$$

the theorem therefore follows. □

As a special case, Theorem 5.12 yields the conclusion that the set of all coprime pairs of positive integers has the asymptotic density $6\pi^{-2}$.

A slight generalization. Although the general results of §§ 3 and 5 (particularly Theorems 3.1, 5.1, Lemmas 3.5, 3.6 and 5.2, and Corollary 5.3) are adequate for virtually all applications desired in this book, it may be worth mentioning some slightly generalized forms of the results, that may be proved in essentially the same way. For example, the extended results are convenient in dealing with parts of the next chapter, although they are not essential for this purpose.

The generalizations in question arise by replacing functions such as $f,g \in \text{Dir}(G)$ by functions $f^*,g^* \in \text{Dir}(|G|)$ satisfying corresponding relations in $\text{Dir}(|G|)$, without assuming anything about possible functions $f,g \in \text{Dir}(G)$ such that $f^* = \tilde{f}$ and $g^* = \tilde{g}$, say. For example, we have:

3.1.* Theorem. *Let* $f^*,g^* \in \text{Dir}(|G|)$ *be functions such that in* $\text{Dir}(|G|)$

$$f^*(z) = [\tilde{\zeta}_G(mz)]^k g^*(z),$$

where m and k are positive integers. Suppose that $g^*(z)$ is absolutely convergent whenever Re $z>\nu$, *for some* $\nu<\delta/m$. *Then as* $x\to\infty$

$$N(f^*, x) = \frac{A}{(k-1)!}\left(\frac{\delta A}{m}\right)^{k-1} [g^*(\delta/m)+o(1)]x^{\delta/m}(\log x)^{k-1}.$$

An asterisk * will be used to indicate the extended form of any of the above particular results, like Theorem 3.1, when as in the next chapter it may be convenient to refer to it.

Selected bibliography for Chapter 4

Sections 1 *and* 2: Bateman and Diamond [1], R. S. Hall [1], Knopfmacher [6, 9], Landau [1, 9], Wegmann [1].

Sections 3—5: E. Cohen [1, 3], Hardy and Wright [1], Horadam [3,4, 9], Knopfmacher [6, 9], Knopfmacher and Ridley [1], Landau [1, 7], J. E. Nyman [1], Rieger [1], Wegmann [1].

CHAPTER 5

ASYMPTOTIC ENUMERATION, AND FURTHER "STATISTICAL" PROPERTIES OF ARITHMETICAL FUNCTIONS

This chapter contains further results on the asymptotic properties of arithmetical functions. Firstly, § 1 is concerned largely with questions of asymptotic enumeration regarding the isomorphism classes in various specific arithmetical categories. Certain of the theorems proved then establish the validity of Axiom A for the particular categories considered. Thereafter, some further asymptotic "statistical" properties of arithmetical functions are discussed. The conclusions have applications to both asymptotic enumeration and to more "purely" number-theoretical problems.

§ 1. Asymptotic enumeration in certain categories

In this section, some of the results of the preceding chapter will be applied to the investigation of asymptotic enumeration questions concerning the isomorphism classes of objects in certain of the categories discussed in Chapter 1. In particular, we establish the validity of Axiom A for a variety of natural arithmetical categories.

Firstly, let $\mathfrak{F}$ denote the category of all modules of finite cardinal over the domain D of all algebraic integers in a given algebraic number field K. Recall from § 1 of Chapter 4 that the semigroup G_K of all non-zero integral ideals in K satisfies Axiom A with

$$N_K(x) = A_K x + O(x^{\eta_0}),$$

where A_K is a positive constant depending on K and $\eta_0 = 1-2(1+[K:\mathbf{Q}])^{-1}$.

1.1. Theorem. *The category* $\mathfrak{F}$ *satisfies Axiom* A *with*

$$N_{\mathfrak{F}}(x) = \left[A_K \prod_{r=2}^{\infty} \zeta_K(r)\right] x + \begin{cases} O(x^{1/2}) & \text{if } [K:\mathbf{Q}] < 3, \\ O(x^{1/2}\log x) & \text{if } [K:\mathbf{Q}] = 3, \\ O(x^{\eta_0}) & \text{if } [K:\mathbf{Q}] > 3. \end{cases}$$

1.2. Corollary (Erdös and Szekeres). *The category $\mathscr{A}$ of all finite abelian groups satisfies Axiom* A *with*

$$N_{\mathscr{A}}(x) = \left[\prod_{r=2}^{\infty} \zeta(r)\right] x + O(x^{1/2}).$$

1.3. Corollary. *Let $a_D(n)$ denote the total number of isomorphism classes of modules of cardinal n in $\mathfrak{F}$. Then, as an arithmetical function of n, $a_D(n)$ has the asymptotic mean-value*

$$A_K \prod_{r=2}^{\infty} \zeta_K(r).$$

In particular, the total number $a(n)$ of isomorphism classes of finite abelian groups of order n has the asymptotic mean-value

$$\prod_{r=2}^{\infty} \zeta(r) = 2.29485 \dots .$$

We shall deduce Theorem 1.1 from some more general results which in particular imply that $\mathfrak{F}$ has infinitely many subcategories that also satisfy Axiom A. Although results about the category $\mathscr{A}$ will be deduced from ones about $\mathfrak{F}$, the reader who so desires should have no difficulty in reading off arguments directly applicable to $\mathscr{A}$ from those given for $\mathfrak{F}$.

Now consider a finite or infinite sequence

$$\langle k \rangle = (k_1, k_2, \dots)$$

of positive integers k_i such that $k_i < k_{i+1}$ $(i=1, 2, \dots)$. Let $\mathfrak{F}^{\langle k \rangle}$ denote the subcategory of all modules M in $\mathfrak{F}$ such that every indecomposable direct summand of M is isomorphic to a cyclic module of the form D/P^r, where P is a prime ideal in D and $r = k_i$ for some i. From the discussion of $\mathfrak{F}$ in Chapter 1, it is easy to see that $\mathfrak{F}^{\langle k \rangle}$ forms an arithmetical category. Further, its zeta function (enumerating function) is

$$\begin{aligned} \zeta_{\mathfrak{F}^{\langle k \rangle}}(z) &= \prod \{(1-|P|^{-rz})^{-1} \text{: prime ideals } P \text{ in } D, r \in \langle k \rangle\} \\ &= \prod_{i \geqq 1} \zeta_K(k_i z); \end{aligned}$$

compare the discussion of $\mathfrak{F}$ in Chapter 3.

In the special case when $\langle k \rangle$ consists of exactly *one* positive integer k_1, it follows that

$$\zeta_{\mathfrak{F}^{\langle k \rangle}}(z) = \zeta_K(k_1 z).$$

Hence Theorem 4.5.1 or an obvious direct argument shows that

$$N_{\mathfrak{F}^{\langle k\rangle}}(z) = A_K x^{1/k_1} + O(x^{\eta_0/k_1}).$$

Hence in this case it is obvious that $\mathfrak{F}^{\langle k\rangle}$ satisfies Axiom A.

A simple extension of the preceding case occurs when $\langle k\rangle$ is merely a finite sequence $(k_1, \ldots, k_r)$. Here

$$\zeta_{\mathfrak{F}^{\langle k\rangle}}(z) = \zeta_K(k_1 z) g^*(z),$$

where

$$g^*(z) = \zeta_K(k_2 z) h^*(z), \qquad h^*(z) = \prod_{i=3}^{r} \zeta_K(k_i z).$$

(An empty product may be taken to be 1.) Since G_K satisfies Axiom A with $\delta = 1$, Lemma 4.3.5* (see the final paragraph of Chapter 4, § 5) implies that $h^*(z)$ is absolutely convergent for $\operatorname{Re} z > 1/k_3$. Therefore Theorem 4.3.1* implies that

$$N(|g^*|, x) = N(g^*, x) = O(x^{1/k_2}).$$

Hence Theorem 4.5.1* gives

$$N_{\mathfrak{F}^{\langle k\rangle}}(x) = \left[A_K \prod_{i=2}^{r} \zeta_K(k_i/k_1)\right] x^{1/k_1}$$

$$+ \begin{cases} O(x^{\eta_0/k_1}) & \text{if } k_2 > k_1/\eta_0, \\ O(x^{\varepsilon + k_2^{-1}}) & \text{if } k_2 \leq k_1/\eta_0. \end{cases}$$

Thus $\mathfrak{F}^{\langle k\rangle}$ satisfies Axiom A in this case also. (Note that $g^* = \tilde{g}$, $h^* = \tilde{h}$ for suitable $g, h \in \operatorname{Dir}(G_K)$; hence one could avoid extended theorems like 4.3.1* and so on, if preferred.)

The general case will follow in the same way as above, as soon as the question of convergence has been dealt with:

1.4. Lemma. *For any finite or infinite sequence $\langle k\rangle$ as above, the series $\zeta_{\mathfrak{F}^{\langle k\rangle}}(z)$ and the product $\prod_{i\geq 1} \zeta_K(k_i z)$ converge to the same value, and hence define an analytic function of z, whenever* $\operatorname{Re} z > 1/k_1$.

Proof. Consider the double series

$$\sum_{n=2}^{\infty} \sum_{r=k}^{\infty} n^{\varepsilon - rs},$$

where $\varepsilon > 0$, $s > 0$ and k is a positive integer. This can be written as

$$\sum_{n=2}^{\infty} \frac{n^{\varepsilon - ks}}{1 - n^{-s}},$$

which has the same convergence behaviour as

$$\int_2^\infty \frac{y^{\varepsilon-ks}}{1-y^{-s}}\,dy.$$

By substituting $u=y^s$, one sees that this integral converges when

$$\int_{2^s}^\infty u^{((1+\varepsilon)/s)-k-1}\,du$$

converges, i.e. for $s>(1+\varepsilon)/k$. Therefore the given double series, and hence also $\sum_{r=k}^\infty \sum_{n=2}^\infty n^{\varepsilon-rs}$, converges for $s>(1+\varepsilon)/k$.

Now consider

$$\sum_{i\geqq 1} [\zeta_K(k_i s)-1] = \sum_{i\geqq 1} \sum_{n=2}^\infty K(n) n^{-k_i s}$$

for $s>1/k_1$, where $K(n)$ denotes the total number of ideals of norm n in G_K. In the next section it will be shown that $K(n)=O(n^\varepsilon)$ for every $\varepsilon>0$. Hence, for every $\varepsilon>0$, the last double series is dominated by $\sum_{r=k_1}^\infty \sum_{n=2}^\infty n^{\varepsilon-rs}$. Since ε is arbitrary, the previous conclusion therefore implies that $\sum_{i\geqq 1}[\zeta_K(k_i s)-1]$ converges for $s>1/k_1$. Hence the product $\prod_{i\geqq 1}\zeta_K(k_i z)$ converges absolutely when $\operatorname{Re} z>1/k_1$.

Now arrange the prime ideals of G_K in some convenient sequence $P_1, P_2, \ldots$. Then, for real $s>1/k_1$ and $\langle k\rangle$ infinite,

$$\begin{aligned}\prod_{i\geqq 1} \zeta_K(k_i s) &= \prod_{i\geqq 1} \prod_{j=1}^\infty (1-|P_j|^{-k_i s})^{-1} \\ &= \lim_{M\to\infty}\left\{\lim_{N\to\infty} \prod_{i=1}^M \prod_{j=1}^N (1-|P_j|^{-k_i s})^{-1}\right\} \\ &= \lim_{M,N\to\infty}\left\{\prod_{i=1}^M \prod_{j=1}^N (1-|P_j|^{-k_i s})^{-1}\right\},\end{aligned}$$

a *double* limit in the sense of the following standard analytic lemma, which is quoted for the reader's convenience.

1.5. Lemma. *Given a double sequence (X_{mn}) of complex numbers X_{mn} $(m,n=1,2,\ldots)$, define*

$$X = \lim_{m,n\to\infty} X_{mn}$$

if and only if for every $\varepsilon>0$ there exists $n_0(\varepsilon)$ such that

$$|X_{mn}-X| < \varepsilon \quad \textit{whenever} \quad m,n \geqq n_0(\varepsilon)$$

Then suppose that each X_{mn} is real and that, for each m, X_{mn} increases (decreases) monotonically with n, while, for each n, X_{mn} increases (decreases) monotonically with m. In that case, if one of the limits

$$\lim_{m,n\to\infty} X_{mn}, \qquad \lim_{m\to\infty}\{\lim_{n\to\infty} X_{mn}\}, \qquad \lim_{n\to\infty}\{\lim_{m\to\infty} X_{mn}\}$$

exists, then all of them exist and coincide. □

It follows now that, for real $s>1/k_1$,

$$\prod_{i\geq 1} \zeta_K(k_i s) = \lim_{M\to\infty}\left\{\prod_{i=1}^{M}\prod_{j=1}^{M}(1-|P_j|^{-k_i s})^{-1}\right\}$$

$$= \lim_{M\to\infty}\left\{\prod_{n=1}^{M}(1+Y_n(s))\right\} = \prod_{n=1}^{\infty}(1+Y_n(s))$$

for certain positive numbers $Y_n(s)$ such that $\sum_{n=1}^{\infty} Y_n(s)$ converges and dominates the series $\sum_{p\in \boldsymbol{P}}|p|^{-s}$, where here $\boldsymbol{P}$ denotes the set of all $\mathfrak{F}^{\langle k\rangle}$-primes. By Proposition 4.2.1, it then follows that the series $\zeta_{\mathfrak{F}^{\langle k\rangle}}(z)$ is absolutely convergent for $\operatorname{Re} z>1/k_1$.

Finally, we note that, for $\operatorname{Re} z>1/k_1$, the Euler product formula gives

$$\zeta_{\mathfrak{F}^{\langle k\rangle}}(z) = \prod_{p\in\boldsymbol{P}}(1+|p|^{-z}+|p|^{-2z}+\ldots).$$

Since the right-hand product is absolutely convergent, so is the series

$$L(z) = \sum_{p\in\boldsymbol{P}}\log(1+|p|^{-z}+|p|^{-2z}+\ldots),$$

where log denotes the principal value of the logarithm. Then

$$\zeta_{\mathfrak{F}^{\langle k\rangle}}(z) = \exp L(z).$$

Since absolutely convergent series may be bracketed arbitrarily into sub-series so as to give a series of sub-series with the same sum, the earlier discussion of the primes of $\mathfrak{F}^{\langle k\rangle}$ shows that

$$L(z) = \sum_{i\geq 1} L_i(z)$$

where $\exp L_i(z)=\zeta_K(k_i z)$. Hence Lemma 1.4 follows. □

1.6. Theorem. *For any finite or infinite sequence $\langle k\rangle$, the category $\mathfrak{F}^{\langle k\rangle}$ satisfies Axiom* A *with*

$$N_{\mathfrak{F}^{\langle k\rangle}}(x) = \left[A_K \prod_{i\geq 2}\zeta_K(k_i/k_1)\right] x^{1/k_1} + O(x^{\eta_K}),$$

where

$$\eta_k = \begin{cases} \eta_0/k_1 & \textit{if } [K:\mathrm{Q}] > (k_2+k_1)/(k_2-k_1), \\ \varepsilon + k_2^{-1} & \textit{if } [K:\mathrm{Q}] \leqq (k_2+k_1)/(k_2-k_1) \ \ (\varepsilon > 0 \text{ arbitrary}). \end{cases}$$

If $k_1=1$, *the error term* $\mathrm{O}(x^{\eta_k})$ *may be written as*

$$\begin{cases} \mathrm{O}(x^{\eta_0}) & \textit{if } [K:\mathrm{Q}] > (k_2+1)/(k_2-1), \\ \mathrm{O}(x^{\eta_0} \log x) & \textit{if } [K:\mathrm{Q}] = (k_2+1)/(k_2-1), \\ \mathrm{O}(x^{1/k_2}) & \textit{if } [K:\mathrm{Q}] < (k_2+1)/(k_2-1), \end{cases}$$

and in this case the total number of isomorphism classes of modules of cardinal n *in* $\mathfrak{F}^{\langle k \rangle}$ *has the asymptotic mean-value*

$$A_K \prod_{i \geqq 2} \zeta_K(k_i).$$

Proof. The argument is now directly similar to that used above for the case when $\langle k \rangle$ is finite. For,

$$\zeta_{\mathfrak{F}^{\langle k \rangle}}(z) = \zeta_K(k_1 z) g^*(z),$$

where

$$g^*(z) = \zeta_K(k_2 z) h^*(z), \qquad h^*(z) = \prod_{i \geqq 3} \zeta_K(k_i z),$$

and now Lemma 1.4 implies that $h^*(z)$ is absolutely convergent for $\mathrm{Re}\, z > 1/k_3$ since

$$h^*(z) = \zeta_{\mathfrak{F}^{\langle k' \rangle}}(z),$$

where $\langle k' \rangle = (k_3, k_4, \ldots)$. This leads to the first conclusion, and the second one follows with the aid of Corollary 4.5.3*. □

Next consider the category $\mathfrak{S}$ of all semisimple finite rings. It was noted in Chapter 3 that the subcategory $\mathfrak{S}_c$ of all commutative rings in $\mathfrak{S}$ is arithmetically equivalent to the category $\mathscr{A}$ of finite abelian groups. Hence, for every finite or infinite sequence $\langle k \rangle$ as above, there is a subcategory $\mathfrak{S}_c^{\langle k \rangle}$ of $\mathfrak{S}_c$ with exactly the same enumeration properties as the subcategory $\mathscr{A}^{\langle k \rangle}$ of $\mathscr{A}$. We shall now show that the entire category $\mathfrak{S}$ also satisfies Axiom A.

1.7. Theorem. *The category* $\mathfrak{S}$ *satisfies Axiom* A *with*

$$N_{\mathfrak{S}}(x) = \Big[\prod_{rm^2 > 1} \zeta(rm^2)\Big]\, x + \mathrm{O}(x^{1/2}).$$

In particular, the total number $S(n)$ of isomorphism classes of rings of cardinal n in $\mathfrak{S}$ has the asymptotic mean-value

$$\prod_{rm^2>1} \zeta(rm^2) = 2.49961 \ldots .$$

Proof. The discussion of $\mathfrak{S}$ in Chapter 3 shows that in Dir $(G_{\mathbf{Z}})$

$$\zeta_{\mathfrak{S}}(z) = \zeta(z)g(z),$$

where

$$g(z) = \zeta(2z)h(z),$$

$$h(z) = \left\{\prod_{n=3}^{\infty} \zeta(nz)\right\}\left\{\prod_{r=1}^{\infty} \prod_{m=2}^{\infty} \zeta(rm^2 z)\right\}.$$

Now let

$$g_k(z) = \left\{\prod_{n=k}^{\infty} \zeta(nz)\right\}\left\{\prod_{r=1}^{\infty} \prod_{m=2}^{\infty} \zeta(rm^2 z)\right\}.$$

1.8. Lemma. *The series and product for $g_k(z)$ converge to the same value whenever* Re $z>\max\{\frac{1}{4}, k^{-1}\}$.

Proof. We have

$$g_k(z) = \left\{\prod_{n=k}^{\infty} \zeta(nz)\right\} F(z),$$

where

$$F(z) = \prod_{r=1}^{\infty} \prod_{m=2}^{\infty} \zeta(rm^2 z).$$

Therefore, by Lemmas 1.4 and 4.3.5, the required conclusion will follow as soon as it has been shown that the series and product for $F(z)$ converge to the same value whenever Re $z>\frac{1}{4}$.

For this purpose, first consider a real $s>\frac{1}{4}$ and

$$\sum_{m=2}^{\infty} \sum_{n=2}^{\infty} \sum_{r=1}^{\infty} n^{-rm^2 s} = \sum_{m=2}^{\infty} \sum_{n=2}^{\infty} (n^{m^2 s}-1)^{-1}.$$

Now, for $m \geqq 2, n \geqq 2$ and $s>\frac{1}{4}$,

$$n^{m^2 s} \geqq 2^{4s} > 2$$

and so

$$n^{m^2 s} - 1 > \tfrac{1}{2} n^{m^2 s}.$$

Therefore the last double series is dominated by

$$2 \sum_{m=2}^{\infty} \sum_{n=2}^{\infty} n^{-m^2 s} = 2 \sum_{m=2}^{\infty} [\zeta(m^2 s) - 1].$$

The latter series converges, by the proof of Lemma 1.4; hence

$$\sum_{m=2}^{\infty}\sum_{n=2}^{\infty}\sum_{r=1}^{\infty} n^{-rm^2 z}$$

is absolutely convergent for Re $z>\frac{1}{4}$, and one may interchange the orders of summation as convenient.

It follows now that

$$\sum_{r=1}^{\infty}\sum_{m=2}^{\infty}[\zeta(rm^2 z)-1] = \sum_{r=1}^{\infty}\sum_{m=2}^{\infty}\sum_{n=2}^{\infty} n^{-rm^2 z}$$

is absolutely convergent for Re $z>\frac{1}{4}$. Then, in essentially the same way as in dealing with $\prod_{i\geqq 1}\zeta_K(k_i s)$ in the proof of Lemma 1.4, one sees that, for real $s>\frac{1}{4}$,

$$\sum_{r=1}^{\infty}\sum_{m=2}^{\infty}[\zeta(rm^2 s)-1] = \sum_{n=1}^{\infty} W_n(s),$$

where each $W_n(s)$ is a finite sum of terms of the form $\zeta(rm^2 s)-1$. Since the latter terms are all positive, one may remove brackets and deduce that

$$\sum_{r=1}^{\infty}\sum_{m=2}^{\infty}[\zeta(rm^2 s)-1] = \sum_{i=1}^{\infty}[\zeta(k_i s)-1]$$

for a suitable arrangement $k_1\leqq k_2\leqq k_3\leqq\ldots$ of the values rm^2 ($r=1,2,\ldots$; $m=2,3,\ldots$). It then follows that the product $\prod_{i=1}^{\infty}\zeta(k_i z)$ is absolutely convergent for Re $z>\frac{1}{4}$.

In order to complete the proof, one may note that $F(z)$ can be interpreted as the enumerating function $\zeta_{\mathfrak{S}'}(z)$ of a certain subcategory $\mathfrak{S}'$ of $\mathfrak{S}$, and, using the Euler product formula for $\zeta(k_i z)$, continue with the discussion of $\prod_{i\geqq 1}\zeta(k_i z)$ in essentially the same way as for $\prod_i \zeta_K(k_i z)$ in Lemma 1.4. □

Theorem 1.7 may now be deduced by essentially the same argument as used earlier. □

It was mentioned in Chapter 3 that, just as the category $\mathscr{A}$ is the simplest member $\mathfrak{F}_{\mathbf{Z}}$ of the family of categories $\mathfrak{F}_D$, where D ranges over the domains of all algebraic integers in the various non-isomorphic algebraic number fields K, so too is $\mathfrak{S}$ the simplest case $\mathfrak{S}_{\mathbf{Z}}$ of a similar family $\mathfrak{S}_D$ of categories of finite algebras of a certain kind where D ranges over the same set of domains. Further, each category $\mathfrak{S}_D$ is also an arithmetical category satisfying Axiom A; for more details, see the author [7].

Next we shall show that not only do the functions $a(n)$ and $S(n)$ above possess asymptotic mean-values, but they also have asymptotic k^{th} moments for every $k=1, 2, \ldots$. A similar fact will be proved for the category $\mathfrak{F}=\mathfrak{F}_D$. (For the category $\mathfrak{S}_D$ just mentioned, see the author and Ridley [1].)

1.9. Theorem. *Let $a(I)$ denote the total number of isomorphism classes of modules M in $\mathfrak{F}$ such that $\Phi(M)=I\in G_K$, where $\Phi: G_{\mathfrak{F}}\to G_K$ is the homomorphism that sends the class of D/P^r to P^r (P any prime ideal in D, any $r\geqq 1$). Then the function $a(I)$ of $I\in G_K$ possesses an asymptotic k^{th} moment for every $k=1, 2, \ldots$.*

Proof. The case $k=1$ of this theorem follows from Theorem 1.1, since

$$N_{\mathfrak{F}}(x) = \sum_{n\leqq x} a_D(n) = \sum_{\substack{I\in G_K\\ |I|\leqq x}} a(I).$$

(Note that the asymptotic mean-value of $a_D(n)$ differs from that of $a(I)$ by the factor A_K.) Now, the discussion in Chapter 3, § 2, shows that $a(I)$ is a PIM-function on G_K, and, for every prime ideal P in G_K, $a(P^r)=p(r)$ — the classical partition function. Therefore, if f denotes the pointwise k^{th} power of $a(I)$, then

$$\hat{f} = 1+\sum_{r=1}^{\infty} [p(r)]^k t^r.$$

Since $p(1)=1$ and the next lemma shows that $p(r)\leqq e^{B\sqrt{r}}$ for some constant $B>0$, it follows that

$$[p(r)]^k = O(c^r) \quad \text{as} \quad r\to\infty,$$

for every $c>1$. Therefore Proposition 4.3.7 implies that

$$f(z) = \zeta_{G_K}(z)g(z),$$

where $\tilde{g}(z)$ is absolutely convergent whenever $\operatorname{Re} z>\nu$, for some $\nu<1$. Therefore Theorem 1.9 follows from Theorem 4.3.1. □

Before considering the lemma mentioned, we also state:

1.10. Corollary. *Let $a(n)$ denote the total number of isomorphism classes of finite abelian groups of order n. Then $a(n)$ possesses an asymptotic k^{th} moment for every $k=1, 2, \ldots$.*

The next lemma, Corollary 1.12 and Lemma 1.14 below are all consequences of more general (and sharper) results to be proved in Chapter 7, and a similar comment applies to the intermediate estimates given below for certain special zeta or generating functions. However, it may perhaps be interesting to provide direct (and more elementary) proofs of the particular conclusions needed at this stage. Further, since *in this book* functions like $p(r)$ first arise from the consideration of arithmetical categories, the discussion will be phrased in terms of certain categories. Although this procedure is certainly not essential, it has the convenient additional feature that it allows direct use to be made of some earlier results about arithmetical semigroups.

1.11. Lemma. *There exists a positive constant B such that*

$$p(r) \leqq e^{B\sqrt{r}} \qquad (r = 1, 2, \ldots).$$

Proof. Consider any additive arithmetical category $\mathfrak{C}$ which (like the category $\mathscr{A}(p)$, say) has the property that its primes may be arranged in a sequence $P_1, P_2, \ldots$ with $\partial(P_n)=n$ $(n=1, 2, \ldots)$. Then regard $\mathfrak{C}$ as an ordinary arithmetical category by associating with $\mathfrak{C}$ the particular norm $|a|=\exp[\partial(a)]$. In that case,

$$\zeta_{\mathfrak{C}}(z) = \sum_{n=0}^{\infty} p(n)e^{-nz},$$

and the Euler product formula is

$$\zeta_{\mathfrak{C}}(z) = \prod_{r=1}^{\infty} (1-e^{-rz})^{-1}.$$

Since $\sum_{r=1}^{\infty} e^{-rz}$ is absolutely convergent for $\operatorname{Re} z>0$, it follows from Proposition 4.2.1 that $\zeta_{\mathfrak{C}}(z)$ is absolutely convergent and the Euler product formula is valid analytically, for $\operatorname{Re} z>0$.

In particular, for real $s>0$,

$$\log \zeta_{\mathfrak{C}}(s) = -\sum_{r=1}^{\infty}{}' \log(1-e^{-rs}) = \sum_{r=1}^{\infty}\sum_{m=1}^{\infty} m^{-1}e^{-rms}$$
$$= \sum_{m=1}^{\infty} \frac{e^{-ms}}{m(1-e^{-ms})}.$$

Now

$$me^{-(m-1)s}(1-e^{-s}) < 1-e^{-ms} < m(1-e^{-s}),$$

and so

$$\frac{e^{-ms}}{m^2(1-e^{-s})} < \frac{e^{-ms}}{m(1-e^{-ms})} < \frac{e^{-s}}{m^2(1-e^{-s})}.$$

Therefore

$$(1-e^{-s})^{-1}\sum_{m=1}^{\infty} m^{-2}e^{-ms} < \log \zeta_{\mathfrak{C}}(s) < e^{-s}(1-e^{-s})^{-1}\sum_{m=1}^{\infty} m^{-2}.$$

Since $1-e^{-s} \sim s$ as $s \to 0$, it follows now from Abel's limit theorem for power series that

$$\log \zeta_{\mathfrak{C}}(s) \sim \tfrac{1}{6}\pi^2 s^{-1} \quad \text{as } s \to 0+.$$

By using this conclusion, and a 'Tauberian' theorem of Hardy and Ramanujan discussed in Part III below, it is now possible to deduce that

$$\log p(n) \sim \pi \sqrt{\tfrac{2}{3}n} \quad \text{as } n \to \infty.$$

However, it is sufficient for present purposes to conclude from the previous statement that, for sufficiently small $s>0$ and any $r \geqq 1$,

$$\log p(r)e^{-rs} < \log \zeta_{\mathfrak{C}}(s) \leqq 2s^{-1};$$

this gives

$$\log p(r) \leqq 2s^{-1} + rs,$$

and so, if $s = r^{-1/2}$ and r is sufficiently large,

$$\log p(r) \leqq 3\sqrt{r}.$$

Hence there is a positive constant B such that the lemma is true. □

For later purposes, we also note:

1.12. Corollary. *Let $a_D(n)$ denote the total number of isomorphism classes of modules of cardinal n in the category $\mathfrak{F}$. Then, for any rational prime p, there is a constant $B'>0$ such that*

$$a_D(p^r) \leqq e^{B'\sqrt{r}} \qquad (r = 1, 2, \ldots).$$

Proof. Let $\mathfrak{F}(p)$ denote the subcategory of $\mathfrak{F}$ consisting of all modules of cardinal some power of p. It was shown in Chapter 3, § 2 that the generating function

$$Z_{\mathfrak{F}(p)}(y) = \sum_{n=0}^{\infty} a_D(p^n)y^n$$

satisfies

$$Z_{\mathfrak{F}(p)}(y) = Z(y^{\alpha_1})Z(y^{\alpha_2}) \ldots Z(y^{\alpha_m})$$

for certain positive integers α_i, where $Z(y) = \sum_{n=0}^{\infty} p(n)y^n$. Therefore, for real $s>0$ and $\mathfrak{C}$ as above,

$$\log Z_{\mathfrak{F}(p)}(e^{-s}) = \sum_{i=1}^{m} \log Z(e^{-\alpha_i s}) = \sum_{i=1}^{m} \log \zeta_{\mathfrak{C}}(\alpha_i s)$$
$$\sim \tfrac{1}{6}\pi^2(\alpha_1^{-1} + \ldots + \alpha_m^{-1})s^{-1} \quad \text{as } s \to 0+.$$

Hence the proof may be completed in essentially the same way as above. □

1.13. Theorem. *Let* $S(n)$ *denote the total number of isomorphism classes of semisimple finite rings of cardinal* n. *Then* $S(n)$ *possesses an asymptotic* k^{th} *moment for every* $k=1, 2, \ldots$.

Proof. The case $k=1$ of this theorem was included in Theorem 1.7. In general, by Chapter 3, § 2, $S(n)$ is a PIM-function on G_Z and, if f denotes its pointwise k^{th} power, then

$$\hat{f} = 1 + \sum_{r=1}^{\infty} [s(r)]^k t^r,$$

where $s(n)$ is a certain generalized partition function with $s(1)=1$. By the next lemma, there is a constant $B''>0$ such that $s(r) \leqq e^{B''\sqrt{r}}$, and so in this case Proposition 4.3.7 implies that $f(z)=\zeta(z)g(z)$, where $g(z)$ converges absolutely whenever $\operatorname{Re} z<\nu$, for some $\nu<1$. This leads to the conclusion. □

1.14. Lemma. *There is a positive constant* B'' *such that*

$$s(r) \leqq e^{B''\sqrt{r}} \qquad (r = 1, 2, \ldots).$$

Proof. Let $\mathfrak{S}(p)$ denote the category of all semisimple finite rings of cardinal some power of p, where p is a fixed rational prime. Then

$$Z_{\mathfrak{S}(p)}(y) = \sum_{n=0}^{\infty} s(n) y^n,$$

and it was shown in Chapter 3, § 2 that algebraically,

$$Z_{\mathfrak{S}(p)}(y) = \prod_{m=1}^{\infty} Z(y^{m^2}),$$

where $Z(y)$ is the partition generating function above. This and the previous discussion of $Z(y)$ suggests the *formal* deduction (as $s \to 0+$)

$$\log Z_{\mathfrak{S}(p)}(e^{-s}) = \sum_{m=1}^{\infty} \log Z(e^{-m^2 s})$$

$$\sim \sum_{m=1}^{\infty} \tfrac{1}{6}\pi^2 (m^2 s)^{-1} = \tfrac{1}{36}\pi^4 s^{-1}.$$

The last conclusion leads to the desired result by the same type of argument as before. One way of justifying it is as follows:

We have

$$Z_{\mathfrak{S}(p)}(y) = \prod_{m=1}^{\infty} \prod_{r=1}^{\infty} (1 - y^{rm^2})^{-1},$$

and, for complex y with $|y|<1$,

$$\sum_{m=1}^{\infty}\sum_{r=1}^{\infty} y^{rm^2} = \sum_{m=1}^{\infty} y^{m^2}(1-y^{m^2})^{-1},$$

which is a so-called 'Lambert' series. The theory of such series (see Knopp [1], say) implies that the present series converges because its "associated" power series $\sum_{m=1}^{\infty} y^{m^2}$ converges for $|y|<1$. As in a discussion of double series in the proof of Lemma 1.8, it follows that $\sum_{i=1}^{\infty} y^{k_i}$ converges for real y with $0<y<1$, where $k_1 \leqq k_2 \leqq k_3 \leqq \ldots$ is some ascending arrangement of the values rm^2 for $r,m=1,2,\ldots$.

Therefore the product $\prod_{i=1}^{\infty}(1-y^{k_i})$ is absolutely convergent for complex y with $|y|<1$, and hence the same is true of $\prod_{i=1}^{\infty}(1-y^{k_i})^{-1}$ and thus also of $\sum_{i=1}^{\infty}\log(1-y^{k_i})^{-1}$, where log denotes the principal value of the logarithm. This implies that $\sum_{m=1}^{\infty} L_m(y)$ is absolutely convergent for $|y|<1$, where

$$L_m(y) = \sum_{r=1}^{\infty} \log(1-y^{rm^2})^{-1}.$$

Now

$$\mathrm{e}^{L_m(y)} = Z(y^{m^2}) \quad \text{for } |y|<1,$$

since the proof of Lemma 1.11 implies that

$$Z(y) = \prod_{r=1}^{\infty}(1-y^r)^{-1} \quad \text{for } |y|<1.$$

Hence

$$\prod_{m=1}^{\infty} Z(y^{m^2}) = \prod_{m=1}^{\infty}\left(1+\sum_{n=1}^{\infty} p(n)y^{nm^2}\right)$$

converges for $|y|<1$.

It follows easily that the series

$$\sum_{m=1}^{\infty}\left|\sum_{n=1}^{\infty} p(n)y^{nm^2}\right|$$

is uniformly convergent in every disc $|y|\leqq\varrho$, where $\varrho<1$. Therefore standard theorems on infinite products of analytic functions (see Knopp [1], say) at last imply that analytically

$$\sum_{n=0}^{\infty} s(n)y^n = Z_{\mathfrak{S}(p)}(y) = \prod_{m=1}^{\infty} Z(y^{m^2}) \quad \text{for } |y|<1.$$

By referring to the proof of Lemma 1.11, it now follows for real $s>0$ that

$$\log Z_{\mathfrak{S}(p)}(\mathrm{e}^{-s}) = \sum_{m=1}^{\infty} \log Z(\mathrm{e}^{-m^2 s}) = \sum_{m=1}^{\infty}\sum_{r=1}^{\infty} \frac{\mathrm{e}^{-rm^2 s}}{r(1-\mathrm{e}^{-rm^2 s})},$$

and also that

$$\frac{\mathrm{e}^{-rm^2 s}}{r^2 m^4(1-\mathrm{e}^{-s})} < \frac{\mathrm{e}^{-rm^2 s}}{rm^2(1-\mathrm{e}^{-rm^2 s})} < \frac{\mathrm{e}^{-s}}{r^2 m^4(1-\mathrm{e}^{-s})}.$$

Therefore

$$(1-\mathrm{e}^{-s})^{-1} \sum_{m=1}^{\infty}\sum_{r=1}^{\infty} \frac{\mathrm{e}^{-rm^2 s}}{r^2 m^2} < \log Z_{\mathfrak{S}(p)}(\mathrm{e}^{-s})$$

$$< \mathrm{e}^{-s}(1-\mathrm{e}^{-s})^{-1} \sum_{m=1}^{\infty}\sum_{r=1}^{\infty} \frac{1}{r^2 m^2}.$$

Now, if

$$f_m(s) = \sum_{r=1}^{\infty} \frac{\mathrm{e}^{-rm^2 s}}{r^2 m^2},$$

then $|f_m(s)| \leqq \frac{1}{6}\pi^2 m^{-2}$ $(s \geqq 0)$, and so $\sum_{m=1}^{\infty} f_m(s)$ is uniformly convergent for $s \geqq 0$. It therefore follows finally that

$$\log Z_{\mathfrak{S}(p)}(\mathrm{e}^{-s}) \sim \tfrac{1}{36}\pi^4 s^{-1} \quad \text{as } s \to 0+.$$

This is the conclusion indicated earlier, and so the lemma is proved. □

§ 2. Maximum orders of magnitude

Up to now, most of the arithmetical functions considered were functions f whose values $f(a)$ fluctuate wildly as $|a|$ increases. Consequently, most of the emphasis has been on investigating the average values or related properties of such functions. By contrast, it is sometimes interesting or useful to obtain some information about the "local" orders of magnitude, and particularly the 'maximum' and 'minimum' orders of magnitude, of a given function f on an arithmetical semigroup G. For example, it will be shown in a moment that many of the functions f considered earlier have the "regular" property that, for *any* $\varepsilon>0$,

$$f(a) = \mathrm{O}(|a|^{\varepsilon}) \quad \text{as } |a| \to \infty.$$

Results of this type, and more precise ones about the maximum and minimum orders of magnitude of f, are also often of interest in connection with questions of asymptotic enumeration concerning specific arithmetical categories, regardless of a particular reader's interest in number theory *per se*. In fact, enumerative functions such as $a(n)$, $a_D(n)$ and $S(n)$ will receive special attention below.

Firstly, although its present applications may also be derived from later considerations in many cases, it may be of interest to consider the following elementary result which is directly related to the "regular" property just referred to. (Here, and elsewhere in this section, the given arithmetical semigroup G need *not* necessarily satisfy Axiom A.)

2.1. Proposition. *Let f denote a multiplicative function in* Dir (G) *such that, for every prime $p \in P$,*

$$f(p^r) \to 0 \quad \text{as } |p|^r \to \infty.$$

Then $f(a) \to 0$ as $|a| \to \infty$.

Proof. By hypothesis, given any $\varepsilon > 0$, there is a number X_ε such that $|f(p^r)| < \varepsilon$ whenever $p \in P$ and $|p|^r > X_\varepsilon$. Since $\pi_G(x)$ is finite for every $x > 0$, it follows that there is a constant B such that $|f(p^r)| \leqq B$ for all $p \in P$ and $r \geqq 1$. Further, there are at most a finite number of elements $b \in G$ with the property that $b = p_1^{\alpha_1} p_2^{\alpha_2} \ldots p_m^{\alpha_m}$ for some m and distinct $p_i \in P$ such that $|p_i|^{\alpha_i} \leqq X_\varepsilon$; hence there is a number Y_ε such that every element b of this kind has $|b| \leqq Y_\varepsilon$.

Now let N_ε denote the total number of prime-powers $p^r \in G$ such that $|p|^r \leqq X_\varepsilon$, and, given $\varepsilon' = \varepsilon B^{-N_1}$, consider an element $a \in G$ such that $|a| > Y_{\varepsilon'}$. In that case, if $a = p_1^{\alpha_1} p_2^{\alpha_2} \ldots p_m^{\alpha_m}$, where the $p_i \in P$ are distinct, then $|p_i|^{\alpha_i} > X_{\varepsilon'}$ for at least one i. Hence $|f(p_i^{\alpha_i})| < \varepsilon'$ for such i. Therefore

$$|f(a)| = |f(p_1^{\alpha_1})| \, |f(p_2^{\alpha_2})| \ldots |f(p_m^{\alpha_m})| < \varepsilon' B^{N_1} = \varepsilon,$$

since N_1 prime powers p^r have $|p|^r \leqq X_1$ (so that $|f(p^r)| \leqq B$), while every other prime-power q^s has $|q|^s > X_1$ (giving $|f(q^s)| < 1$). This shows that $|f(a)| < \varepsilon$ whenever $|a| > Y_{\varepsilon'}$. Therefore $f(a) \to 0$ as $|a| \to \infty$. □

2.2. Corollary. *Let f denote a multiplicative function in* Dir (G) *such that, for any $\varepsilon > 0$ and any prime-power p^r,*

$$f(p^r) = O(|p|^{r\varepsilon}) \quad \text{as } |p|^r \to \infty.$$

Then, for every $\varepsilon > 0$,

$$f(a) = O(|a|^\varepsilon) \quad \text{as } |a| \to \infty.$$

Proof. Given $\varepsilon > 0$, define $g(a) = |a|^{-\varepsilon} f(a)$ for $a \in G$. Then $g(p^r) = o(1)$ as $|p|^r \to \infty$ for $p \in P$, since one of the hypotheses implies that $f(p^r) = O(|p|^{r\varepsilon'})$

for $\varepsilon'<\varepsilon$, so that $f(p^r)=o(|p|^{r\varepsilon})$. Therefore Proposition 2.1 implies that $g(a)=o(1)$, and hence $f(a)=o(|a|^{\varepsilon})$ as $|a|\to\infty$. The last conclusion establishes the corollary. □

(Note that the last conclusion is not stronger than the required one!)

Now consider some applications of Corollary 2.2 to special arithmetical functions and enumeration questions.

2.3. Proposition. *Consider any* $\varepsilon>0$. *Then:*

(i) $d(a)=O(|a|^{\varepsilon})$ *as* $|a|\to\infty$;

(ii) $d_*(a)=O(|a|^{\varepsilon})$ *as* $|a|\to\infty$;

(iii) $\beta(a)=O(|a|^{\varepsilon})$ *as* $|a|\to\infty$;

(iv) $\sigma_w(a)=O(|a|^{w+\varepsilon})$ *as* $|a|\to\infty$ $(w>0)$;

(v) $\varphi_w(a)|a|^{-w+\varepsilon}\to\infty$ *as* $|a|\to\infty$ $(w>0)$.

Proof. For $p\in P$ and $q_0=\min\{|p|:p\in P\}$,

$$\frac{d(p^r)}{|p|^{r\varepsilon}}=\frac{r+1}{|p|^{r\varepsilon}}\leqq\frac{2r}{|p|^{r\varepsilon}}=\frac{2\log|p|^r}{|p|^{r\varepsilon}\log|p|}\leqq\frac{2\log|p|^r}{(\log q_0)|p|^{r\varepsilon}}$$
$$\to 0 \quad \text{as } |p|^r\to\infty.$$

Therefore part (i) follows from Corollary 2.2, and parts (ii) and (iii) then follow from the inequalities

$$d_*(a)\leqq d(a), \qquad \beta(a)\leqq d(a).$$

Next, the definition of σ_w shows that it is multiplicative and, for $p\in P$,

$$\sigma_w(p^r)=1+|p|^w+\ldots+|p|^{rw}=|p|^{rw}\left[\frac{1-|p|^{-(r+1)w}}{1-|p|^{-w}}\right]$$
$$=O(|p|^{rw}) \quad \text{as } |p|^r\to\infty \quad (\text{if } w>0).$$

Therefore, by Corollary 2.2,

$$\bar{\sigma}_w(a)=|a|^{-w}\sigma_w(a)=O(|a|^{\varepsilon}) \quad \text{as } |a|\to\infty,$$

and part (iv) follows.

Finally, note that φ_w is multiplicative and, for $p\in P$,

$$\varphi_w(p^r)=|p|^{rw}-|p|^{(r-1)w}=|p|^{rw}[1-|p|^{-w}].$$

It follows that $|p^r|^{w-\varepsilon}/\varphi_w(p^r)=$

$$|p^r|^{-\varepsilon}(1-|p|^{-w})^{-1}\leqq|p|^{-r\varepsilon}(1-q_0^{-w})^{-1}.$$ So $|a|^{w-\varepsilon}/\varphi_w(a)=o(1)$,

which implies part (v). □

2.4. Theorem. *Let $\mathfrak{C}$ denote one of the arithmetical categories $\mathscr{A}$, $\mathfrak{F}$, $\mathfrak{F}^{\langle k\rangle}$ or $\mathfrak{S}$. Then, for any $\varepsilon>0$,*

$$G_{\mathfrak{C}}(n) = O(n^{\varepsilon}) \quad \textit{as } n \to \infty.$$

Proof. It is only necessary to consider $\mathfrak{F}$ and $\mathfrak{S}$, and the argument for $\mathfrak{S}$ is directly parallel to that for $\mathfrak{F}$. Therefore consider $\mathfrak{F}$, and note that, by the discussion of Chapter 3, $G_{\mathfrak{F}}(n)=a_D(n)$ is a multiplicative function of the positive integer n. Further, $a_D(p^r) \leqq e^{B'\sqrt{r}}$ for every rational prime-power p^r, by Corollary 1.12. Hence, for any given $\varepsilon>0$,

$$p^{-r\varepsilon}a_D(p^r) \leqq 2^{-r\varepsilon}\exp[B'\sqrt{r}] = \exp[B'\sqrt{r}-r\varepsilon\log 2].$$

Now, for any real numbers x and C,

$$0 \leqq (\varepsilon\log 2)(x-C(2\varepsilon\log 2)^{-1})^2 = x^2\varepsilon\log 2 - Cx + C^2(4\varepsilon\log 2)^{-1},$$

i.e.,

$$Cx - x^2\varepsilon\log 2 \leqq C^2(4\varepsilon\log 2)^{-1} = C_{\varepsilon},$$

say. Therefore, if $x=\sqrt{r}$ and $C=B'$, we obtain

$$p^{-r\varepsilon}a_D(p^r) \leqq \exp[B'_{\varepsilon}],$$

which shows that the required conditions for Corollary 2.2 are fulfilled. □

2.5. Corollary. *Let $K(n)$ denote the total number of integral ideals of norm n in a given algebraic number field K. Then, for any $\varepsilon>0$,*

$$K(n) = O(n^{\varepsilon}) \quad \textit{as } n \to \infty.$$

Proof. $K(n)=G_K(n)$ and G_K is arithmetically equivalent to the particular arithmetical semigroup $G_{\mathfrak{F}^{\langle k\rangle}}$ for which $\langle k\rangle$ is the finite sequence consisting of 1 alone. Therefore the conclusion follows. □

Now consider the problem of determining the 'maximum' and 'minimum' orders of magnitude of a given function $f\in \mathrm{Dir}\,(G)$, which for convenience may be supposed to take only non-negative real values. Since the values taken by f may not be bounded, this question must be interpreted in a special way:

Let H denote an infinite subset of G, and suppose that there is a positive real-valued function $F(x)$ of $x>x_0$ such that

$$\limsup_{\substack{|a|\to\infty \\ a\in H}} \{f(a)/F(|a|)\} = 1,$$

i.e.,

$$f(a) < (1+\varepsilon)F(|a|)$$

whenever $|a|$ is sufficiently large ($a \in H$), while

$$f(a) > (1-\varepsilon)F(|a|)$$

for infinitely many $a \in H$. In such a case, we shall say that $f(a)$ has the **maximum** order of magnitude $F(|a|)$ for $a \in H$. If, alternatively,

$$\liminf_{\substack{|a| \to \infty \\ a \in H}} \{f(a)/F(|a|)\} = 1,$$

we shall say that $f(a)$ has the **minimum** order of magnitude $F(|a|)$ for $a \in H$. (When $H = G$, as is usually the case below, reference to it may be omitted.)

Although these definitions are not designed to cover all special cases (such as when $\lim_{|a| \to \infty} f(a) = 0$, for example), they are adequate for the particular natural arithmetical functions studied below. Frequently, one of these orders of magnitude is relatively easy or trivial to determine, while the other is fairly subtle to investigate. In particular, for most of the functions treated below, the minimum order of magnitude is easy to find, while the maximum order of magnitude is known only in a weak sense which will be described shortly.

Firstly, consider some of the easy cases: Let $f \in \mathrm{Dir}\,(G)$ be a PIM-function such that

$$\hat{f} = 1 + \sum_{r=1}^{\infty} c_r t^r,$$

where $1 \leqq c_1 \leqq c_2 \leqq \ldots$, and suppose that the set P of primes for G is infinite. In that case, $f(a) \geqq c_1$ for $a \neq 1$, and $f(p) = c_1$ for $p \in P$. Therefore

$$\liminf_{|a| \to \infty} f(a) = c_1,$$

i.e. the minimum order of magnitude of $f(a)$ (for $a \in G$) is c_1. For example,

$$\liminf_{|a| \to \infty} d(a) = \liminf_{|a| \to \infty} d_*(a) = 2,$$

$$\liminf_{|a| \to \infty} \beta(a) = 1,$$

$$\liminf_{n \to \infty} a(n) = \liminf_{n \to \infty} S(n) = 1.$$

Other types of functions are sometimes also easy to deal with. For example, it is also easy to see that

$$\liminf_{|a| \to \infty} \omega(a) = \liminf_{|a| \to \infty} \Omega(a) = 1.$$

Further, the following argument shows that $\Omega(a)$ *has the maximum order of magnitude* $\log|a|/\log q_0$, where $q_0=\min\{|p|:p\in P\}$.

Firstly,

$$\Omega(a) = \log|a|/\log q_0$$

whenever a is a prime-power p_0^r with $|p_0|=q_0$. Secondly, for any $p\in P$ and $r\geqq 1$,

$$\Omega(p^r) = r = \log|p^r|/\log|p| \leqq \log|p^r|/\log q_0,$$

which implies that $\Omega(a)\leqq\log|a|/\log q_0$ for every $a\in G$. Hence

$$\limsup_{|a|\to\infty} \frac{\Omega(a)}{\log|a|/\log q_0} = 1. \qquad \square$$

The maximum order of magnitude of $\omega(a)$ is much less obvious, and will be dealt with later, subject to Axiom A.

For two further simple examples, we note that the discussion of σ_w and φ_w ($w>0$) in the proof of Proposition 2.3 leads directly to the conclusions

$$\liminf_{|a|\to\infty} |a|^{-w}\sigma_w(a) = 1, \qquad \limsup_{|a|\to\infty} |a|^{-w}\varphi_w(a) = 1.$$

Turning to the less obvious cases, it is a surprising fact that many natural arithmetical functions f have the following apparently special property: Relative to some subset H of G, there exists a constant $\tau=\tau_H(f)\geqq 1$ such that

$$\limsup_{\substack{|a|\to\infty \\ a\in H}} \frac{\log f(a)}{\log|a|/\log\log|a|} = \log\tau.$$

In such a case, the constant τ will be called the **indicator** of f over H; if $H=G$, reference to H may be omitted.

If f has the indicator τ over H, $\log f(a)$ has the maximum order of magnitude $\log|a|\log\tau/\log\log|a|$ for $a\in H$, and then it may be reasonable to say that $f(a)$ has the **logarithmic** maximum order of magnitude $\tau^{\log|a|/\log\log|a|}$ for $a\in H$. However, in general this appears to provide only partial information about the "true" maximum order of magnitude of f over H, in the sense defined previously. In virtually all non-trivial cases for which a given function f has been shown to possess a finite indicator τ, even in that of the divisor function d over $G_{\mathbf{Z}}$, the problem of determining the true maximum remains open!

Before establishing the existence of a finite indicator and deriving information about its value, in each of a variety of specific cases, consider some auxiliary results leading in this direction:

2.6. Lemma. *Let G denote an arithmetical semigroup such that* $N_G(x)=$ $=O(x^{\delta})$ *as* $x\to\infty$, *where* δ *is a positive constant. Let f denote a non-negative real-valued multiplicative function on G such that* $f(p^r)\leqq e^{B\sqrt{r}}$ *for all primes* $p\in P$ *and* $r\geqq 1$, *where B is a positive constant. Then*

$$\limsup_{|a|\to\infty}\frac{\log f(a)}{\log|a|/\log\log|a|}\leqq\delta\log\varrho$$

for any $\varrho\geqq 1$ *such that* $\varrho^r\geqq f(p^r)$ *for all* $p\in P$ *and* $r\geqq 1$.

Proof. Consider any $p\in P$, $r\geqq 1$ and $v>0$. Then

$$|p|^{-rv}f(p^r)\leqq q_0{}^{-rv}\exp[B\sqrt{r}]=\exp[B\sqrt{r}-rv\log q_0],$$

where $q_0=\min\{|p|:p\in P\}$. Then, by the same type of argument as in the proof of Theorem 2.4,

$$B\sqrt{r}-rv\log q_0\leqq C/v,$$

and hence

$$|p|^{-rv}f(p^r)\leqq e^{C/v},$$

where

$$C=B^2(4\log q_0)^{-1}.$$

Now consider any number $\varrho\geqq 1$ such that $\varrho^r\geqq f(p^r)$ for all $p\in P$ and $r\geqq 1$. If $v>0$ and $|p|\geqq\varrho^{1/v}$, then $|p|^{rv}\geqq\varrho^r\geqq f(p^r)$, and so $|p|^{-rv}f(p^r)\leqq 1$. Therefore, if $a=\prod_{p\in P}p^r$, where $r=r(p)$,

$$|a|^{-v}f(a)=\prod_{\substack{p\in P\\ p|a}}|p|^{-rv}f(p^r)\leqq\prod_{|p|<\varrho^{1/v}}|p|^{-rv}f(p^r)$$

$$\leqq\prod_{|p|<\varrho^{1/v}}e^{C/v}\leqq(e^{C/v})^{N_G(\varrho^{1/v})},$$

assuming that $\varrho>1$, since otherwise the lemma is trivial.

It follows now that $f(a)=O(|a|^v)$ as $|a|\to\infty$. In order to obtain the sharper conclusion of the lemma, now let $\varepsilon>0$ be given and put

$$v=\delta(1+\tfrac{1}{2}\varepsilon)\log\varrho/\log\log|a|.$$

Then

$$\varrho^{1/v}=(\log|a|)^{1/\delta(1+\varepsilon/2)},$$

and since $N_G(x) \leqq A' x^\delta$ for some constant $A' > 0$, it follows that

$$\begin{aligned} \log f(a) &\leqq \nu \log |a| + (C/\nu) N_G(\varrho^{1/\nu}) \\ &\leqq \frac{\delta(1+\frac{1}{2}\varepsilon) \log \varrho \cdot \log |a|}{\log \log |a|} + \frac{A' C \log \log |a| \cdot (\log |a|)^{1/(1+\varepsilon/2)}}{\delta(1+\frac{1}{2}\varepsilon) \log \varrho} \\ &< \frac{\delta(1+\varepsilon) \log \varrho \cdot \log |a|}{\log \log |a|} \quad \text{for } |a| \text{ sufficiently large,} \end{aligned}$$

since

$$\frac{\log \log y \cdot (\log y)^{1/(1+\varepsilon/2)}}{\log y / \log \log y} = (\log \log y)^2 (\log y)^{-\varepsilon/2(1+\varepsilon/2)} \to 0 \quad \text{as } y \to \infty.$$

This implies the lemma. □

2.7. Lemma. *Let f denote a non-negative real-valued multiplicative function on some arithmetical semigroup G. Suppose that G contains a subset H such that:*

(i) $\theta_H(x) =_{\text{def.}} \sum \{\log |p| : p \in P \cap H, |p| \leqq x\} > B' x^\delta$ *for some positive constants B' and δ;*

(ii) *for some positive integer t, there is a constant $c_t \geqq 1$ such that $f(p^t) \geqq c_t$ whenever $p \in P \cap H$.*

Then

$$(\delta/t) \log c_t \leqq \limsup_{|a| \to \infty} \frac{\log f(a)}{\log |a| / \log \log |a|},$$

and if H is also a sub-semigroup of G, then

$$(\delta/t) \log c_t \leqq \limsup_{\substack{|a| \to \infty \\ a \in H}} \frac{\log f(a)}{\log |a| / \log \log |a|}.$$

Proof. Given $x > 0$ sufficiently large for the product to be non-vacuous, define

$$a = a(x) = \prod_{\substack{p \in P \cap H \\ |p| \leqq x}} p^t.$$

If H is a sub-semigroup of G, then this element a will lie in H. Also,

$$f(a) = \prod_{\substack{p \in P \cap H \\ |p| \leqq x}} f(p^t) \geqq c_t^{\pi_H(x)},$$

where

$$\pi_H(x) = \sum_{\substack{p \in P \cap H \\ |p| \leqq x}} 1 \geqq \theta_H(x) / \log x.$$

Since

$$\log|a| = t\theta_H(x) > tB'x^\delta,$$

if follows that

$$\log\log|a| > \log tB' + \delta\log x > \delta(1-\varepsilon)\log x$$

for x sufficiently large (depending on the given $\varepsilon>0$). Therefore

$$\log f(a) \geqq \frac{\theta_H(x)\log c_t}{\log x} > \frac{\delta(1-\varepsilon)\log|a|\cdot\log c_t}{t\log\log|a|}$$

for x sufficiently large. The lemma therefore follows. □

A combination of Lemmas 2.6 and 2.7 yields:

2.8. Corollary. *Let G denote an arithmetical semigroup such that $N_G(x)=O(x^\delta)$ as $x\to\infty$, where δ is a positive constant. Let f denote a non-negative real-valued multiplicative function on G such that $f(p^r)\leqq e^{B\sqrt{r}}$ for all primes $p\in P$ and $r\geqq 1$, where B is a positive constant. Suppose that G contains a subset H such that*

(i) $\theta_H(x)=\sum\{\log|p|:p\in P\cap H,\ |p|\leqq x\}>B'x^\delta$ *for some constant* $B'>0$,

(ii) *for some positive integer t, there is a constant $c_t\geqq 1$ such that $f(p^t)=c_t$ whenever $p\in P\cap H$, and $c_t^{r/t}\geqq f(p^r)$ for all $p\in P$ and $r\geqq 1$.*

Then f has the indicator $c_t^{\delta/t}$ over G, and, if H is also a sub-semigroup of G, then f also has the indicator $c_t^{\delta/t}$ over H. □

In order to apply the above auxiliary results to some particular functions of interest, we shall *now assume Axiom* A again. Then it will follow from the proof of the abstract prime number theorem in the next chapter that

$$\theta_G(x) = \sum_{\substack{p\in P\\ |p|\leqq x}} \log|p| \sim \delta^{-1}x^\delta \quad \text{as } x\to\infty.$$

Accepting this for the moment, we may now prove:

2.9. Theorem. *Let G denote an arithmetical semigroup satisfying Axiom* A. *Then:*

(i) *d and d_* both have the indicator 2^δ over G,*

(ii) *β has the indicator $3^{\delta/3}$ over G,*

(iii) *$\omega(a)$ has the (true) maximum order of magnitude*

$$\delta\log|a|/\log\log|a| \quad \textit{for } a\in G.$$

Proof. For $p \in P$ and $r \geqq 1$,

$$d_*(p^r) = 2 \leqq 2^r, \qquad d(p^r) = r+1 \leqq (1+1)^r = 2^r.$$

Also, $\beta(p^r) = r$, and so each of these functions f has the property that $f(p^r) \leqq e^{B\sqrt{r}}$ for some constant $B > 0$. Choose $H = G$ and $t = 1$ for d and d_*. Then the above inequalities show that all the conditions of Corollary 2.8 are satisfied for these two functions. Hence part (i) follows. Since $d_*(a) = 2^{\omega(a)}$, the conclusion for d_* then implies that

$$\limsup_{|a| \to \infty} \frac{\omega(a) \log 2}{\log |a| / \log \log |a|} = \delta \log 2.$$

This implies part (iii).

Finally, part (ii) will follow once it has been shown that $\beta(p^r) \leqq 3^{r/3}$ (so that one may take $t = 3$ and $c_t = 3$ in this case). Now, this inequality is certainly satisfied for $r \leqq 3$. Also, if $r \leqq 3^{r/3}$, then

$$r+1 \leqq 3^{r/3} + 1 \leqq 3^{(r+1)/3}$$

provided that

$$1 + 3^{-r/3} \leqq 3^{1/3}.$$

But, for $r \geqq 3$,

$$1 + 3^{-r/3} \leqq 1 + 3^{-1} = \tfrac{4}{3} < 3^{1/3}.$$

Hence the conclusion follows. □

Now let $\mathfrak{C}$ denote some arithmetical category whose objects are finite sets possessing some additional structure. If the function $G_{\mathfrak{C}}(n)$ possesses a finite indicator τ over $G_{\mathbf{Z}}$, we shall refer to $\tau = \tau(\mathfrak{C})$ as the **indicator** of $\mathfrak{C}$.

2.10. Theorem (Krätzel). *The category $\mathscr{A}$ of all finite abelian groups possesses the indicator $\tau(\mathscr{A}) = 5^{1/4}$.*

Proof. The function $a(n) = G_{\mathscr{A}}(n)$ is a PIM-function such that $a(p^r) \leqq e^{B\sqrt{r}}$ for a positive constant B, if p is prime and $r \geqq 1$. Since $G_{\mathbf{Z}}$ is the prototype of Axiom A semigroups, the theorem will then follow from Corollary 2.8 as soon as an integer t and constant c_t with appropriate properties have been determined. Now, a little calculation with the first ten or so values of $a(p^r) = \boldsymbol{p}(r)$ (see Appendix 2) soon makes it conceivable that $t = 4$ and $c_4 = \boldsymbol{p}(4) = 5$ have the required properties. We shall now show that this is indeed the case.

For this purpose, consider the generating function

$$Z(y) = \sum_{n=0}^{\infty} p(n)y^n = \prod_{r=1}^{\infty} (1-y^r)^{-1},$$

and the operation D_L of logarithmic differentiation used in Chapter 3, § 2. Then

$$D_L(Z(y)) = \frac{Z'(y)}{Z(y)} = \sum_{r=1}^{\infty} r y^{r-1}(1-y^r)^{-1}$$

$$= \sum_{r=1}^{\infty} r y^{r-1}(1+y^r+y^{2r}+\dots) = \sum_{n=1}^{\infty} y^{n-1} \sum_{r|n} r$$

$$= \sum_{n=1}^{\infty} \sigma(n) y^{n-1}.$$

Therefore

$$\sum_{n=1}^{\infty} n p(n) y^{n-1} = Z(y) \sum_{n=1}^{\infty} \sigma(n) y^{n-1},$$

which shows that

$$n p(n) = \sum_{k=1}^{n} \sigma(k) p(n-k) = \sum_{k=1}^{n} p(n-k) \sum_{r|k} r$$

$$= \sum_{rs \leqq n} r p(n-rs).$$

Now suppose as an inductive hypothesis that $p(r) \leqq 5^{r/4}$ for $r \leqq n-1$. Then

$$n p(n) \leqq \sum_{s \leqq n} \sum_{r \leqq n/s} r \cdot 5^{(n-rs)/4}$$

$$< 5^{n/4} \sum_{s=1}^{\infty} \sum_{r=1}^{\infty} r \cdot 5^{-rs/4} = 5^{n/4} \sum_{s=1}^{\infty} \frac{5^{-s/4}}{(1-5^{-s/4})^2}$$

$$< 5^{n/4} \sum_{s=1}^{\infty} \frac{16}{s^2 (\log 5)^2},$$

by the inequality

$$x^{-1}(1-x^{-1})^{-2} < (\log x)^{-2} \qquad [x > 1],$$

which may be verified as follows: Consider the function

$$g(x) = x^{1/2} - x^{-1/2} - \log x \qquad (x > 0).$$

Then

$$g'(x) = x^{-3/2}(\tfrac{1}{2}x+\tfrac{1}{2}-x^{1/2})\begin{cases} > 0 & \text{when } x < 1, \\ = 0 & \text{when } x = 1, \\ > 0 & \text{when } x > 1, \end{cases}$$

since $h(x)=\frac{1}{2}x+\frac{1}{2}-x^{1/2}$ has a minimum at $x=1$. Since $g(1)=0$, it follows that $g(x)>0$ for $x>1$. This implies the stated inequality.

It now follows that

$$p(n) < 5^{n/4}\frac{8\pi^2}{3n(\log 5)^2} \leqq 5^{n/4}$$

when

$$n \geqq \frac{8\pi^2}{3(\log 5)^2}.$$

The last inequality is satisfied when $n \geqq 11$, and since one may verify the inductive hypothesis by direct calculation for $r \leqq 10$, it follows that

$$p(r) \leqq 5^{r/4} \quad \text{for } r \geqq 1.$$

This proves the theorem. □

The reader may perhaps care to prove the next theorem as an exercise. (For low values of $s(n)=S(p^n)$, see Appendix 2.)

2.11. Theorem. *The category* $\mathfrak{S}$ *of all semisimple finite rings possesses the indicator* $\tau(\mathfrak{S})=6^{1/4}$. □

This theorem and various others concerning the indicators of particular arithmetical functions, including some functions treated earlier, may be found in the author's papers [8, 10]. For example, it turns out that the category $\mathfrak{F}$ has the indicator $5^{1/2}$ if $[K:\mathbf{Q}]=2$, it has some indicator $\tau(\mathfrak{F}) \leqq [K:\mathbf{Q}]$ if $[K:\mathbf{Q}] \geqq 3$, and for an 'absolutely abelian' number field K with $[K:\mathbf{Q}] \geqq 3$ the indicator is precisely $[K:\mathbf{Q}]$. (An absolutely abelian number field is one whose Galois group over $\mathbf{Q}$ is abelian.) In order to avoid making this section unduly long, we shall prove only one further result, which provides partial information about the total number $K(n)$ of integral ideals of norm n in the number field K. Further results in this direction, depending on theorems outside the scope of this book, are included in the author's papers just referred to. For example, the function $K(n)$ possesses a finite indicator $\tau(K)$ (called the 'indicator of K') with the property that $\tau(K) \leqq [K:\mathbf{Q}]$ in all cases, and $\tau(K)=[K:\mathbf{Q}]$ whenever K is absolutely abelian.

2.12. Theorem. *Let K be an algebraic number field such that there exists a set H of rational primes p that 'split completely' in K (*i.e., *the ideal (p) factorizes into $[K:\mathbf{Q}]$ distinct prime ideals P_i, for each $p \in H$), such that $x = O(\theta_H(x))$. Then $K(n)$ has the indicator $\tau(K) = [K:\mathbf{Q}]$.*

Proof. For any rational prime p and $r \geqq 1$, the proof of Corollary 2.5 implies that $K(p^r) \leqq e^{B\sqrt{r}}$ for some positive constant B. Also, the hypothesis about H implies that $K(p) = [K:\mathbf{Q}]$ for every $p \in H$. Hence, if one lets $t=1$ and $c_1 = [K:\mathbf{Q}]$ in Corollary 2.8, the theorem will follow as soon as it is proved that $K(p^r) \leqq [K:\mathbf{Q}]^r$ for every rational prime p and $r \geqq 1$.

For this purpose, consider an arbitrary rational prime p and suppose that $(p) = P_1^{e_1} P_2^{e_2} \dots P_k^{e_k}$, where the P_i are distinct prime ideals in K, $e_i > 0$ and $|P_i| = p^{\alpha_i}$. Then the earlier discussion of G_K shows that, in terms of formal power series,

$$\begin{aligned} 1 + \sum_{r=1}^{\infty} K(p^r) y^r &= \left(\sum_{r_1=0}^{\infty} y^{r_1 \alpha_1} \right) \dots \left(\sum_{r_k=0}^{\infty} y^{r_k \alpha_k} \right) \\ &= (1 - y^{\alpha_1})^{-1} \dots (1 - y^{\alpha_k})^{-1}. \end{aligned}$$

Now consider any formal rational function of the form

$$\sum_{r=0}^{\infty} a_r y^r = (1 - y^{t_1})^{-1} \dots (1 - y^{t_n})^{-1},$$

where the t_i are positive integers. Suppose as an inductive hypothesis that $a_r \leqq (t_1 + \dots + t_n)^r$ for each r, whenever $n = N-1$; this is certainly true when $n=1$. Then consider any such product of the form

$$\sum_{r=0}^{\infty} b_r y^r = (1 - y^{t_1})^{-1} \dots (1 - y^{t_N})^{-1},$$

and let

$$\sum_{r=0}^{\infty} c_r y^r = (1 - y^{t_2})^{-1} \dots (1 - y^{t_N})^{-1}.$$

Then it follows that

$$b_r = c_r + c_{r-t_1} + \dots + c_{r-st_1},$$

where

$$st_1 \leqq r \leqq (s+1) t_1.$$

If $T = t_2 + \dots + t_N$, then the inductive hypothesis implies that

$$\begin{aligned} b_r &\leqq T^r + T^{r-t_1} + \dots + T^{r-st_1} \\ &\leqq T^r (1 + T^{-1} + \dots + T^{-r}) \\ &= T^r (1 + t_1/T)^r = (t_1 + T)^r. \end{aligned}$$

This proves the inductive statement, and hence

$$K(p^r) \leqq (\alpha_1 + \ldots + \alpha_k)^r \leqq [K:\mathbf{Q}]^r,$$

as required. □

2.13. Corollary. *The field* $\mathbf{Q}(\sqrt{-1})$ *possesses the indicator* 2. *Hence the function* $r(n)$ *(the total number of integral lattice points* (a, b) *such that* $a^2+b^2=n$*) also possesses the indicator* 2.*

Proof. If $K=\mathbf{Q}(\sqrt{-1})$, then $K(n)=\frac{1}{4}r(n)$, and so the corollary will follow once it has been shown that K satisfies the hypothesis of Theorem 2.12. Now, the description of the primes in $G_{\mathbf{Z}[\sqrt{-1}]}$ given in Chapter 1 shows that for this particular arithmetical semigroup G

$$\pi_G(x) = 1+\pi_{4,3}(\sqrt{x})+2\pi_{4,1}(x) \quad \text{for } x \geqq 2,$$

where $\pi_{k,r}(x)$ denotes the total number of rational primes $p \equiv r \pmod{k}$ such that $p \leqq x$. Also,

$$\theta_G(x) = \log 2+2\theta_{4,3}(\sqrt{x})+2\theta_{4,1}(x) \quad \text{for } x \geqq 2,$$

where $\theta_{k,r}(x)=\sum \log p$, summing over all rational primes $p \equiv r \pmod{k}$ such that $p \leqq x$. By the temporarily assumed result about $\theta_G(x)$ when G satisfies Axiom A, and the inequality $\theta_{4,3}(\sqrt{x}) \leqq (\sqrt{x}) \log \sqrt{x}$, it follows here that $\theta_{4,1}(x) \sim \frac{1}{2}x$ as $x \to \infty$. Therefore H may be chosen to be the set of all rational primes $p \equiv 1 \pmod{4}$. □

It may be noted that the above special set H is actually a sub-semigroup of $G_{\mathbf{Z}}$. Hence the preceding conclusion about $\theta_{4,1}(x)$ and Corollary 2.8 also imply that $d(n)$, $d_*(n)$, $\beta(n)$, $a(n)$ and $S(n)$ $(n=1, 2, \ldots)$ possess finite indicators over H, and the values of these indicators are the same as the corresponding ones over $G_{\mathbf{Z}}$. In fact, these conclusion are merely examples of more general results that can be derived with the partial aid of the theory of 'arithmetical formations' discussed in Chapter 9; see the author [9, 10]. In another direction, Corollary 2.13 is of course merely a special case of the earlier mentioned conclusion about $\tau(K)$ when K is absolutely abelian; for such number fields, there always exist sets H satisfying the hypothesis of Theorem 2.12.

* This statement contradicts one that appears without proof in Hardy and Wright [1] §18.7. The discrepancy is presumably due to some minor miscalculation in the unpublished argument of the latter authors.

§ 3. Distribution functions of prime-independent functions

In Chapter 4 and in § 1 of this chapter, we have seen various examples of arithmetical functions f which not only have finite asymptotic mean-values but also possess finite asymptotic k^{th} moments for each $k=1, 2, \ldots$. This phenomenon suggests the question as to whether those functions also possess statistical 'distribution' functions in some reasonable sense. In this section, some results in this direction will be established with the aid of standard theorems about mathematical 'distribution' functions. The latter theorems will not be proved here, and so this section is less self-contained than the rest of the book. However, for the reader's convenience, various basic definitions and relevant theorems will be quoted in detail. (For this purpose, use will be made of the book of Lukacs [1], and quotations of theorems will be given under headings of the form [3.1; L22], where L22 indicates a reference to page 22 of Lukacs [1], for example.)

It is worth pointing out that there exists an extensive theory of the "statistical" or "probabilistic" properties of arithmetical functions on G_Z (with isolated cases of extensions to more general semigroups); see the book of Kubilius [1] and the survey by Galambos [1]. The discussion below is not intended to provide more than a very brief introduction to this interesting field.

Now consider some of the relevant concepts. Firstly, a **distribution** function is defined to be any non-decreasing real-valued function $F(x)$ of a real variable x such that:

(i) F is right continuous, i.e. $F(x)=\lim_{y\to x+} F(y)$ for all x,

(ii) $\lim_{y\to -\infty} F(y)=0$ and $\lim_{y\to +\infty} F(y)=1$.

[3.1; L3]. *Every distribution function F can be decomposed into the form*

$$F(x) = aF_{\mathrm{d}}(x)+bF_{\mathrm{c}}(x),$$

where a and b are real numbers with $0\leqq a\leqq 1$, $0\leqq b\leqq 1$ and $a+b=1$, F_{c} is a continuous distribution function and F_{d} is a distribution function which is a step function with a finite or countable number of discontinuity points. (A function of type F_{d} is called **discrete**.)

Now let f denote a real-valued function on some given arithmetical semigroup G. Given any $q\geqq 1$, one may associate with f and q a discrete distribution function $F_q(f)$ such that

$$F_q(f)(x) = \left(N_G(q)\right)^{-1} \operatorname{card}\left\{a\in G\colon\ |a| \leqq q,\ f(a) \leqq x\right\}.$$

In the language of probability theory (which is suggestive but *not* essential to the present treatment), $F_q(f)$ may be regarded as the distribution function of the "random variable" f on the discrete probability space $G[q]=$ $=\{a\in G: |a|\leqq q\}$ when each point of $G[q]$ is assigned the same "probability" $1/N_G(q)$. Thus $F_q(f)$ is the probability distribution function corresponding to f on $G[q]$ when the "probability measure" of any subset E of $G[q]$ is defined to be its density

$$(1/N_G(q))\,\text{card}\,E = d_{G[q]}(E) = \lim_{x\to\infty} d_{G[q]}(E, x),$$

in the notation of Chapter 4, § 3.

The last comment immediately suggests that it may be fruitful to approach statistical questions over G itself by taking as the "probability" of any set E its asymptotic density $d(E)=d_G(E)$, whenever this exists. Unfortunately, apart from difficulties connected with its existence in general, as was pointed out earlier, asymptotic density lacks the technically vital property of being countably additive. (For example, every single element of G has "measure" zero, while the (countable) semigroup G has "measure" one!) Thus the naturally desirable "measure" $d(E)$ is not always suitable for the application of probabilistic methods to G, and it is largely because of this that more subtle or alternative ideas need to be invoked.

In the present brief discussion of "statistical" questions, despite the undoubted appeal of a more "probabilistic" approach, it will be convenient to proceed in terms of concepts from the mathematical theory of distribution functions which do not depend explicitly on the theory of probability. Firstly, if $1<q_0<q_1<q_2<\ldots$ denote the distinct values taken by the norm $|\ |$ on G, we note that the distribution functions $F_q(f)$ above all arise from the sequence $F_1(f), F_{q_0}(f), F_{q_1}(f), \ldots$, and, clearly if one wishes to associate a suitable function with f and the entire semigroup G, it is reasonable to consider the "limit" as q (or q_n) tends to infinity. Here, we shall use the standard concept of 'limit' of the theory of distribution functions:

Let $F_1, F_2, \ldots$ be an arbitrary sequence of distribution functions. Such a sequence is called **(weakly) convergent** if and only if there exists a non-decreasing function F such that

$$\lim_{n\to\infty} F_n(x) = F(x)$$

at every continuity point x of F; then F is called the **(weak) limit** of the functions F_n, and one writes

$$F = \operatorname*{Lim}_{n\to\infty} F_n$$

(using a capital L).

In a case like this, F is bounded but not necessarily a distribution function. Given a real-valued arithmetical function f on a given (but otherwise arbitrary) arithmetical semigroup G, we shall say that f possesses an **asymptotic distribution function** $F(f)$ if there exists a distribution function

$$F(f) = \operatorname*{Lim}_{n\to\infty} F_{q_n}(f).$$

(For some purposes, it is of interest to consider this concept relative to some given subset H of G, by first defining 'relative' distribution functions $F_{H[q]}(f)$ in the above manner but with $G[q]$ now replaced by $H[q]$, and then saying that f possesses a **relative** asymptotic distribution function $F_H(f)$ if there exists a distribution function

$$F_H(f) = \operatorname*{Lim}_{n\to\infty} F_{H[q_n]}(f);$$

when $H=G$, the suffix and reference to H may as usual be omitted.)

In order to establish the existence of asymptotic distribution functions in certain cases, and in fact for the entire theory of distribution functions in general, a basic tool is provided by the **characteristic** function φ of a given distribution function F. This is defined for real t as the 'Fourier–Stieltjes Transform'

$$\varphi(t) = \int_{-\infty}^{\infty} e^{itx}\, dF(x).$$

[3.1; L15]. *The characteristic function φ of F is uniformly continuous over the whole real line and has*

$$|\varphi(t)| \leqq \varphi(0) = \int_{-\infty}^{\infty} dF(x) = 1.$$

[3.1; L28]. *If F_1 and F_2 are distribution functions with the same characteristic function φ, then $F_1 = F_2$.*

If F is a discrete distribution function with steps of height A_r at its discontinuity points x_r $(r=1, 2, \ldots)$, then its characteristic function reduces to

$$\varphi(t) = \sum_r A_r e^{itx_r},$$

the series being absolutely and uniformly convergent over the whole real line. In particular, if F is one of the distribution functions $F_q(f)$ associated with a real-valued function f on an arithmetical semigroup G and some $q>0$, suppose that f takes the distinct values $x_1<\ldots<x_m$ on $G[q]$ and that exactly n_r elements $a\in G[q]$ have $f(a)=x_r$. Then

$$\varphi(t)=\sum_{r=1}^{m}\frac{n_r}{N_G(q)}\mathrm{e}^{\mathrm{i}tx_r}=\frac{1}{N_G(q)}\sum_{a\in G[q]}\mathrm{e}^{\mathrm{i}tf(a)}.$$

The basic theorem about weak convergence of distribution functions is

[3.1; L48]. **Continuity Theorem.** *A sequence $F_1, F_2, \ldots$ of distribution functions F_n converges weakly to a distribution function F if and only if the corresponding sequence $\varphi_1, \varphi_2, \ldots$ of characteristic functions φ_n is pointwise convergent to a function φ which is continuous at $t=0$. In that case, φ is the characteristic function of F.*

By applying this theorem and the previous computation of φ for a distribution function of the type $F_q(f)$, we deduce:

3.2. Proposition. *Let f denote a real-valued function on an arithmetical semigroup G, and for real t let $\mathrm{e}_t(f)$ denote the arithmetical function on G such that*

$$\mathrm{e}_t(f)(a)=\mathrm{e}^{\mathrm{i}tf(a)}\quad \text{for } a\in G.$$

Then f possesses an asymptotic distribution function $F(f)$ if and only if for every real t the function $\mathrm{e}_t(f)$ possesses an asymptotic mean-value $m(\mathrm{e}_t(f))=\varphi(f)(t)$, say, and

$$\varphi(f)(t)\to 1 \quad \textit{as } t\to 0.$$

In that case, $\varphi(f)$ is the characteristic function of $F(f)$. □

Now *suppose that G satisfies Axiom* A, and let f be a real-valued prime-independent additive (PIA-) function on G such that $f(p)=0$ for every prime $p\in P$. Then $\mathrm{e}_t(f)$ is a PIM-function such that $\mathrm{e}_t(f)(p)=1$ and

$$\mathrm{e}_t(f)(p^r)=O(1)\quad \text{as } r\to\infty,$$

for every prime $p\in P$. Therefore Proposition 4.3.7, Lemma 4.3.8 and Theorem 4.3.1 imply that $\mathrm{e}_t(f)$ has the asymptotic mean-value

$$\varphi(f)(t)=\prod_{p\in P}(1-|p|^{-\delta})\left(\sum_{r=0}^{\infty}|p|^{-r\delta}\exp\left[\mathrm{i}tf(p^r)\right]\right).$$

By some remarks made at the end of Chapter 4, § 3, the same conclusion and formula are valid if f is not necessarily prime-independent but it remains true that $f(p)=0$ for every prime $p\in P$.

Then

$$\varphi(f)(t) = \prod_{p\in P} \{1+g_p(t)\},$$

where

$$g_p(t) = \sum_{r=2}^{\infty} \{\exp[itf(p^r)] - \exp[itf(p^{r-1})]\}|p|^{-r\delta}.$$

Therefore

$$\begin{aligned}|g_p(t)| &\leqq 2p^{-2\delta}(1+|p|^{-\delta}+|p|^{-2\delta}+\dots)\\ &\leqq 2|p|^{-2\delta}(1+q_0{}^{-\delta}+q_0{}^{-2\delta}+\dots)\\ &= 2(1-q_0{}^{-\delta})^{-1}|p|^{-2\delta}.\end{aligned}$$

It follows that $\sum_{p\in P} g_p(t)$ and hence the product for $\varphi(f)(t)$ is uniformly convergent relative to t, and similar inequalities show that the above series for $g_p(t)$ is also uniformly convergent relative to t. Hence $\varphi(f)(t)$ is a continuous function of t, and so Proposition 3.2 implies

3.3. Proposition. *Suppose that G is an arithmetical semigroup satisfying Axiom A, and let f denote a real-valued additive function on G such that $f(p)=0$ for every prime $p\in P$. Then f possesses an asymptotic distribution function $F(f)$, and the characteristic function of $F(f)$ has the form*

$$\varphi(f)(t) = \prod_{p\in P}\left\{1+\sum_{r=2}^{\infty} \{\exp[itf(p^r)] - \exp[itf(p^{r-1})]\}|p|^{-r\delta}\right\}. \qquad \square$$

3.4. Corollary. *Suppose that G is an arithmetical semigroup satisfying Axiom A, and let h denote a positive real-valued multiplicative function on G such that $h(p)=1$ for every prime $p\in P$. Then h possesses an asymptotic distribution function $F(h)$ and $F(h)(x)=0$ for $x\leqq 0$.*

Proof. The function $\log h$ has an asymptotic distribution function $F_*(\log h)$, say, and this implies that $F(h)$ is an asymptotic distribution function for h, where

$$F(h)(x) = \begin{cases} F_*(\log h)(\log x) & \text{for } x>0,\\ 0 & \text{for } x\leqq 0.\end{cases} \qquad \square$$

Before applying these results to special arithmetical functions of interest, consider some general corollaries. Let f denote a real-valued additive function on an arithmetical semigroup G satisfying Axiom A, such that $f(p)=0$ for every $p\in P$. Then the above comments about uniform convergence in connection with $\varphi(f)(t)$ show that each function $g_p(t)$ and hence also $\varphi(f)(t)$ is the limit of a uniformly convergent sequence of almost periodic functions of t. (Recall that an *almost-periodic* function $\psi(x)$ of real x is a complex-valued function ψ such that, given any $\varepsilon>0$, there exists a trigonometric polynomial

$$T(x) = \sum_{k=1}^{n} c_k \exp[i\alpha_k x]$$

(depending on ε) with the property that $|\psi(x)-T(x)|<\varepsilon$ for all x.) By a well-known result about almost periodic functions (see Corduneanu [1] or Maak [1], say), it follows that $\varphi(f)(t)$ is also an almost periodic function of t. Therefore $F(f)$ is a discrete distribution function, by:

[3.1; L36]. *A distribution function F is discrete if and only if its characteristic function φ is almost periodic.*

Now note that

$$F(f) = \operatorname*{Lim}_{n\to\infty} F_{q_n}(f)$$

and that the discontinuity points of $F_{q_n}(f)$ are the various values $f(a)$ for $a \in G[q_n]$. By the next statement, this implies that the discontinuity points of $F(f)$ are the various values $f(a)$ for $a \in G$ or else limit points of the set of all such values.

[3.1; L63]. *Let $F, F_1, F_2, \ldots$ be distribution functions such that*

$$F = \operatorname*{Lim}_{n\to\infty} F_n.$$

Then $S_F \subseteq \text{closure}(\liminf S_{F_n})$, *where $S_{F'}$ denotes the set of all points of increase of a given distribution function F' and* $\liminf S_{F_n}$ *denotes the set of all points x that belong to S_{F_n} for all except perhaps a finite number of integers n.*

Now suppose that the set of all values $f(a)$ for $a \in G$ has *no* limit points; for example, this will happen if f takes only rational integer values, of if $f=\log h$, where h is a positive integer-valued and multiplicative function on G with $h(p)=1$ for every $p \in P$. In this case S_F is the set of values $f(a)$ for $a \in G$. By an earlier remark about discrete distribution functions, if $S_F=\{x_1, x_2, \ldots\}$, then $\varphi(f)(t)$ may be expressed as the sum of an absolutely and uniformly convergent series:

$$\varphi(f)(t) = \sum_r A_r(f) \exp[itx_r],$$

where $A_r(f)$ denotes the jump of $F(f)$ at x_r. The numbers $A_r(f)$ are then the 'Fourier coefficients' of the function $\varphi(f)$.

Given any sufficiently small $\varepsilon>0$, $F(f)$ is continuous at the points $x_r \pm \varepsilon$, and therefore

$$\lim_{n\to\infty} \{F_{q_n}(f)(x_r \pm \varepsilon)\} = F(f)(x_r \pm \varepsilon).$$

Therefore, for sufficiently small $\varepsilon>0$,

$$\begin{aligned} A_r(f) &= F(f)(x_r+\varepsilon) - F(f)(x_r-\varepsilon) \\ &= \lim_{n\to\infty} \{F_{q_n}(f)(x_r+\varepsilon) - F_{q_n}(f)(x_r-\varepsilon)\} \\ &= \lim_{n\to\infty} A_{q_n,r}(f), \end{aligned}$$

where $A_{q_n,r}(f)$ denotes the jump of $F_{q_n}(f)$ at x_r. If exactly n_r elements $a \in G[q_n]$ have $f(a)=x_r$, then

$$A_{q_n,r}(f) = n_r/N_G(q_n),$$

which may be called the **frequency** of the elements $a \in G[q_n]$ such that $f(a)=x_r$. It follows that

$$A_r(f) = \lim_{q\to\infty} \left\{(N_G(q))^{-1} \operatorname{card}\{a \in G[q]\colon f(a) = x_r\}\right\}.$$

This proves that the set of all elements $a \in G$ such that $f(a)=x_r$ possesses the asymptotic density $A_r(f)$. This density may also be called the **asymptotic** frequency of the elements $a \in G$ such that $f(a)=x_r$.

The asymptotic frequencies $A_r(f)$ have the property that

$$\sum_r A_r(f) = \varphi(f)(0) = 1,$$

which, though intuitively plausible, is not an *a priori* conclusion. In general, we note as above that the frequency $A_r(f)$ may be represented as the 'Fourier coefficient'

$$\lim_{T\to\infty} \left\{T^{-1} \int_0^T \varphi(f)(t) \exp[-itx_r]\, dt\right\}.$$

However, if f is also integer-valued, then $\varphi(f)(t)$ is an ordinary periodic function with the more familiar type of Fourier coefficients

$$a_n(f) = \frac{1}{2\pi}\int_0^{2\pi} \varphi(f)(t)\exp[-int]\,dt,$$

where $a_n(f)$ may be interpreted as the asymptotic frequency of the set of elements $a \in G$ such that $f(a)=n$.

All these conclusions may be collected in the form of:

3.5. Proposition. *Let G be an arithmetical semigroup satisfying Axiom* A, *and let f denote a real-valued additive function on G such that $f(p)=0$ for every $p \in P$. Then $F(f)$ is a discrete distribution function, with an almost periodic characteristic function $\varphi(f)$. If the set of all values $f(a)$ $[a \in G]$ has no limit points, the Fourier coefficients of $\varphi(f)$ coincide with the various asymptotic frequencies with which f takes on particular values. If, in addition, f is integer-valued then $\varphi(f)$ is a periodic function of period 2π and these frequencies are its Fourier coefficients in the more common sense of this term. In either case, the sum of the asymptotic frequencies is $\sum_r A_r(f)=1$ (if the values $f(a)$ $(a \in G)$ have no limit points).* □

Amongst special functions satisfying the above conditions, we have firstly $\log h$, where h denotes any one of the functions listed below; Corollary 3.4 then provides a way of carrying the resulting conclusions over to the functions h themselves.

(i) The enumerative functions $a(n)$ and $S(n)$ of a positive integer n;

(ii) the functions β, d/d_* and d_*/d on any given semigroup G satisfying Axiom A;

(iii) the enumerative function $a(I)$ of $I \in G_K$, where K is an arbitrary algebraic number field.

There is a more detailed discussion of the asymptotic frequencies for $a(n)$ in the paper of Kendall and Rankin [1]; very similar conclusions may be deduced for other functions, such as $S(n)$ and $\beta(n)$ for example. (In their paper, Kendall and Rankin also point out a peculiarity that can arise with asymptotic frequencies in certain cases: they observe that the function $r(n)$ of $n \in G_{\mathbf{Z}}$, which was discussed in Chapter 4, § 1, has asymptotic frequencies $a_0, a_1, a_2, \ldots$ such that $\sum a_m m=0$, while the asymptotic mean-value of $r(n)$ is π. It is then natural to ask how the earlier results about asymptotic k^{th} moments of functions like $a(n)$, $S(n)$ and $\beta(n)$ are related to those specifying the existence of distribution functions in these cases, i.e., one may ask whether the asymptotic moments coincide with the corresponding moments of those distribution functions. Although the example just mentioned shows that such conclusions are not automatic, affirmative answers for $a(n)$, $S(n)$ and certain other functions of $n \in G_{\mathbf{Z}}$ are discussed in the author's note [14].)

An interesting case of an integer-valued additive function taking the value zero on all primes is provided by the pointwise difference $\Omega-\omega$:

$$(\Omega-\omega)(a) = \Omega(a)-\omega(a) \quad \text{for } a \in G.$$

3.6. Proposition. (Rényi's formula). *Let G be an arithmetical semigroup satisfying Axiom* A. *Then the asymptotic frequency $a_n(\Omega-\omega)$ of the elements $a \in G$ such that $\Omega(a)-\omega(a)=n$ exists for each $n=0,1,2,\ldots$, and may be calculated from the power series formula*

$$\sum_{n=0}^{\infty} a_n(\Omega-\omega)z^n = \prod_{p \in P} (1-|p|^{-\delta})(1+(|p|^{\delta}-z)^{-1}).$$

Proof. The characteristic function of the asymptotic distribution function of $\Omega-\omega$ is

$$\begin{aligned}\varphi(\Omega-\omega)(t) &= \prod_{p \in P} (1-|p|^{-\delta}) \left(1+\sum_{r=1}^{\infty} |p|^{-r\delta} \exp\,[\mathrm{i}t(r-1)]\right) \\ &= \prod_{p \in P} (1-|p|^{-\delta})(1+(|p|^{\delta}-\mathrm{e}^{\mathrm{i}t})^{-1}).\end{aligned}$$

Now, if

$$\begin{aligned}g(z) &= \prod_{p \in P} (1-|p|^{-\delta})(1+(|p|^{\delta}-z)^{-1}) \\ &= \prod_{p \in P} (1-|p|^{-\delta}) \left(1+|p|^{-\delta} \sum_{r=0}^{\infty} z^r |p|^{-r\delta}\right),\end{aligned}$$

then a modification of the earlier discussion of the infinite product for a function $\varphi(f)(t)$ shows that the product for $g(z)$ is uniformly convergent for $|z| \leq c_0{}^{\delta}$ whenever $c_0 < q_0$. Hence $g(z)$ is an analytic function of z in the region $|z| < q_0{}^{\delta} = \min\{|p|^{\delta} : p \in P\}$. The nth coefficient of the Taylor expansion of $g(z)$ about the origin can then be expressed in the form

$$\begin{aligned}a_n &= \frac{1}{2\pi \mathrm{i}} \int_{|z|=1} g(z) z^{-n-1}\, \mathrm{d}z = \frac{1}{2\pi} \int_0^{2\pi} g(\mathrm{e}^{\mathrm{i}t}) \mathrm{e}^{-\mathrm{i}tn}\, \mathrm{d}t \\ &= \frac{1}{2\pi} \int_0^{2\pi} \varphi(\Omega-\omega)(t) \mathrm{e}^{-\mathrm{i}nt}\, \mathrm{d}t,\end{aligned}$$

which is the nth Fourier coefficient of $\varphi(\Omega-\omega)$. Thus $a_n = a_n(\Omega-\omega)$. □

It is interesting to note that the above proposition gives

$$a_0(\Omega-\omega) = \prod_{p\in P}(1-|p|^{-2\delta}) = 1/\zeta_G(2\delta),$$

which (as it should be) is the asymptotic density of the set G_2 of all square-free elements in G as given by Proposition 4.4.5.

Proposition 3.6 can also be used to determine the asymptotic behaviour of $a_n(\Omega-\omega)$ as $n\to\infty$. For the sake of simplicity, attention will be confined to the case when G has only one prime of norm q_0; for example, $G_{\mathbf{Z}}$, $G_{\mathscr{A}}$ and $G_{\mathfrak{S}}$ each have only one prime of norm 2.

3.7. Proposition. *Let G denote an arithmetical semigroup satisfying Axiom* A *which has only one prime $p\in P$ of minimum norm q_0. Then as $n\to\infty$,*

$$a_n(\Omega-\omega) \sim q_0^{-(n+2)\delta}(q_0^{\delta}-1)\prod_{\substack{p\in P\\ |p|>q_0}}(1-|p|^{-\delta})(1+(|p|^{\delta}-q_0^{\delta})^{-1}).$$

Proof. In general, if

$$q_1 = \min\{|p| : p\in P, |p|\neq q_0\},$$

then the previous function

$$g(z) = \{(1-q_0^{-\delta})(1+(q_0^{\delta}-z)^{-1})\}^{P(q_0)}h(z),$$

say, where a similar technique to that for $g(z)$ above shows that $h(z)$ is an analytic function of z for $|z|<q_1^{\delta}$. (By iterating this argument, one sees that $g(z)$ is actually a meromorphic function of z with poles at the various values $|p|^{\delta}$ $(p\in P)$.) If $P(q_0)=1$, then it follows that

$$g(z) = (1-q_0^{-\delta})(q_0^{\delta}-z)^{-1}h(q_0^{\delta})+h_1(z),$$

where $h_1(z)$ is analytic for $|z|<q_1^{\delta}$. Then, for $|z|<q_0^{\delta}$,

$$g(z) = (1-q_0^{-\delta})h(q_0^{\delta})q_0^{-\delta}\sum_{n=0}^{\infty}(q_0^{-\delta}z)^n+\sum_{n=0}^{\infty}c_nz^n,$$

where the second power series has radius of convergence q_1^{δ}. Then

$$\limsup_{n\to\infty}|c_n|^{1/n} = q_1^{-\delta},$$

and so

$$a_n(\Omega-\omega) \sim (1-q_0^{-\delta})h(q_0^{\delta})q_0^{-(1+n)\delta} \quad \text{as } n\to\infty. \qquad \square$$

The reader may perhaps like to consider the case when $P(q_0)\neq 1$, as an exercise.

Selected bibliography for Chapter 5

Section 1: Duttlinger [1], Erdös and Szekeres [1], Kendall and Rankin [1], Knopfmacher [5, 7], Knopfmacher and Ridley [1], Richert [1], Šapiro–Pyateckiĭ [1], P. G. Schmidt [1], Schwarz [1], Srinivasan [1].

Section 2: Chandrasekharan and Narasimhan [1], Hardy and Wright [1], Heppner [1, 2], Horadam [16], Knopfmacher [8, 10, 11], Krätzel [1], Ramanujan [1], Schwarz and Wirsing [1], Wegmann [1].

Section 3: Baĭbulatov [1], Danilov [1], de Kroon [1], Galambos [1], Juškis [1–3], Kac [1], Knopfmacher [14], Kubilius [1], Rieger [7], Schoenberg [1].

CHAPTER 6

THE ABSTRACT PRIME NUMBER THEOREM

The main theorem proved in this chapter is the abstract prime number theorem

$$\pi_G(x) \sim \frac{x^\delta}{\delta \log x} \quad \text{as } x \to \infty,$$

which is true for any arithmetical semigroup G satisfying Axiom A. In addition, the chapter contains some related theorems about the approximate densities of certain sets and the asymptotic behaviours of various special arithmetical functions. Throughout the chapter, *G denotes an arithmetical semigroup satisfying Axiom* A.

§ 1. The fundamental theorem

Our method of proof will follow one of the standard patterns used for the ordinary integers or ideals in an algebraic number field. Firstly, consider the von Mangoldt function Λ on G such that

$$\Lambda(a) = \begin{cases} \log|p| & \text{if } a = p^r \text{ for some } p \in P,\ r > 0, \\ 0 & \text{otherwise.} \end{cases}$$

The relevance of Λ to the present problem is shown by the following lemma, which implies that the abstract prime number theorem follows from the conclusion

$$\psi(x) \sim \delta^{-1} x^\delta \quad \text{as } x \to \infty,$$

where $\psi(x) = N(\Lambda, x)$.

1.1. Lemma. *If $\psi(x) \sim \delta^{-1} x^\delta$ as $x \to \infty$, then*

$$\pi_G(x) \sim x^\delta/(\delta \log x) \quad \textit{as } x \to \infty.$$

Proof. By definition,

$$\psi(x) = \sum_{|a| \leqq x} \Lambda(a) = \sum_{\substack{p \in P \\ |p^r| \leqq x}} \log|p|$$

$$= \sum_{p \in P} \log|p| \sum_{r \leqq \log x/\log|p|} 1$$

$$\leqq \sum_{\substack{p \in P \\ |p| \leqq x}} \log x = \pi_G(x) \log x.$$

Also, if $1 < y \leqq x$, then

$$\pi_G(x) = \pi_G(y) + \sum_{\substack{p \in P \\ y < |p| \leqq x}}' 1$$

$$\leqq \pi_G(y) + \sum_{y < |p| \leqq x} \log|p|/\log y$$

$$\leqq N_G(y) + \psi(x)/\log y = \mathrm{O}(y^\delta) + \psi(x)/\log y.$$

If x is sufficiently large and $y = x(\log x)^{-2/\delta}$, it follows that

$$\psi(x) \leqq \pi_G(x) \log x \leqq \mathrm{O}(x^\delta/\log x) + \psi(x) \log x/\log y.$$

Since

$$\log y = \log x - 2\delta^{-1} \log\log x,$$

this implies that, if $\psi(x) \sim \delta^{-1} x^\delta$, then

$$\delta^{-1} \sim \psi(x) x^{-\delta} \sim \pi_G(x) x^{-\delta} \log x \quad \text{as } x \to \infty,$$

and the lemma follows. □

In order to prove the required result about $\psi(x)$, now consider the formal Dirichlet series

$$\Lambda(z) = -\zeta_G'(z)/\zeta_G(z);$$

cf. Theorem 2.6.1. Then, earlier facts about the zeta function and Lemma 4.3.5 imply that

$$\tilde{\Lambda}(z) = -\zeta_G'(z)/\zeta_G(z)$$

for all complex z with $\operatorname{Re} z > \delta$.

1.2. Lemma. *$\tilde{\Lambda}(z)$ can be extended to an analytic function of z for all $z \neq \delta$ with $\operatorname{Re} z \geqq \delta$, in such a way that:*

(i) $\tilde{\Lambda}(z) = \delta^{-1} z \int_0^\infty e^{-zy/\delta} \psi(e^{y/\delta})\, dy$ *for* $\operatorname{Re} z > \delta$,

(ii) $\delta z^{-1} \tilde{\Lambda}(z)$ *has a simple pole with residue* 1 *at* $z = \delta$.

Proof. For Re $z=\sigma>\delta$,

$$\psi(x)x^{-z} = O(\pi_G(x)(\log x)x^{-\sigma}) = O(x^{\delta-\sigma}\log x)$$
$$= o(1) \quad \text{as } x \to \infty,$$

and so Lemma 4.2.5 implies that for such z

$$\tilde{\Lambda}(z) = \sum_{a\in G} \Lambda(a)|a|^{-z} = z\int_1^\infty \psi(t)t^{-z-1}\,dt$$
$$= \delta^{-1}z\int_0^\infty e^{-zy/\delta}\psi(e^{y/\delta})\,dy.$$

Next, Proposition 4.2.13 implies that $\tilde{\Lambda}(z)$ may be extended analytically into the entire region Re $z>\eta$, $z\neq\delta$, with the exception of poles arising from zeros of $\zeta_G(z)$. Now, for $z\neq\delta$ sufficiently close to δ, Proposition 4.2.13 shows that

$$\zeta_G(z) = \frac{\delta A}{z-\delta} + g(z),$$

where $g(z)$ is analytic. Therefore, for such z,

$$\frac{\delta}{z}\tilde{\Lambda}(z) - \frac{1}{z-\delta} = -\left\{\frac{\delta}{z}\frac{\zeta_G'(z)}{\zeta_G(z)} + \frac{1}{z-\delta}\right\}$$
$$= -\left\{\frac{\delta[-\delta A(z-\delta)^{-2}+g'(z)]}{z[\delta A(z-\delta)^{-1}+g(z)]} + \frac{1}{z-\delta}\right\}$$
$$= -\frac{\delta A + \delta(z-\delta)g'(z) + zg(z)}{z[\delta A + (z-\delta)g(z)]}$$

which is a quotient of analytic functions, the denominator being non-zero for z sufficiently close to δ. Therefore (ii) is true, and the initial statement of the lemma will now follow from:

1.3. Lemma. $\zeta_G(z)\neq 0$ *for all* $z\neq\delta$ *with* Re $z\geqq\delta$.

Proof. By Proposition 4.2.6, it is only necessary to consider Re $z=\delta$. For this purpose, if Re $w>\delta$ and log denotes the principal value of the logarithm once again, note that the Euler product formula implies that

$$\zeta_G(w) = \exp\left[\sum_{p\in P}\log(1-|p|^{-w})^{-1}\right] = \exp\left[\sum_{p\in P}\sum_{m=1}^{\infty} m^{-1}|p|^{-mw}\right]$$
$$= \exp\left[\sum_{a\in G}\frac{\Lambda(a)}{\log|a|}|a|^{-w}\right]$$

(if one lets $\Lambda(1)/\log 1=0$). Hence, for $\sigma>\delta$ and any real t,

$$|\zeta_G{}^3(\sigma)\zeta_G{}^4(\sigma+\mathrm{i}t)\zeta_G(\sigma+2\mathrm{i}t)| = \left|\exp\left[\sum_{a\in G}\frac{\Lambda(a)}{\log|a|}|a|^{-\sigma}(3+4|a|^{-\mathrm{i}t}+|a|^{-2\mathrm{i}t})\right]\right|$$

$$= \exp\left[\sum_{a\in G}\frac{\Lambda(a)}{\log|a|}|a|^{-\sigma}\operatorname{Re}(3+4|a|^{-\mathrm{i}t}+|a|^{-2\mathrm{i}t})\right]$$

$$\geqq \exp 0 = 1,$$

since it is easy to verify that $\operatorname{Re}(3+4w+w^2)\geqq 0$ for any complex w with $|w|=1$. It follows that, for $\sigma>\delta$,

$$|(\sigma-\delta)\zeta_G(\sigma)|^3|(\sigma-\delta)^{-1}\zeta_G(\sigma+\mathrm{i}t)|^4|\zeta_G(\sigma+2\mathrm{i}t)| \geqq (\sigma-\delta)^{-1}.$$

Since $\zeta_G(z)$ has a simple pole with residue δA at $z=\delta$, if $\zeta_G(\delta+\mathrm{i}t)=0$ for $t\neq 0$, then, as $\sigma\to\delta+$, the right-hand side of the last inequality tends to infinity, while the left-hand side tends to the finite quantity

$$(\delta A)^3|\zeta_G{}'(\delta+\mathrm{i}t)|^4|\zeta_G(\delta+2\mathrm{i}t)|.$$

This contradiction implies that $\zeta_G(\delta+\mathrm{i}t)\neq 0$ for $t\neq 0$. □

For the purpose of deducing information about $\psi(x)$ from the above facts about $\tilde{\Lambda}(z)$, now consider the following 'Tauberian' theorem.

1.4. Theorem (Ikehara). *Let φ be a non-negative and non-decreasing real-valued function of $y\geqq 0$. Suppose that*

$$F(z) = \int_0^\infty \mathrm{e}^{-zy}\varphi(y)\,\mathrm{d}y$$

is a convergent integral for every complex z with $\operatorname{Re} z>1$, *and that $F(z)$ can be extended to an analytic function of z for all $z\neq 1$ with* $\operatorname{Re} z\geqq 1$ *in such a way that $F(z)$ has a simple pole with residue* 1 *at $z=1$. Then*

$$\varphi(y) \sim \mathrm{e}^y \quad \textit{as } y\to\infty.$$

Proof. Strictly speaking, despite its special interest for number theory, this is a theorem of pure analysis. However, we include a proof for the reader's convenience, the argument being on lines due largely to Bochner and Landau; see also Chandrasekharan [1].

First consider:

1.5. Lemma. *Let*

$$V(y) = \mathrm{e}^{-y}\varphi(y).$$

Then, for any $\alpha>0$,

$$\lim_{x\to\infty}\int_{-\infty}^{x} V(x-y)\frac{\sin^2 \alpha y}{\alpha y^2}\,\mathrm{d}y = \int_{-\infty}^{\infty}\frac{\sin^2 y}{y^2}\,\mathrm{d}y.$$

Proof. It is clear that

$$\beta = \int_{-\infty}^{\infty}\frac{\sin^2 y}{y^2}\,\mathrm{d}y > 0$$

is a convergent integral. (In fact, $\beta=\pi$, but this will not be relevant here.)

Now let

$$f(z) = F(z)-(z-1)^{-1}.$$

Then the assumptions on $F(z)$ imply that $f(z)$ is analytic for $\operatorname{Re} z \geqq 1$, and, since

$$(z-1)^{-1} = \int_0^{\infty} \mathrm{e}^{-(z-1)y}\,\mathrm{d}y$$

for $\operatorname{Re} z>1$, it follows that, for such z,

$$f(z) = \int_0^\infty [V(y)-1]\mathrm{e}^{-(z-1)y}\,\mathrm{d}y.$$

Further, given any $\alpha>0$, we note as follows that this integral is uniformly convergent whenever $z=\sigma+it$ for a fixed $\sigma>1$ and $|t|\leqq\alpha$: Since φ is a non-negative and non-decreasing function, for any $s>1$ and $x>0$, we have

$$F(s) \geqq \int_x^\infty \mathrm{e}^{-sy}\varphi(y)\,\mathrm{d}y \geqq \varphi(x)\int_x^\infty \mathrm{e}^{-sy}\,\mathrm{d}y = s^{-1}\varphi(x)\mathrm{e}^{-sx}.$$

Therefore

$$\varphi(x) \leqq sF(s)\mathrm{e}^{sx},$$

and so for any fixed $\varepsilon>0$

$$V(y)\mathrm{e}^{-\varepsilon y} = \varphi(y)\mathrm{e}^{-(1+\varepsilon)y} = \mathrm{o}(1) \quad \text{as } y\to\infty,$$

since one may choose the previous $s=1+\frac{1}{2}\varepsilon$, say. In particular, the uniform convergence follows if one chooses $\varepsilon=\frac{1}{2}(\sigma-1)$, say.

This conclusion implies that, for $x>0$,

$$\int_{-\alpha}^{\alpha} f(\sigma+it)(1-|t|/\alpha)\,\mathrm{e}^{itx}\,\mathrm{d}t =$$

$$= \int_{-\alpha}^{\alpha} (1-|t|/\alpha)\,\mathrm{e}^{itx}\left[\int_{0}^{\infty} [V(y)-1]\,\mathrm{e}^{-(\sigma+it-1)y}\,\mathrm{d}y\right]\mathrm{d}t$$

$$= \int_{0}^{\infty} [V(y)-1]\,\mathrm{e}^{-(\sigma-1)y}\left[\int_{-\alpha}^{\alpha} (1-|t|/\alpha)\,\mathrm{e}^{it(x-y)}\,\mathrm{d}t\right]\mathrm{d}y$$

$$= \int_{0}^{\infty} [V(y)-1]\,\mathrm{e}^{-(\sigma-1)y}\,\frac{\sin^2 \frac{1}{2}\alpha(x-y)}{\frac{1}{2}\alpha(x-y)^2}\,\mathrm{d}y.$$

Now, since $f(z)$ is analytic and hence continuous in z for $\operatorname{Re} z \geqq 1$, $f(\sigma+it) \to f(1+it)$ uniformly as $\sigma \to 1+$ with $|t| \leqq \alpha$. Therefore

$$\int_{-\alpha}^{\alpha} f(1+it)(1-|t|/\alpha)\,\mathrm{e}^{itx}\,\mathrm{d}t =$$

$$= \lim_{\sigma\to 1+}\left\{\int_{0}^{\infty} [V(y)-1]\,\mathrm{e}^{-(\sigma-1)y}\,\frac{\sin^2 \frac{1}{2}\alpha(x-y)}{\frac{1}{2}\alpha(x-y)^2}\,\mathrm{d}y\right\}$$

$$= \lim_{\sigma\to 1+}\left\{\int_{0}^{\infty} V(y)\,\mathrm{e}^{-(\sigma-1)y}\,\frac{\sin^2 \frac{1}{2}\alpha(x-y)}{\frac{1}{2}\alpha(x-y)^2}\,\mathrm{d}y\right\} - \beta(x),$$

where

$$\beta(x) = \int_{0}^{\infty} \frac{\sin^2 \frac{1}{2}\alpha(x-y)}{\frac{1}{2}\alpha(x-y)^2}\,\mathrm{d}y = \lim_{\sigma\to 1+}\left\{\int_{0}^{\infty} \mathrm{e}^{-(\sigma-1)y}\,\frac{\sin^2 \frac{1}{2}\alpha(x-y)}{\frac{1}{2}\alpha(x-y)^2}\,\mathrm{d}y\right\}$$

since the last integral is uniformly convergent over any interval $1 \leqq \sigma \leqq \sigma_0$. This shows that the other right-hand limit does exist, and since the related integrand is non-negative and non-decreasing, the limit must be

$$\int_{0}^{\infty} V(y)\,\frac{\sin^2 \frac{1}{2}\alpha(x-y)}{\frac{1}{2}\alpha(x-y)^2}\,\mathrm{d}y = L(x),$$

say. Thus

$$L(x) = \beta(x) + \int_{-\alpha}^{\alpha} f(1+it)(1-|t|/\alpha)\,\mathrm{e}^{itx}\,\mathrm{d}t.$$

Now

$$\lim_{x\to\infty} \beta(x) = \lim_{x\to\infty}\left\{\int_{-\infty}^{\frac{1}{2}\alpha x} \frac{\sin^2 u}{u^2}\,\mathrm{d}u\right\} = \int_{-\infty}^{\infty} \frac{\sin^2 u}{u^2}\,\mathrm{d}u = \beta,$$

while the well-known Riemann–Lebesgue lemma of Fourier analysis implies that the second integral in the expression for $L(x)$ tends to 0 as $x \to \infty$. Therefore

$$\beta = \lim_{x\to\infty} L(x) = \lim_{x\to\infty} \left\{ \int_{-\infty}^{x} V(x-u) \frac{\sin^2 \frac{1}{2}\alpha u}{\frac{1}{2}\alpha u^2} \, du \right\},$$

which proves the lemma since $\alpha > 0$ is arbitrary. □

It now follows from the lemma that, for any fixed positive numbers X and α,

$$\beta \geqq \limsup_{x\to\infty} \left\{ \int_{-X}^{X} V(x-y) \frac{\sin^2 \alpha y}{\alpha y^2} \, dy \right\},$$

since the integrand is a non-negative function. Also, for $|y| \leqq X$,

$$V(x-y) \geqq V(x-X) e^{y-X} \geqq V(x-X) e^{-2X}$$

since $V(y)e^y = \varphi(y)$ is a monotonic increasing function of y. Therefore

$$\beta \geqq \limsup_{x\to\infty} \left\{ V(x-X) e^{-2X} \int_{-X}^{X} \frac{\sin^2 \alpha y}{\alpha y^2} \, dy \right\}$$

$$= \limsup_{x\to\infty} \left\{ V(x) e^{-2X} \int_{-X}^{X} \frac{\sin^2 \alpha y}{\alpha y^2} \, dy \right\}$$

If Y is positive and $X = 1/Y$, $\alpha = Y^2$, it follows then that

$$\beta \geqq \limsup_{x\to\infty} \left\{ V(x) e^{-2/Y} \int_{-Y}^{Y} \frac{\sin^2 u}{u^2} \, du \right\}$$

$$\to \limsup_{x\to\infty} \left\{ V(x) \int_{-\infty}^{\infty} \frac{\sin^2 u}{u^2} \, du \right\} \quad \text{as } Y \to \infty.$$

Thus $1 \geqq \limsup_{x\to\infty} V(x)$, and so the required conclusion $\lim_{x\to\infty} V(x) = 1$ will follow as soon as it has been shown that $1 \leqq \liminf_{x\to\infty} V(x)$.

For this purpose, let X and α be fixed positive numbers, and use the fact that $V(y)\, e^y = \varphi(y)$ is monotonic increasing in order to now deduce that

$$\int_{-X}^{X} V(x-y) \frac{\sin^2 \alpha y}{\alpha y^2} \, dy \leqq V(x+X) e^{2X} \int_{-X}^{X} \frac{\sin^2 \alpha y}{\alpha y^2} \, dy.$$

Then note that, since $1 \geqq \limsup_{x\to\infty} V(x)$,

$$V(x) = O(1) \quad \text{as } x \to \infty.$$

Therefore Lemma 1.5 implies that

$$\begin{aligned}\beta &= \lim_{x\to\infty}\left\{\int_{-X}^{X} V(x-y)\frac{\sin^2\alpha y}{\alpha y^2}\,\mathrm{d}y + \left(\int_{-\infty}^{-X}+\int_{X}^{\infty}\right)\mathrm{O}(1)\frac{\sin^2\alpha y}{\alpha y^2}\,\mathrm{d}y\right\}\\ &\leqq \liminf_{x\to\infty}\left\{V(x+X)\,\mathrm{e}^{2X}\int_{-X}^{X}\frac{\sin^2\alpha y}{\alpha y^2}\,\mathrm{d}y\right\} + \left(\int_{-\infty}^{-X}+\int_{X}^{\infty}\right)\mathrm{O}(1)\frac{\sin^2\alpha y}{\alpha y^2}\,\mathrm{d}y\\ &= \liminf_{x\to\infty}\left\{V(x)\,\mathrm{e}^{2/Y}\int_{-Y}^{Y}\frac{\sin^2 u}{u^2}\,\mathrm{d}u\right\} + \left(\int_{-\infty}^{-Y}+\int_{Y}^{\infty}\right)\mathrm{O}(1)\frac{\sin^2 u}{u^2}\,\mathrm{d}u,\end{aligned}$$

if $X=1/Y$ and $\alpha=Y^2$ for $Y>0$ again. By letting $Y\to\infty$, one therefore sees that

$$\beta \leqq \liminf_{x\to\infty} V(x)\,\beta,$$

and so Theorem 1.4 follows. □

In order to complete the proof of the abstract prime number theorem, now put

$$\varphi(y) = \delta\psi(\mathrm{e}^{y/\delta}), \qquad F(z) = \delta z^{-1}\tilde{\Lambda}(\delta z).$$

Then

$$F(z) = \int_0^\infty \mathrm{e}^{-zy}\varphi(y)\,\mathrm{d}y \quad \text{for } \operatorname{Re} z > 1,$$

by Lemma 1.2, and that lemma implies that $F(z)$ satisfies all the required conditions of Ikehara's theorem. Therefore

$$\delta\psi(\mathrm{e}^{y/\delta}) = \varphi(y) \sim \mathrm{e}^{y} \quad \text{as } y\to\infty,$$

which shows that $\psi(x)\sim\delta^{-1}x^{\delta}$ as $x\to\infty$. □

Finally, in view of the use made of this conclusion in Chapter 5, we note that it now follows that $\theta(x)\sim\delta^{-1}x^{\delta}$ as $x\to\infty$, where

$$\theta(x) = \theta_G(x) = \sum_{\substack{p\in P\\ |p|\leqq x}} \log|p|.$$

For,

$$\begin{aligned}\psi(x) = N(\Lambda, x) &= \sum_{\substack{p\in P,\, r\geqq 1\\ |p^r|\leqq x}} \log|p|\\ &= \sum_{\substack{p\in P,\, r\geqq 1\\ |p|\leqq x^{1/r}}} \log|p| = \sum_{r\geqq 1}\theta(x^{1/r})\\ &= \theta(x) + \mathrm{O}\left(x^{\delta/2}\log x\right),\end{aligned}$$

since

$$\theta(x) \leqq \pi_G(x)\log x = \mathrm{O}(x^{\delta})$$

and since the sum over r ends when $x^{1/r} < q_0$, i.e., when $r > \log x / \log q_0$, so that only $O(\log x)$ terms are involved. This proves the statement about $\theta(x)$.

It is worth mentioning that various authors have investigated the difference $\pi_G(x) - \mathrm{li}\,(x^\delta)$ as $x \to \infty$, where $\mathrm{li}\,(x)$ is the 'logarithmic integral'

$$\int_{q_0}^{x} \frac{1}{\log u}\, du.$$

In particular, Wegmann [1] has proved a theorem which implies that

$$\pi_G(x) - \mathrm{li}\,(x^\delta) = O\big(x^\delta (\log x)^{-\alpha}\big)$$

for every $\alpha > 0$. The proof depends on subtle "elementary" methods; see also Segal [1]. By using deeper tools of analysis, Malliavin [1], R. S. Hall [1] and Müller [1] have obtained sharper estimates, and still sharper ones are known for the particular semigroups $G_{\mathbf{Z}}$ and G_K; see Mitsui [1] and Walfisz [1]. (The fact that, for example, the above result of Wegmann implies the abstract prime number theorem may be seen by considering $\mathrm{li}\,(x)$ and integrating by parts, expressing $\int_{q_0}^{x}$ as $\int_{q_0}^{\sqrt{x}} + \int_{\sqrt{x}}^{x}$ in order to estimate the error term that arises; compare Lemma 2.6 below.)

§ 2. Asymptotic properties of prime-divisor functions

This section is concerned largely with applications of some results of § 1 to asymptotic questions involving the functions ω, Ω, μ, λ and λ_k defined in Chapter 2.

Firstly, consider the subsets G_{even} and G_{odd} of G consisting of all elements a such that $\Omega(a)$ is even or odd, respectively. Although the conclusions of the next theorem may appear to be as expected, they are not necessarily true without Axiom A.

2.1. Theorem. *The sets G_{even} and G_{odd} both have asymptotic density $\frac{1}{2}$ in G. In addition, they both have asymptotic relative density $\frac{1}{2}$ in the set G_k of all k-free elements of G.*

For the proof, first consider the following consequence of Ikehara's theorem 1.4.

2.2. Lemma. *Let $f \in \mathrm{Dir}\,(G)$ be a real-valued function which is dominated by a non-negative real-valued function $f_1 \in \mathrm{Dir}\,(G)$ in the sense that $f(a) =$*

$=O(f_1(a))$ *as* $|a|\to\infty$. *Suppose that* $\tilde{f}_1(z)$ *converges for* $\operatorname{Re} z>\delta$, *and that it can be extended to an analytic function of* z *for* $\operatorname{Re} z\geqq\delta$, $z\neq\delta$, *in such a way that* $\tilde{f}_1(z)$ *has a simple pole with residue* 1 *at* $z=\delta$. *Suppose also that* $\tilde{f}(z)$ *can be extended to an analytic function of* z *for* $\operatorname{Re} z\geqq\delta$, *with the possible exception of a simple pole with real residue at* $z=\delta$. *Then*

$$N(f,x)=\alpha\delta^{-1}x^{\delta}+o(x^{\delta})\quad \text{as } x\to\infty,$$

where α *denotes the residue of* $\tilde{f}(z)$ *at* $z=\delta$ *if* $\tilde{f}(z)$ *has a pole there, and* $\alpha=0$ *otherwise.*

Proof. Since f is dominated by f_1, $\tilde{f}(z)$ converges absolutely for $\operatorname{Re} z>\delta$. Therefore, for $z=\sigma>\delta$,

$$|N(f,x)x^{-z}|\leqq x^{-\sigma}\sum_{|a|\leqq x}|f(a)|\leqq x^{-(\sigma-\delta)/2}\sum_{|a|\leqq x}|f(a)|\,|a|^{-\delta-(\sigma-\delta)/2}$$

$$=O(x^{-(\sigma-\delta)/2})=o(1)\quad \text{as } x\to\infty.$$

Thus the Partial Summation Lemma 4.2.3 implies that, for $\operatorname{Re} z>\delta$,

$$\tilde{f}(z)=z\int_1^{\infty}N(f,t)t^{-z-1}\,dt=\frac{z}{\delta}\int_0^{\infty}e^{-zy/\delta}N(f,e^{y/\delta})\,dy.$$

A similar formula holds for f_1, and, by the assumption on f_1, $\delta z^{-1}\tilde{f}_1(z)$ defines an analytic function for $\operatorname{Re} z\geqq\delta$, $z\neq\delta$, which has a simple pole with residue 1 at $z=\delta$. As in the application of Ikehara's theorem to $\tilde{\Lambda}(z)$ in § 1, now let

$$\varphi(y)=\delta N(f_1,e^{y/\delta}),\qquad F(z)=\delta z^{-1}\tilde{f}_1(\delta z).$$

Then, in the same way, Ikehara's theorem implies that $N(f_1,x)\sim\delta^{-1}x^{\delta}$ as $x\to\infty$, which is the required conclusion for f_1.

Next define

$$g(z)=(\alpha+C)^{-1}[f(z)+Cf_1(z)]$$

in Dir (G), where $C>0$ is a constant such that $\alpha+C>0$ and $|f(a)|\leqq$ $\leqq Cf_1(a)$ for $a\in G$. Then g is a non-negative real-valued function on G with the same analytic properties as f_1 above. Hence

$$N(f,x)+CN(f_1,x)\sim\delta^{-1}(\alpha+C)x^{\delta}\quad \text{as } x\to\infty,$$

and the lemma follows. □

Now consider

2.3. Lemma. *The functions* μ, λ *and* λ_k *all have the asymptotic mean-value* 0 *on* G.

Proof. By definition of μ and Theorem 2.6.1,

$$\mu(z) = 1/\zeta_G(z), \qquad \lambda(z) = \zeta_G(2z)/\zeta_G(z),$$

and

$$\lambda_k(z) = \begin{cases} \zeta_G(2z)/\zeta_G(z)\zeta_G(kz) & \text{if } k \text{ is even,} \\ \zeta_G(2z)\zeta_G(kz)/\zeta_G(z)\zeta_G(2kz) & \text{if } k \text{ is odd.} \end{cases}$$

Since $\zeta_G(z)$ converges and is non-zero for $\operatorname{Re} z>\delta$, and may be continued analytically into the entire region $\operatorname{Re} z>\eta$, $z\neq\delta$, in such a way that it is non-zero for $\operatorname{Re} z\geqq\delta$, $z\neq\delta$, it follows that each of the preceding functions f defines an absolutely convergent series $\hat{f}(z)$ for $\operatorname{Re} z>\delta$ that may be continued analytically into the entire region $\operatorname{Re} z\geqq\delta$. The present lemma therefore follows from Lemma 2.2, since each of these functions f is dominated by $f_1(z)=(\delta A)^{-1}\zeta_G(z)$. □

In order to deduce Theorem 2.1, now define

$$N_{\text{even}}(x) = \sum_{\substack{a\in G_{\text{even}} \\ |a|\leqq x}} 1, \qquad N_{\text{odd}}(x) = \sum_{\substack{a\in G_{\text{odd}} \\ |a|\leqq x}} 1,$$

$$N^{(k)}_{\text{even}}(x) = \sum_{\substack{a\in G_k\cap G_{\text{even}} \\ |a|\leqq x}} 1, \qquad N^{(k)}_{\text{odd}}(x) = \sum_{\substack{a\in G_k\cap G_{\text{odd}} \\ |a|\leqq x}} 1.$$

Then

$$N_{\text{even}}(x)+N_{\text{odd}}(x) = N_G(x) \sim Ax^{\delta} \quad \text{as } x\to\infty,$$

and, by Proposition 4.5.5,

$$N^{(k)}_{\text{even}}(x)+N^{(k)}_{\text{odd}}(x) = Q_k(x) \sim \frac{Ax^{\delta}}{\zeta_G(k\delta)} \quad \text{as } x\to\infty.$$

On the other hand,

$$N_{\text{even}}(x)-N_{\text{odd}}(x) = \sum_{|a|\leqq x} (-1)^{\Omega(a)} = N(\lambda, x),$$

while

$$N^{(k)}_{\text{even}}(x)-N^{(k)}_{\text{odd}}(x) = \sum_{\substack{a\in G_k \\ |a|\leqq x}} (-1)^{\Omega(a)} = N(\lambda_k, x).$$

Therefore Theorem 2.1 follows from Lemma 2.3. □

Next, in contrast with the fact that the divisor function d has the approximate average value $\delta A \log x$ on $G[x]$, we find that the prime-divisor functions ω and Ω both have the approximate average value $\log\log x$ on $G[x]$:

2.4. Theorem. *As* $x \to \infty$,

$$N(\omega, x) = Ax^{\delta}(\log\log x + B_G) + O(x^{\delta}/\log x),$$

$$N(\Omega, x) = Ax^{\delta}(\log\log x + B_G') + o(x^{\delta}),$$

where

$$B_G = \lim_{x\to\infty} \Big\{ \sum_{\substack{p\in P\\ |p|\leqq x}} |p|^{-\delta} - \log\log x \Big\},$$

$$B_G' = B_G + \sum_{p\in P} \{|p|^{\delta}(|p|^{\delta}-1)\}^{-1}.$$

The proof requires:

2.5. Lemma. *There exists a constant* B_G *such that*

$$\sum_{\substack{p\in P\\ |p|\leqq x}} |p|^{-\delta} = \log\log x + B_G + O(1/\log x).$$

Proof. Consider

$$\begin{aligned}\sum_{|a|\leqq x} |a|^{-\delta}\Lambda(a) - \sum_{\substack{p\in P\\ |p|\leqq x}} |p|^{-\delta}\log|p| &= \sum_{\substack{p\in P,\, r\geqq 2\\ |p^r|\leqq x}} |p|^{-m\delta}\log|p| \\ &< \sum_{p\in P} (|p|^{-2\delta} + |p|^{-3\delta} + \ldots)\log|p| \\ &= \sum_{p\in P} |p|^{-\delta}(|p|^{\delta}-1)^{-1}\log|p|.\end{aligned}$$

The last series is dominated by

$$\sum_{a\in G} |a|^{-\delta}(|a|^{\delta}-1)^{-1}\log|a| < \infty.$$

and so

$$\sum_{p\in P} |p|^{-\delta}\log|p| = \sum_{|a|\leqq x} |a|^{-\delta}\Lambda(a) + O(1).$$

Now, partial summation and the conclusion about $\psi(x)$ of § 1 gives

$$\begin{aligned}\sum_{|a|\leqq x} |a|^{-\delta}\Lambda(a) &= (Ax^{\delta})^{-1} \sum_{|a|\leqq x} \Lambda(a)\{N_G(|a|^{-1}x) + O(|a|^{-\eta}x^{\eta})\} \\ &= (Ax^{\delta})^{-1}\{ \sum_{|a|\leqq x} \log|a| + O(x^{\eta} \sum_{|a|\leqq x} |a|^{-\eta}\Lambda(a)\} \\ &= \log x + O(1).\end{aligned}$$

Thus

$$\sum_{\substack{p\in P\\ |p|\leqq x}} |p|^{-\delta}\log|p| = \log x + S(x),$$

where $S(x)=O(1)$. Partial summation then gives

$$\sum_{\substack{p\in P\\|p|\leqq x}} |p|^{-\delta} = \frac{\log x+S(x)}{\log x}+\int_{q_0}^{x}\frac{\log t+S(t)}{t(\log t)^2}\,dt$$

$$= 1+\frac{S(x)}{\log x}+\log\log x-\log\log q_0$$

$$+\int_{q_0}^{\infty}\frac{S(t)\,dt}{t(\log t)^2}-\int_{x}^{\infty}\frac{S(t)\,dt}{t(\log t)^2}.$$

Since $S(x)=O(1)$, the conclusion of the lemma follows with B_G a constant depending explicitly on the function $S(x)$ (but not on x).

Now note that

$$N(\omega, x) = \sum_{\substack{|a|\leqq x\\ p\in P,\, p|a}} 1 = \sum_{\substack{p\in P\\|p|\leqq x}} N_G(x/|p|)$$

$$= \sum_{\substack{p\in P\\|p|\leqq x}} \{Ax^{\delta}|p|^{-\delta}+O(x^{\eta}|p|^{-\eta})\}.$$

Then partial summation and Lemma 2.5 imply that, if

$$T(x) = \sum_{\substack{p\in P\\|p|\leqq x}} |p|^{-\delta},$$

then

$$\sum_{\substack{p\in P\\|p|\leqq x}} |p|^{-\eta} = T(x)x^{\delta-\eta}-(\delta-\eta)\int_{q_0}^{x} T(t)t^{\delta-\eta-1}\,dt$$

$$= O(x^{\delta-\eta}/\log x),$$

by use of the asymptotic estimate for $T(x)$, integration by parts, and use of the following elementary but frequently useful estimate.

2.6. Lemma. *For any* $\alpha>0$, β *and* $c>1$,

$$\int_c^x t^{\alpha-1}(\log t)^{\beta}\,dt = \alpha^{-1}x^{\alpha}(\log x)^{\beta}+O\big(x^{\alpha}(\log x)^{\beta-1}\big).$$

Proof. For $x\geqq c^2$ and $\beta'>0$,

$$\int_c^x t^{\alpha-1}(\log t)^{-\beta'}\,dt = \Big(\int_c^{\sqrt{x}}+\int_{\sqrt{x}}^{x}\Big)\,t^{\alpha-1}(\log t)^{-\beta'}\,dt$$

$$< \frac{x^{\alpha/2}-c^{\alpha}}{\alpha(\log c)^{\beta'}}+\frac{x^{\alpha}-x^{\alpha/2}}{\alpha(\frac{1}{2}\log x)^{\beta'}} = O\big(x^{\alpha}(\log x)^{-\beta'}\big).$$

The lemma therefore follows, by integrating by parts. □

The conclusion for ω now follows from Lemma 2.5 and the estimate for

$$\sum_{\substack{p \in P \\ |p| \leqq x}} |p|^{-\eta}.$$

For Ω we have

$$N(\Omega, x) = \sum_{\substack{|a| \leqq x,\, p^m | a \\ p \in P,\, m \geqq 1}} 1$$

$$= N(\omega, x) + \sum_{\substack{p \in P \\ m \geqq 2}} N_G(x/|p|^m)$$

$$= N(\omega, x) + Ax^\delta \sum_{\substack{p \in P \\ M(x,p) \geqq 2}} (|p|^{-2\delta} + \ldots + |p|^{-M(x,p)\delta})$$

$$+ \mathrm{O}\Big(x^\eta \sum_{\substack{p \in P \\ M(x,p) \geqq 2}} (|p|^{-2\eta} + \ldots + |p|^{-M(x,p)\eta})\Big),$$

where $M(x, p) = [\log x / \log |p|]$, the integral part of $\log x / \log |p|$. Therefore

$$N(\Omega, x) = N(\omega, x) + Ax^\delta \left\{ \sum_{p \in P} \sum_{r=2}^{\infty} |p|^{-r\delta} + \mathrm{o}(1) \right\}$$

$$+ \mathrm{O}(x^\eta \sum_{p \in P} M(x, p) |p|^{-\eta}).$$

Now

$$\psi(x) = N(\Lambda, x) = \sum_{p \in P} M(x, p) \log |p|;$$

so partial summation yields

$$\sum_{p \in P} M(x, p) |p|^{-\eta} = (x^\eta \log x)^{-1} \psi(x) - \int_{q_0}^{x} \psi(t) g'(t)\, \mathrm{d}t$$

where $g(t) = (t^\eta \log t)^{-1}$. Since $\psi(x) = \mathrm{O}(x^\delta)$, Lemma 2.6 then implies that

$$\sum_{p \in P} M(x, p) |p|^{-\eta} = \mathrm{O}(x^{\delta - \eta} / \log x),$$

and so the estimate for Ω follows from that for ω. □

Lastly, we consider the frequency with which the functions ω and Ω take on a given value $k \geqq 1$. For this purpose, consider the subsets T_k, R_k and P_k of G consisting of all elements $a \in G$ such that (i) $\Omega(a) = k$, or (ii) $\omega(a) = k$, or (iii) $\omega(a) = k$ and a is square-free, respectively.

2.7. Theorem (Landau's Formulae). *The subsets T_k, R_k and P_k each have the approximate density*

$$\frac{(\log\log x)^{k-1}}{(k-1)!\,\delta A \log x}$$

in $G[x]$.

Proof. Let $\tau_k(x)$, $\varrho_k(x)$ and $\pi_k(x)$ denote the total numbers of elements $a \in G$ with $|a| \leqq x$ such that a lies in T_k, R_k or P_k, respectively. Consider the functions

$$L_k(x) = \sum |p_1 p_2 \dots p_k|^{-\delta}, \qquad \Pi_k(x) = \sum 1,$$

$$\theta_k(x) = \sum \log |p_1 p_2 \dots p_k|,$$

where each sum is over all ordered k-tuples $(p_1, \dots, p_k)$ of primes $p_i \in P$ such that $|p_1 p_2 \dots p_k| \leqq x$. Then it is not difficult to verify that

$$k!\,\pi_k(x) \leqq \Pi_k(x) \leqq k!\,\tau_k(x),$$

and, for $k \geqq 2$,

$$\tau_k(x) - \pi_k(x) \leqq \sum \{1:\ |p_1 p_2 \dots p_{k-1} p_{k-1}| \leqq x\}$$

$$\leqq \sum \{1:\ |p_1 p_2 \dots p_{k-1}| \leqq x\} = \Pi_{k-1}(x).$$

Therefore, for $k \geqq 2$,

$$(k!)^{-1}\Pi_k(x) - \Pi_{k-1}(x) \leqq \pi_k(x) \leqq (k!)^{-1}\Pi_k(x),$$

$$(k!)^{-1}\Pi_k(x) \leqq \tau_k(x) \leqq \pi_k(x) + \Pi_{k-1}(x).$$

Now,

$$\tau_1(x) = \pi_1(x) = \pi_G(x)$$

$$\sim x^\delta/\delta \log x \quad \text{as } x \to \infty.$$

Also

$$\varrho_1(x) = \sum \{1:\ p \in P,\ m \geqq 1,\ |p^m| \leqq x\}$$

$$= \sum \{\pi_G(x^{1/m}):\ m \leqq \log x/\log q_0\}$$

$$= \pi_G(x) + O\big(\pi_G(x^{1/2}) \log x\big)$$

$$\sim x^\delta/\delta \log x \quad \text{as } x \to \infty.$$

Hence it is only necessary to consider the case when $k \geqq 2$, and the above inequalities for $\pi_k(x)$ and $\tau_k(x)$ show that the results for these functions will follow from the conclusion

$$\Pi_k(x) \sim k(\delta \log x)^{-1} x^\delta (\log\log x)^{k-1} \quad \text{as } x \to \infty.$$

For $\varrho_k(x)$, consider the inequality

$$\begin{aligned} 0 &\leqq \varrho_k(x) - \pi_k(x) \\ &= \sum \{1\colon\ p \in P,\ m \geqq 2, \omega(c) = k-1,\ |p^m c| \leqq x\} \\ &= \sum \{\pi_G((x/|c|)^{1/m})\colon\ \omega(c) = k-1,\ |c| \leqq x,\ 2 \leqq m \leqq \log(x/|c|)/\log q_0\} \\ &= \sum \{O((x/|c|)^{\delta/2})\colon\ \omega(c) = k-1,\ |c| \leqq x\}. \end{aligned}$$

Then suppose, as an inductive hypothesis, that

$$\varrho_r(x) = O(x^\delta (\log x)^{-1} (\log\log x)^{r-1});$$

for $r=1$, this was verified above. In that case, partial summation yields

$$\begin{aligned} \sum \{|c|^{-\delta/2}\colon\ \omega(c) = r,\ |c| \leqq x\} &= \varrho_r(x) x^{-1/2} + \tfrac{1}{2}\delta \int_1^x \varrho_r(t) t^{-\delta/2-1}\,dt \\ &= O(x^{\delta/2} (\log x)^{-1} (\log\log x)^{r-1}) + O(x^{\delta/2} (\log x)^{\varepsilon-2}), \end{aligned}$$

after integrating by parts and using Lemma 2.6 again (with $\varepsilon > 0$ arbitrary). Therefore

$$0 \leqq \varrho_{r+1}(x) - \pi_{r+1}(x) = O(x^\delta (\log x)^{-1} (\log\log x)^{r-1}).$$

If one assumes that the asymptotic estimate for $\pi_k(x)$ has already been proved, then the inductive conclusion about $\varrho_{r+1}(x)$ follows. Hence the required asymptotic estimate for $\varrho_k(x)$ will follow from that for $\pi_k(x)$ and the theorem will be proved, as soon as the above asymptotic statement about $\Pi_k(x)$ has been established.

For this purpose, note that

$$\Pi_k(x) = \sum_{|a| \leqq x} n(a), \qquad \theta_k(x) = \sum_{|a| \leqq x} n(a) \log |a|,$$

where $n(a)$ is the total number of ordered k-tuples $(p_1, \ldots, p_k)$ of primes $p_i \in P$ such that $p_1 p_2 \ldots p_k = a$. Therefore, by partial summation,

$$\begin{aligned} \theta_k(x) &= \Pi_k(x) \log x - \int_1^x \Pi_k(t) t^{-1}\,dt \\ &= \Pi_k(x) \log x + O(x^\delta), \end{aligned}$$

since

$$\Pi_k(t) \leqq k!\, \tau_k(t) = O(N_G(t)) = O(t^\delta).$$

Hence the stated asymptotic formula for $\Pi_k(x)$ will follow when it has been proved that

$$\theta_k(x) \sim k\delta^{-1} (\log\log x)^{k-1} \quad \text{as } x \to \infty.$$

In order to do this, consider

$$k\theta_{k+1}(x) = \sum_{|p_1 p_2 \cdots p_{k+1}| \leq x} \log |p_1 p_2 \cdots p_{k+1}|^k$$
$$= \sum_{|p_1 p_2 \cdots p_{k+1}| \leq x} \{\log |p_2 p_3 \cdots p_{k+1}| +$$
$$+ \log |p_1 p_3 \cdots p_{k+1}| + \ldots + \log |p_1 p_2 \cdots p_k|\}$$
$$= (k+1) \sum_{|p_1| \leq x} \theta_k(x/|p_1|).$$

Also, if $L_0(x)=1$, consider

$$L_k(x) = \sum_{|p_1 p_2 \cdots p_k| \leq x} |p_1 p_2 \cdots p_k|^{-\delta}$$
$$= \sum_{|p_1| \leq x} |p_1|^{-\delta} L_{k-1}(x/|p_1|).$$

If one writes

$$f_k(x) = \theta_k(x) - k\delta^{-1} x^\delta L_{k-1}(x),$$

the above equations then yield

$$k f_{k+1}(x) = (k+1) \sum_{\substack{p \in P \\ |p| \leq x}} f_k(x/|p|).$$

Now, by § 1,

$$f_1(x) = \theta_G(x) - \delta^{-1} x^\delta = o(x^\delta) \quad \text{as } x \to \infty.$$

Then suppose as an inductive hypothesis that

$$f_k(x) = o\big(x^\delta (\log \log x)^{k-1}\big) \quad \text{as } x \to \infty.$$

In that case, given any $\varepsilon > 0$, there is a number $x_0 = x_0(k, \varepsilon)$ such that

$$|f_k(x)| < \varepsilon x^\delta (\log \log x)^{k-1}$$

whenever $x \geqq x_0$, while there is a constant $B = B(k, \varepsilon)$ such that

$$|f_k(x)| < B$$

whenever $x < x_0$. Therefore

$$|f_{k+1}(x)| \leqq \frac{k+1}{k} \Big(\sum_{|p| \leqq x/x_0} + \sum_{x/x_0 < |p| \leqq x} \Big) |f_k(x/|p|)$$

$$< 2\varepsilon (\log \log x)^{k-1} \sum_{|p| \leqq x/x_0} (x/|p|)^\delta + 2B \sum_{x/x_0 < |p| \leqq x} 1$$

$$< 4\varepsilon x^\delta (\log \log x)^k + 4ABx^\delta < 5\varepsilon x^\delta (\log \log x)^k$$

for x sufficiently large. Hence the inductive statement follows, and so

$$\theta_k(x) = k\delta^{-1} x L_{k-1}(x) + o\big(x^\delta (\log\log x)^{k-1}\big) \quad \text{as } x \to \infty.$$

Finally,

$$\Big(\sum_{|p| \leqq x^{1/k}} |p|^{-\delta}\Big)^k \leqq L_k(x) \leqq \Big(\sum_{|p| \leqq x} |p|^{-\delta}\Big)^k,$$

and therefore Lemma 2.5 now implies that

$$L_k(x) \sim (\log\log x)^k \quad \text{as } x \to \infty.$$

This gives the required result for $\theta_k(x)$, and hence the theorem is proved. □

§ 3. Maximum and minimum orders of magnitude of certain functions

This section contains some applications of a corollary of the abstract prime number theorem to the determination of the 'true' maximum and minimum orders of magnitude of the divisor-sum function

$$\sigma_w(a) = \sum_{d|a} |d|^w$$

and the Euler-type function

$$\varphi_w(a) = \sum_{d|a} \mu(d)\, |a/d|^w.$$

As a preliminary, we need asymptotic information about the Euler product for G at the critical value $z = \delta$, in the form of

3.1. Lemma (Mertens's Formula). *As* $x \to \infty$,

$$\prod_{\substack{p \in P \\ |p| \leqq x}} (1 - |p|^\delta) = e^{-\gamma}/(\delta A \log x) + O((\log x)^{-2}),$$

where $\gamma = 0.57721\ldots$ *is the classical Euler constant.*

Proof. Consider the series

$$-\sum_{p \in P} \{|p|^{-\delta} + \log(1 - |p|^{-\delta})\} = \sum_{p \in P} \sum_{m=2}^{\infty} m^{-1} |p|^{-m\delta}$$

$$< \tfrac{1}{2} \sum_{p \in P} \sum_{m=2}^{\infty} |p|^{-m\delta}$$

$$= \tfrac{1}{2} \sum_{p \in P} |p|^{-\delta} (|p|^\delta - 1)^{-1},$$

which converges (compare the proof of Lemma 2.5). Therefore

$$-\sum_{\substack{p\in P\\|p|\leqq x}} \log(1-|p|^{-\delta}) = \sum_{|p|\leqq x}\sum_{m=1}^{\infty} m^{-1}|p|^{-m\delta}$$

$$= H(\delta)+r(x)+\sum_{|p|\leqq x}|p|^{-\delta},$$

say, where $H(\delta)$ denotes the sum of the first series above. By Lemma 2.5, it follows that

$$-\sum_{|p|\leqq x}\log(1-|p|^{-\delta}) = \log\log x + B_G + H(\delta) + O((\log x)^{-1}),$$

since the preceding discussion of $H(\delta)$ shows that

$$r(x) = \sum_{|p|\leqq x}\sum_{m=2} m^{-1}|p|^{-m\delta} = O\left(\sum_{|p|>x}|p|^{-\delta}(|p|^{\delta}-1)^{-1}\right)$$

$$= O\left(\sum_{|a|>x}|a|^{-2\delta}\right) = O(x^{-\delta}),$$

by Proposition 4.2.6. Therefore

$$\prod_{|p|\leqq x}(1-|p|^{-\delta}) = (e^{-C}/\log x)\exp[O(1/\log x)]$$

$$= (e^{-C}/\log x)\{1+O(1/\log x)\},$$

where $C=B_G+H(\delta)$.

In order to evaluate C, now consider any real s and

$$H(s) = -\sum_{p\in P}\{|p|^{-s}+\log(1-|p|^{-s})\} = \sum_{p\in P}\sum_{m=2}^{\infty} m^{-1}|p|^{-ms}.$$

By slightly modifying the above discussion of $H(\delta)$ and $r(x)$, one sees that $H(s)$ is convergent for $s>\frac{1}{2}\delta$, and uniformly convergent in $s\geqq s_0$ for any fixed $s_0>\frac{1}{2}\delta$. In particular, $\lim_{s\to\delta} H(s)=H(\delta)$. Also, for real $s>\delta$, the Euler product formula implies that

$$\log\zeta_G(s) = -\sum_{p\in P}\log(1-|p|^{-s}) = \sum_{p\in P}\sum_{m=1}^{\infty} m^{-1}|p|^{-ms}$$

$$= P_G(s)+H(s),$$

where

$$P_G(z) = \sum_{p\in P}|p|^{-z}.$$

Now the fact that $\zeta_G(z)$ has a simple pole with residue δA at $z=\delta$ implies that

$$\lim_{z\to\delta}\{(z-\delta)\zeta_G(z)\} = \delta A.$$

Therefore

$$\lim_{s\to\delta+} \{\log(s-\delta)+\log\zeta_G(s)\} = \log\delta A,$$

and this implies that

$$H(\delta) = \lim_{s\to\delta+} H(s) = \log\delta A - \lim_{s\to\delta+} \{\log(s-\delta)+P_G(s)\},$$

assuming that the right-hand limit exists.

In order to show that the limit does exist and to determine its value, consider real $s>\delta$ and write $P_G(s)$ as

$$\sum_{p\in P} |p|^{-\delta}|p|^{\delta-s}.$$

Then the Partial Summation Lemma 4.2.3 implies that

$$P_G(s) = (s-\delta)\int_{q_0}^{\infty} T(t)\,t^{\delta-s-1}\,dt,$$

where

$$T(x) = \sum_{|p|\leqq x} |p|^{-\delta},$$

since Lemma 2.5 implies that

$$T(x)x^{\delta-s} = o(1) \quad \text{as } x\to\infty.$$

The asymptotic formula of Lemma 2.5 then implies that

$$P_G(s) = B_G q_0{}^{\delta-s}+(s-\delta)\int_{q_0}^{\infty} [\log\log t+O(1/\log t)]\,t^{\delta-s-1}\,dt.$$

Substituting $u=(s-\delta)\log t$, one then has

$$(s-\delta)\int_{q_0}^{\infty} (\log\log t)\,t^{\delta-s-1}\,dt = \int_{(s-\delta)\log q_0}^{\infty} e^{-u}\log u\,du - q_0{}^{\delta-s}\log(s-\delta).$$

Now, by a standard formula from the theory of the Gamma function $\Gamma(z)$, the classical Euler constant (of $G_{\mathbf{Z}}$) can be expressed as

$$\gamma = -\Gamma'(1) = -\int_0^{\infty} e^{-u}\log u\,du.$$

Therefore

$$(s-\delta)\int_{q_0}^{\infty} (\log\log t)\,t^{\delta-s-1}\,dt = -\gamma-\log(s-\delta)+o(1) \quad \text{as } s\to\delta+.$$

Also, the above term $O(1/\log t)$ will be less than any given $\varepsilon>0$ in modulus whenever $t\geqq x_0=x_0(\varepsilon)$, say. Therefore

$$\left|(s-\delta)\int_{q_0}^{\infty} O(1/\log t)\,t^{\delta-s-1}\,dt\right| < \varepsilon x_0{}^{\delta-s}+(s-\delta)\int_{q_0}^{x_0},$$

for short, where it may be assumed that $x_0\geqq q_0$. Hence the right-hand quantity will be less than 2ε for $\delta<s<s_0=s_0(\varepsilon, x_0)=s_0(\varepsilon)$.

It therefore follows that

$$\log(s-\delta)+P_G(s) \to B_G-\gamma \quad \text{as } s \to \delta+,$$

and so

$$C = B_G+H(\delta) = \gamma+\log\delta A.$$

This proves Lemma 3.1. □

Now consider the special functions φ_w and σ_w. The discussion of Chapter 4, § 1 shows that for an arithmetical semigroup satisfying Axiom A the function φ_δ is of particular importance. On the other hand, $d=\sigma_0$ and $\sigma=\sigma_1$ are perhaps the most interesting cases arising from the functions σ_w. For these reasons, the present discussion will be confined to φ_δ and σ, and in view of some statements in Chapter 5, § 2 it will only be necessary to determine their minimum and maximum orders of magnitude, respectively.

3.2. Theorem. *The (true) minimum order of magnitude of $\varphi_\delta(a)$ for $a\in G$* is

$$e^{-\gamma}|a|^\delta/(A\log\log|a|),$$

where γ is the classical Euler constant.

Proof. For $x\geqq q_0$, consider

$$a = a(x) = \prod_{\substack{p\in P\\ |p|\leqq x}} p.$$

Then

$$\varphi_\delta(a) = |a|^\delta \prod_{|p|\leqq x}(1-|p|^{-\delta}) \sim |a|^\delta e^{-\gamma}/(\delta A\log x) \quad \text{as } x\to\infty,$$

by Lemma 3.1. Also, by § 1,

$$\log|a| = \theta_G(x) = \delta^{-1}x^\delta[1+o(1)],$$

and so

$$\log\log|a| \sim \delta\log x \quad \text{as } x\to\infty.$$

Thus

$$\begin{aligned}\varphi_\delta(a) &\sim |a|^\delta e^{-\gamma}/(A\log\log|a|) \quad \text{as } x\to\infty,\\ &< (1+\varepsilon)|a|^\delta e^{-\gamma}/(A\log\log|a|) \quad \text{whenever} \quad x\geqq x_0=x_0(\varepsilon).\end{aligned}$$

Now consider an arbitrary element $c\in G$. Then

$$\begin{aligned}\varphi_\delta(c) &= |c|^\delta \prod\{(1-|p|^{-\delta}):\ p\in P,\ p|c,\ |p|^\delta\leqq\log|c|\}\\ &\quad\times\prod\{(1-|p|^{-\delta}):\ p\in P,\ p|c,\ |p|^\delta>\log|c|\}\\ &= |c|^\delta\Pi_1\Pi_2,\end{aligned}$$

say. Further

$$\begin{aligned}\Pi_1 &\geqq \prod \{(1-|p|^{-\delta}):\ p\in P,\ |p|^\delta \leqq \log|c|\}\\ &\sim e^{-\gamma}/(A\log\log|c|) \quad \text{as } |c|\to\infty,\\ &> (1-\varepsilon)e^{-\gamma}/\{(1-\tfrac{1}{2}\varepsilon)A\log\log|c|\}\end{aligned}$$

for $|c|\geqq x_1(\varepsilon)$, say. On the other hand, if Π_2 involves $n=n(c)$ factors then

$$|c|^\delta \geqq \prod \{|p|^\delta:\ p\in P,\ p|c,\ |p|^\delta > \log|c|\} \geqq (\log|c|)^n,$$

and so $n\leqq\delta\log|c|/\log\log|c|$. Therefore

$$\begin{aligned}\Pi_2 &\geqq (1-1/\log|c|)^n\\ &\geqq (1-1/\log|c|)^{\delta\log|c|/\log\log|c|}\\ &= \exp\{(\delta\log|c|/\log\log|c|)\log(1-1/\log|c|)\}\\ &= \exp[O(1/\log\log|c|)]\\ &> 1-\tfrac{1}{2}\varepsilon\end{aligned}$$

for $|c|\geqq x_2(\varepsilon)$, say. Hence the theorem follows. □

In considering the maximum order of magnitude of σ, the case in which $\delta\leqq 1$ appears to be more amenable than the general case. Since all particular natural arithmetical semigroups satisfying Axiom A that were discussed earlier have $\delta\leqq 1$, we shall restrict attention to this case.

3.3. Theorem. *Let G denote an arithmetical semigroup satisfying Axiom* A *with $\delta\leqq 1$. Then the (true) maximum order of magnitude of $\sigma(a)$ for $a\in G$ is*

$$Ae^\gamma|a|\log\log|a| \quad \text{if } \delta=1,$$

and

$$\zeta_G(1)|a| \quad \text{if } \delta<1.$$

Proof. It follows from the proof of Proposition 4.7.3 that

$$\sigma_w(a)\varphi_w(a) = |a|^{2w}\prod_{i=1}^{m}(1-|p_i|^{-(r_i+1)w}) \quad \text{if } a=\prod_{i=1}^{m}p_i^{r_i},$$

where the p_i are distinct primes and $r_i\geqq 1$. Hence, for real $w>0$,

$$1 \geqq |a|^{-2w}\sigma_w(a)\varphi_w(a)$$

$$\Big[\geqq \prod_{\substack{p\in P\\ p|a}}(1-|p|^{-2w}) > (\zeta_G(2w))^{-1} \quad \text{if } w>\tfrac{1}{2}\delta\Big].$$

If $\delta=1$, it follows that

$$\frac{\sigma(a)}{|a|} \leqq \frac{|a|}{\varphi_1(a)} < \frac{(1+\varepsilon)A \log\log|a|}{(1+\frac{1}{2}\varepsilon)e^{-\gamma}} < \frac{(1+\varepsilon)A \log\log|a|}{e^{-\gamma}}$$

whenever $|a|$ is sufficiently large. On the other hand, if

$$b = b(x, k) = \prod_{\substack{p \in P \\ |p| \leqq x}} p^k,$$

where k is a positive integer, then

$$\sigma(b) = |b| \prod_{\substack{p \in P \\ |p| \leqq x}} \left[\frac{1-|p|^{-(k+1)}}{1-|p|^{-1}}\right] > \frac{(1-\varepsilon)|b| A \log x}{(1-\frac{1}{4}\varepsilon^2)\zeta_G(k+1)e^{-\gamma}}$$

for $x \geqq x_0(\varepsilon, k)$, by Lemma 3.1 and the Euler product formula for $\zeta_G(z)$. This is true for any fixed k, and by Lemma 3.4 below one may choose $k=k(\varepsilon)$ so that $\zeta_G(k+1)<1+\frac{1}{2}\varepsilon$. Fixing this k, choose $x \geqq x_0(\varepsilon, k)$ so that

$$\log\log|b| = \log k + \log\theta_G(x) < \frac{\log x}{1-\frac{1}{2}\varepsilon}.$$

Then the resulting element $b=b(x,k)=b(\varepsilon)$ satisfies

$$\sigma(b) > (1-\varepsilon)Ae^{\gamma}|b| \log\log|b|.$$

This proves the theorem when $\delta=1$.

3.4. Lemma. *For real* w,

$$\lim_{w\to\infty} \zeta_G(w) = 1.$$

Proof. For real $s>0$, let

$$F(s) = \zeta_G(1/s) = \sum_{a \in G} |a|^{-1/s}.$$

Then $F(s)$ is uniformly convergent in $s \leqq s_0$ whenever $1/s_0 > \delta$, by Lemma 4.2.7. Since $|a|^{-1/s} \to 0$ as $s \to 0+$ when $a \neq 1$, a standard 'extended continuity' property of uniformly convergent series (see Knopp [1] § 47, say) then implies that

$$\lim_{s\to 0+} F(s) = 1.$$

This proves the lemma (and here δ of Axiom A may be an arbitrary positive number). □

Finally let $\delta<1$. Then in this case

$$\frac{\sigma(a)}{|a|} \leqq \frac{|a|}{\varphi_1(a)} = \prod_{\substack{p \in P \\ p|a}} (1-|p|^{-1})^{-1} < \zeta_G(1).$$

On the other hand, if $b=b(x, k)$ is defined as before, then

$$\sigma(b) = |b| \prod_{\substack{p \in P \\ |p| \leqq x}} \left[\frac{1-|p|^{-(k+1)}}{1-|p|^{-1}}\right] > \frac{(1-\varepsilon)\zeta_G(1)|b|}{(1+\varepsilon)\zeta_G(k+1)}$$

for fixed k and $x \geqq x_0(\varepsilon, k)$. Then one may choose k so that $\zeta_G(k+1) < 1+\varepsilon$, and this yields an element b for which $\sigma(b) > (1-\varepsilon)\zeta_G(1)|b|$. Therefore this case follows also. □

§ 4. The "law of large numbers" for certain functions

Although much of the preceding discussion has been concerned with different types of "statistical" properties of arithmetical functions, one aspect of such questions which has not been considered so far concerns properties that may not necessarily be universally true over G but are nevertheless valid for 'almost all' elements. Here a property will be said to hold for **almost all** elements of G if and only if it is valid for all elements in some subset of asymptotic density 1 in G. In the first place, this section considers some results about the prime-divisor functions ω and Ω, which are true for almost all elements of G and are analogous to one form of the "law of large numbers" of probability theory.

In order to see the relationship with the probability-theoretical situation, consider initially an arbitrary real-valued function f on G. Given any $q>0$, consider as in Chapter 5, § 3 the discrete probability space consisting of $G[q]$ with the probability measure which assigns to a subset E of $G[q]$ the probability

$$\mathrm{Prob}_q E = \big(N_G(q)\big)^{-1} \operatorname{card} E.$$

The "random variable" f on $G[q]$ has expected value

$$m(f, q) = \big(N_G(q)\big)^{-1} N(f, q)$$

and the variance

$$v(f, q) = \big(N_G(q)\big)^{-1} \sum_{a \in G[q]} [f(a) - m(f, q)]^2.$$

Then the well-known Tchebychev inequality implies that, for any $\varepsilon>0$,

$$\mathrm{Prob}_q \{|f(a) - m(f, q)| \geqq \varepsilon\} \leqq \varepsilon^{-2} v(f, q).$$

If $v(f,q) \neq 0$ and $\xi(q) > 0$, it follows that

$$\mathrm{Prob}_q \{|f(a) - m(f,q)| \geqq \xi(q)\sqrt{v(f,q)}\} \leqq \xi(q)^{-2}.$$

Hence, for any function $\xi(q)$ such that $\xi(q) \to \infty$ as $q \to \infty$,

$$\mathrm{Prob}_q \{|f(a) - m(f,q)| < \xi(q)\sqrt{v(f,q)}\} \to 1 \quad \text{as } q \to \infty.$$

The last conclusion (which does not depend on Axiom A) may be regarded as an analogue of the probabilistic "law of large numbers" for f on G.

In the case when f is known to have the approximate average value $F(q)$ on $G[q]$, so that $m(f,q) \sim F(q)$ as $q \to \infty$, one may ask whether a result similar to the above is true if $m(f,q)$ is replaced by $F(q)$ and $v(f,q)$ is replaced by an estimate for

$$V(f,q) = (N_G(q))^{-1} \sum_{|a| \leqq q} [f(a) - F(q)]^2.$$

The following is a result of this type for the function ω, since ω has the approximate average value $\log\log q$ on $G[q]$ and the argument below shows that $V(\omega, q) = \mathrm{O}(\log\log q)$ as $q \to \infty$.

4.1. Proposition. *Let $\xi(q) > 0$ be any function of q such that $\xi(q) \to \infty$ as $q \to \infty$. Then*

$$\mathrm{Prob}_q \{|\omega(a) - \log\log q| < \xi(q)\sqrt{\log\log q}\} \to 1 \quad \textit{as } q \to \infty.$$

Proof. Consider

$$\sum_{|a| \leqq x} [\omega(a)]^2 = \sum_{|a| \leqq q} \sum_{\substack{p_1, p_2 \in P \\ p_i | a}} 1$$

$$= \sum_{|a| \leqq q} \{ \sum_{\substack{p_1 \neq p_2 \\ p_1 p_2 b = a}} 1 + \sum_{\substack{p \in P \\ pc = a}} 1 \}$$

$$= \sum_{p_1 \neq p_2} N_G(x/|p_1 p_2|) + \sum_p N_G(x/|p|)$$

$$= \sum_{p_1 \neq p_2} \{A(x/|p_1 p_2|)^\delta + \mathrm{O}((x/|p_1 p_2|)^\eta)\}$$

$$+ \sum_p \{A(x/|p|)^\delta + \mathrm{O}((x/|p|)^\eta)\}.$$

Now, for real σ,

$$(\sum_{|p| \leqq \sqrt{x}} |p|^\sigma)^2 \leqq \sum_{|p_1 p_2| \leqq x} |p_1 p_2|^\sigma \leqq (\sum_{|p| \leqq x} |p|^\sigma)^2.$$

Also $\sum_{p \in P} |p|^{-2\delta}$ converges, and, by Lemma 2.5 and the proof of Theorem 2.4,

$$\sum_{|p| \leq x} |p|^{-\delta} = \log\log x + O(1), \qquad \sum_{|p| \leq x} |p|^{-\eta} = O(x^{\delta-\eta}/\log x).$$

Therefore

$$\sum_{|a| \leq x} [\omega(a)]^2 = Ax^\delta[\log\log x + O(1)]^2 + O\Big(\sum_{|p_1| \leq x} (x/|p_1|)^\delta\Big)$$
$$+ Ax^\delta[\log\log x + O(1)] + O(x^\delta)$$
$$= Ax^\delta(\log\log x)^2 + O(x^\delta \log\log x).$$

By using this estimate and Theorem 2.4, it may be deduced that, as $q \to \infty$,

$$V(\omega, q) = (N_G(q))^{-1} \sum_{|a| \leq q} [\omega(a) - \log\log q]^2 = O(\log\log q).$$

Next, let $M(q, \xi)$ denote the total number of elements $a \in G[q]$ such that

$$|\omega(a) - \log\log q| \geqq \xi(q) \log\log q.$$

Then

$$V(\omega, q) \geqq (N_G(q))^{-1} M(q, \xi)(\xi(q))^2 \log\log q,$$

which implies that

$$M(q, \xi) = O((\xi(q))^{-2} N_G(q) = o(N_G(q)) \quad \text{as } q \to \infty.$$

This proves the theorem. □

4.2. Corollary. *Let $\xi(q) > 0$, $\xi(q) \leqq 2\xi(q')$ for $q^c \leqq q'$, sufficiently large q and some positive $c < 1$, and suppose that $\xi(q) \to \infty$ as $q \to \infty$. Then*

$$|\omega(a) - \log\log |a|| < \xi(|a|)(\log\log |a|)^{1/2}$$

for almost all elements $a \in G$, and

$$|\Omega(a) - \log\log |a|| < \xi(|a|)(\log\log |a|)^{1/2}$$

for almost all elements $a \in G$.

Proof. For $x^c \leqq |a| \leqq x$,

$$\log\log x + C \leqq \log\log |a| \leqq \log\log x,$$

where $C = \log c$. Therefore

$$\log\log x + C \leqq \log\log |a| \leqq \log\log x$$

for all elements $a \in G[x]$, except perhaps those $O(N_G(x^c)) = o(N_G(x))$ elements a such that $|a| < x^c$. Therefore, for $x^c \leqq |a| \leqq x$,

$$|\omega(a) - \log\log|a|| \leqq |\omega(a) - \log\log x| + |\log\log x - \log\log|a||$$
$$< \tfrac{1}{4}\xi(x)(\log\log x)^{1/2} + |C|$$

for asymptotically $N_G(x)$ of such elements a, by Proposition 4.1. For such elements,

$$\log\log x \leqq \log\log|a| - C, \qquad \xi(x) \leqq 2\xi(|a|)$$

(when x is sufficiently large); thus, in this case,

$$|\omega(a) - \log\log|a|| < \tfrac{1}{2}\xi(|a|)(\log\log|a| - C)^{1/2} + |C|$$
$$< \xi(|a|)(\log\log|a|)^{1/2}$$

for $|a|$ sufficiently large. Hence, for asymptotically $N_G(x)$ elements $a \in G[x]$,

$$|\omega(a) - \log\log|a|| < \xi(|a|)(\log\log|a|)^{1/2}.$$

This proves the statement about ω.

Next, the total number of elements $a \in G[x]$ such that

$$\Omega(a) - \omega(a) \geqq (\log\log x)^{1/2}$$

cannot exceed

$$(\log\log x)^{-1/2} \sum_{|a| \leqq x} \{\Omega(a) - \omega(a)\} = O((\log\log x)^{-1/2} x^{\delta})$$
$$= o(x^{\delta}),$$

by Theorem 2.4. Therefore the conclusion about Ω is a consequence of that already proved for ω. □

An example of a function ξ satisfying the hypotheses of the above corollary is provided by $\xi(q) = (\log\log q)^{\varepsilon}$, where $\varepsilon > 0$ is a constant; here one may take $c = e^{-1}$.

Following usage first established by Hardy and Ramanujan (who considered functions on G_Z), we shall say that the arithmetical function f on G has the **normal order** $F|(a|)$ for $a \in G$ if, for any fixed $\varepsilon > 0$,

$$(1 - \varepsilon)F(|a|) < |f(a)| < (1 + \varepsilon)F(|a|)$$

for almost all elements $a \in G$. (The set of elements a in question may of course depend on ε.) Then it follows from Corollary 4.2 that *both* ω *and* Ω *have the normal order* $\log\log|a|$ for $a \in G$.

This last conclusion has some interesting implications in relation to certain of the results about maximum orders of magnitude given in Chapter 5, § 2: In the first place, it shows that the normal orders of ω and Ω are of lower order than their maximum orders of magnitude. Secondly, suppose that f is an arithmetical function with the property that there exist positive constants τ_1 and τ_2 such that

$$\tau_1^{\omega(a)} \leqq |f(a)| \leqq \tau_2^{\Omega(a)} \quad [a \in G].$$

Then it follows that, for any given $\varepsilon > 0$,

$$\tau_1^{(1-\varepsilon)\log\log|a|} < |f(a)| < \tau_2^{(1+\varepsilon)\log\log|a|}$$

for almost all elements $a \in G$. If $\tau_1 = \tau_2$, one might then say that the normal order of f is *roughly* $\tau_1^{\log\log|a|}$.

For example, the normal order of the divisor function d is roughly $2^{\log\log|a|}$, since

$$2 \leqq d(p^r) = r+1 \leqq 2^r$$

for any prime-power p^r in G. Thus, the rough normal order of d is of a lower order of magnitude than its logarithmic maximum order of magnitude.

Other inequalities for multiplicative functions which arose in Chapter 5, § 2 lead to the following inequalities and hence conclusions of a partly similar kind:

(i) $\beta(a) \leqq 3^{\Omega(a)/3}$;
(ii) $a(n) \leqq 5^{\Omega(n)/4}$;
(iii) $S(n) \leqq 6^{\Omega(n)/4}$;
(iv) $K(n) \leqq [K:\mathbf{Q}]^{\Omega(n)}$.

(The inequality for $S(p^r)$ is implicit in the statement of Theorem 2.11, in the sense that it is essentially part of the exercise suggested at that point.) These inequalities and the above comments imply that, *given any* $\varepsilon > 0$,

(4.3) $\beta(a) < 3^{((1+\varepsilon)/3)\log\log|a|}$ *for almost all* $a \in G$;

(4.4) $a(n) < 5^{((1+\varepsilon)/4)\log\log n}$ *for almost all* $n \in G_{\mathbf{Z}}$;

(4.5) $S(n) < 6^{((1+\varepsilon)/4)\log\log n}$ *for almost all* $n \in G_{\mathbf{Z}}$;

(4.6) $K(n) < [K:\mathbf{Q}]^{(1+\varepsilon)\log\log n}$ *for almost all* $n \in G_{\mathbf{Z}}$.

These conclusions may be contrasted with those about the corresponding *indicators* which were discussed in Chapter 5.

Selected bibliography for Chapter 6

Section 1: Ahern [1, 2], Amitsur [1, 2], Bateman and Diamond [1], Beurling [1], Bredihin [1, 3, 4, 5], Cibul'skite [1], Davison [1], Diamond [1–4], Forman and Shapiro [1], R. S. Hall [1–3], Horadam [6], Knopfmacher [5–8], Landau [9], Malliavin [1], B. Nyman [1], Rieger [2, 5, 6], Segal [1], Shapiro [1], Wegmann [1].

Section 2: E. Cohen [2], Fluch [1], Hardy and Wright [1], Horadam [16], Kalniń [1], Knopfmacher [6], Landau [3, 5, 10], Prachar [1].

Sections 3/4: Hardy and Wright [1], Heppner [2], Horadam [17], Kalniń [1], Knopfmacher [10], Landau [10].

CHAPTER 7

FOURIER ANALYSIS OF ARITHMETICAL FUNCTIONS

The present chapter is concerned with a kind of analysis of arithmetical functions which differs from that occurring elsewhere in this book. This type of analysis is in many ways parallel to the classical analysis of certain kinds of functions in terms of periodic, or later almost periodic, functions. In fact, it is possible to view it as one special aspect of the wide-ranging program of abstract harmonic analysis over general topological groups and semigroups. However, from this point of view, it is not *a priori* obvious that the application of such analysis to arithmetical functions should lead to conclusions of any great number-theoretical interest.

Although, with hindsight, it is to some extent possible to indicate intrinsic number-theoretical reasons for expecting or seeking some kind of "harmonic" analysis of arithmetical functions, the historical process of discovery occurred differently. In 1918, Ramanujan [2] published a paper containing a number of remarkable formulae expressing some particular functions over G_{Z} as the pointwise convergent sums of certain infinite trigonometrical series. For example, he proved that

$$\frac{\sigma(n)}{n} = \tfrac{1}{6}\pi^2\left\{1+\frac{(-1)^n}{2^2}+\frac{2\cos(\frac{2}{3}\pi n)}{3^2}+\frac{2\cos(\frac{1}{2}\pi n)}{4^2}\right.$$
$$\left.+\frac{2\left[\cos(\frac{2}{5}\pi n)+\cos(\frac{4}{5}\pi n)\right]}{5^2}+\frac{2\cos(\frac{1}{3}\pi n)}{6^2}+\dots\right\},$$

where $n=1, 2, 3, \dots$ and $\sigma(n)$ is the sum of all the divisors of n. If one recalls from Chapter 4, § 5, that $\sigma(n)/n$ has the asymptotic mean-value $\frac{1}{6}\pi^2$ over G_{Z}, then the above formula provides a very striking indication of how the actual values of $\sigma(n)/n$ fluctuate harmonically about their mean-value. In briefer terms, this formula can be written as

$$\frac{\sigma(n)}{n} = \tfrac{1}{6}\pi^2 \sum_{r=1}^{\infty} \frac{c_r(n)}{r^2},$$

where the $c_r(n)$ are certain trigonometrical sums which later became known as *Ramanujan sums*.

By means of *ad hoc* methods, Ramanujan and (slightly thereafter) Hardy established further formulae for particular arithmetical functions over $G_{\mathbf{Z}}$, in terms of the sums $c_r(n)$, some of which again exhibit the fluctuations of a function f with asymptotic mean value $m(f)$ about the value $m(f)$. Then, at a later stage, further authors began investigations into features of harmonic analysis which seemed to explain the existence of at least some of the special formulae. (See particularly the monograph of Wintner [2] and its references.)

At first sight, however interesting such phenomena appear to be for functions of a positive integer, it is not very clear that anything similar is likely to occur for functions over a general arithmetical semigroup. However, although periodicity and trigonometrical sums appear to have no obvious direct counterparts over an arbitrary arithmetical semigroup G, E. Cohen [5], with the aid of a formula for the Ramanujan sums which is readily extended to the general case, was able to establish some initial results for an analytical "Fourier theory" of Ramanujan sums over the particular arithmetical semigroup $G_{\mathscr{A}}$ of all isomorphism classes of finite abelian groups. Further, E. Cohen [4] established a more restricted algebraic theory which applies to certain partial counterparts of periodic functions over any arithmetical semigroup G.

In addition to discussing this latter theory, the present chapter establishes Cohen's basic analytical results about $G_{\mathscr{A}}$ in the context of any arithmetical semigroup G satisfying Axiom A, as well as some further results of abstract harmonic analysis over G. Many of the additional results are modelled along the very elegant lines of Schwarz and Spilker [1], who deal with functions on $G_{\mathbf{Z}}$.

In reading this chapter, a slight familiarity with the theories of periodic and almost periodic functions and with the elements of abstract harmonic analysis might facilitate understanding of the motivation and details of the discussion. However, in principle, such background knowledge may not be strictly essential if definitions are taken at their internal face-values and quoted results of abstract analysis are accepted at least temporarily without proof.

§ 1. Algebraic and topological theory of Ramanujan sums

In this and the next section, G will denote an arbitrary arithmetical semigroup, not necessarily satisfying Axiom A. However, in anticipation of later sections in which Axiom A will again be required, the definition of Ramanujan sums below will be given relative to an arbitrary constant $\delta > 0$ which may be supposed to have been fixed in advance in some unspecified way.

Now consider the complex vector space Dir (G). In this chapter, in place of the previous convolution multiplication, the simpler operation of point-wise multiplication

$$(f \cdot g)(a) = f(a)g(a) \qquad [a \in G]$$

will be used to make the vector space Dir (G) into a new commutative algebra $\mathfrak{D} = \mathfrak{D}(G)$. In this case the constant function ζ_G such that $\zeta_G(a) = 1$ on G acts as identity element, and $\mathfrak{D}$ may be regarded as the unrestricted (complete) direct product

$$\mathfrak{D} = \prod_{a \in G} \mathbf{C}_a$$

of copies $\mathbf{C}_a$ of the complex field $\mathbf{C}$. In addition, it will be convenient to regard $\mathfrak{D}$ as a topological algebra relative to the (cartesian) product topology when each $\mathbf{C}_a$ has the usual topology of the complex numbers.

It will be seen shortly that, within $\mathfrak{D}$, a significant rôle is played by the subset of all **Ramanujan sums** c_r $(r \in G)$. Given $\delta > 0$ and $r \in G$, the function $c_r \in \mathfrak{D}$ is defined by

$$c_r(a) = \sum_{d \mid (r, a)} \mu(r/d) |d|^\delta \qquad [a \in G],$$

where μ denotes the Möbius function on G and (r, a) is the g.c.d. of r and a. (The relationship between this definition and the trigonometrical sums over $G_{\mathbf{Z}}$ mentioned earlier is discussed in § 5 below.)

The following properties of the Ramanujan sums are easy consequences of their definition:

(1.1) $$c_1(a) = 1 \text{ (i.e. } c_1 = \zeta_G), \qquad c_r(1) = \mu(r);$$

(1.2) $$c_r(r) = \sum_{d \mid r} \mu(r/d) |d|^\delta = \varphi_\delta(r),$$

where φ_w is the 'Euler-type' function defined in Chapter 4, § 5;

(1.3) $$|c_r(a)| = \Big|\sum_{d\mid(r,a)} \mu(r/d)|d|^{\delta}\Big| \leq \min\{\sigma_\delta(r), \sigma_\delta(a)\},$$

where σ_w is the divisor-sum function defined in Chapter 4, § 5.

In the present chapter, it will be convenient to denote the particular functions φ_δ and σ_δ simply by φ and σ, respectively. (The Euler function defined in Chapter 2, § 6 will not occur in this chapter except over $G_{\mathbf{Z}}$ — when it coincides with the present function φ!)

The next conclusion is a direct consequence of (1.3), while the one after follows from a simple calculation:

(1.4) For fixed $r \in G$,

$$c_r(a) = O(1) \quad \text{as } |a| \to \infty,$$

while

for fixed $a \in G$,

$$c_r(a) = O(1) \quad \text{as } |r| \to \infty.$$

(1.5) For any prime $p \in P$ and $m > 0$,

$$c_{p^m}(a) = \begin{cases} |p|^{\delta m} - |p|^{\delta(m-1)} & \text{if } p^m \mid a, \\ -|p|^{\delta(m-1)} & \text{if } p^m \nmid a,\ p^{m-1} \mid a, \\ 0 & \text{if } p^{m-1} \nmid a. \end{cases}$$

One further property of the Ramanujan sums worth noting here is given by:

1.6. Proposition. *For any fixed $a \in G$, $c_r(a)$ is a multiplicative function of $r \in G$.*

Proof. Given $a, r \in G$, define

$$\eta_r(a) = \begin{cases} |r|^{\delta} & \text{if } r \mid a, \\ 0 & \text{otherwise.} \end{cases}$$

Then, for fixed $a \in G$, $\eta_r(a)$ is clearly a multiplicative function of $r \in G$. Since

$$c_r(a) = \sum_{d \mid r} \mu(r/d)\eta_d(a)$$

and μ is a multiplicative function, it therefore follows from Proposition 2.4.1 that $c_r(a)$ is a multiplicative function of r. □

Now let $\mathfrak{R} = \mathfrak{R}(G)$ denote the vector subspace of $\mathfrak{D}$ spanned by all the functions c_r ($r \in G$), and, for a given element $k \in G$, let $\mathfrak{R}_k$ denote the further subspace spanned by those functions c_r with $r \mid k$. If $r \mid k$, then $(r, a) =$

$=(r, (k, a))$, and so any function f on G which can be expressed in the form

$$f(a) = \sum_{d \mid (r, a)} g(d, r/d) \qquad (a \in G)$$

will have the property

$$f(a) = f((k, a)) \quad \text{for every } a \in G; \tag{1.7}$$

in particular, every c_r spanning $\mathfrak{R}_k$, and so every $f \in \mathfrak{R}_k$, has this property. In general, any function $f \in \mathfrak{D}$ with the property (1.7) will be said to be **even** (mod k). The set $\mathfrak{D}_k$ of all even functions (mod k) is evidently a subalgebra of $\mathfrak{D}$ containing $\mathfrak{R}_k$.

1.8. Proposition. *The algebra* $\mathfrak{D}_k = \mathfrak{R}_k$.

Proof. Instead of attempting a purely algebraic and computational proof, the argument will make use of the topology of $\mathfrak{D}_k$ induced from that of $\mathfrak{D}$, and the well-known Stone–Weierstrass theorem (see Dugundji [1] Chapter 13, § 3). Firstly, note that every even function f (mod k) is uniquely determined by its values on the set X_k of all divisors of k. Conversely, any complex-valued function on X_k may be extended to a function in $\mathfrak{D}_k$. This implies that the natural projection map

$$\prod_{a \in G} \mathbf{C}_a \to \prod_{d \mid k} \mathbf{C}_d$$

induces a topological algebra isomorphism of $\mathfrak{D}_k$ with

$$\prod_{d \mid k} \mathbf{C}_d = C(X_k),$$

the algebra of all continuous complex-valued functions on X_k when X_k has the discrete topology.

Now consider:

1.9. Lemma. $\mathfrak{R}_k$ *is a subalgebra of* $\mathfrak{D}_k$ *which contains the identity element* c_1, *and also the complex conjugate* $\bar{f}$ *of every function* $f \in \mathfrak{R}_k$.

Proof. It was noted in (1.1) above that c_1 is the identity element ζ_G of the algebra $\mathfrak{D}$; also $\mathfrak{R}_k$ is closed under complex conjugation since the spanning functions c_r are real-valued.

Now note that, by Proposition 1.6,

$$c_r \cdot c_s = c_{rs} \in \mathfrak{R}_k$$

if $r, s|k$ and $(r, s)=1$. Also, for $p\in P$ and $m>n>0$, (1.5) above implies that

$$c_{p^m}\cdot c_{p^n} = (|p|^{\delta n}-|p|^{\delta(n-1)})c_{p^m}.$$

Further, (1.5) implies that

$$[c_{p^n}(a)]^2 = \begin{cases} (|p|^{\delta n}-|p|^{\delta(n-1)})^2 & \text{if } p^n|a, \\ |p|^{2\delta(n-1)} & \text{if } p^n\nmid a,\ p^{n-1}|a, \\ 0 & \text{if } p^{n-1}\nmid a. \end{cases}$$

Hence a little calculation allows one to establish the formula

$$[c_{p^n}]^2 = (|p|^{\delta n}-2|p|^{\delta(n-1)})c_{p^n}+(|p|^{\delta n}-|p|^{\delta(n-1)})\sum_{r=0}^{n-1} c_{p^r}.$$

These formulae show that $c_{p^m}\cdot c_{p^n}\in\mathfrak{R}_k$ whenever $p\in P$ and $p^m,p^n|k$. It follows that $\mathfrak{R}_k$ is a subalgebra of $\mathfrak{D}_k$. □

Next let $\mathfrak{R}_k'$ denote the image of $\mathfrak{R}_k$ in $C(X_k)$ under the above isomorphism $\mathfrak{D}_k\to C(X_k)$. Then $\mathfrak{R}_k'$ is a subalgebra of $C(X_k)$ which contains the identity and is closed under complex conjugation. Also, if d, d' are distinct divisors of k, then there exists a prime-power p^n dividing d but not d', or vice versa; hence

$$c_{p^n}(d) \neq c_{p^n}(d').$$

This implies that the algebra $\mathfrak{R}_k'$ "separates" the points of X_k. Thus $\mathfrak{R}_k'$ satisfies all the requirements of the Stone–Weierstrass theorem, and it follows that it must be dense in $C(X_k)$, relative to the compact-open topology for $C(X_k)$. But the compact–open topology in $C(X_k)$ coincides with the product topology for $\prod_{d|k}\mathbf{C}_d$ (and the same is true for $\mathfrak{D}=$ $=C(G)=\prod_{a\in G}\mathbf{C}_a$ if G takes the discrete topology); compare Dugundji [1] Chapter 12, § 1. Therefore $\mathfrak{R}_k'$ is dense in the topological product $\prod_{d|k}\mathbf{C}_d$. Since a finite-dimensional vector subspace of a topological vector space is closed, by a standard result of functional analysis, it follows that $\mathfrak{R}_k'=C(X_k)$. Hence $\mathfrak{R}_k=\mathfrak{D}_k$. □

It is easy to verify that every even function f (mod k) over $G_{\mathbf{Z}}$ is also periodic (mod k), i.e., $f(m)=f(n)$ whenever $m\equiv n$ (mod k). Thus evenness may be regarded as a restricted form of periodicity in the case of an arbitrary arithmetical semigroup G. The next result shows that the even functions play an important rôle in $\mathfrak{D}$, both generally and in the special case when $G=G_{\mathbf{Z}}$.

1.10. Proposition. *The vector space* $\mathfrak{R}$ *of all even functions in* $\mathfrak{D}$ *forms a dense subalgebra of* $\mathfrak{D}$.

Proof. Since $\mathfrak{R}=\bigcup_{k\in G}\mathfrak{R}_k$, it consists of all the even functions in $\mathfrak{D}$. Further, if one ignores the element $k\in G$, the proof of Lemma 1.9 shows that $\mathfrak{R}$ forms a subalgebra of $\mathfrak{D}$, that it contains the identity element c_1 and that it is closed under complex conjugation. In addition, the previous discussion implies that $\mathfrak{R}$ "separates" arbitrary points of G. Since it was noted above that the present topology on $\mathfrak{D}$ coincides with the compact–open topology on $C(G)$ when G is regarded as a discrete topological space, the Stone–Weierstrass theorem may be applied again so as to now yield Proposition 1.10. □

1.11. Corollary. *Given any* $f\in\mathfrak{D}$, *any* $\varepsilon>0$ *and any number* $x>0$, *there exists a finite linear combination*

$$g = g(\varepsilon, x) = \sum_r \alpha_r c_r \qquad (\alpha_r \textit{ complex})$$

such that $|f(a)-g(a)|<\varepsilon$ *whenever* $|a|\leqq x$.

Proof. This follows from Proposition 1.10 and the definition of the compact–open topology; see Dugundji [1], Chapter 12, § 1. □

The above conclusions show that every arithmetical function is "nearly" even in a weak sense, while every even function coincides with some "Ramanujan polynomial". If one follows the analogy with continuous functions over the real line, it seems reasonable to investigate stronger forms of approximation by means of even functions.

Now every even function c_r is bounded on G, by (1.4). Hence every even function is bounded on G, and $\mathfrak{R}\subseteqq\mathfrak{B}(G)$, the Banach algebra of all bounded complex-valued functions on G under the uniform norm

$$\|f\|_u = \sup_{a\in G} |f(a)|.$$

In analogy with the theory of ordinary almost periodic functions (see Corduneanu [1], for example), it seems reasonable here to define an arithmetical function f to be **(uniformly) almost even** if and only if f belongs to the closure $\mathfrak{D}^*=\mathfrak{D}^*(G)$ of $\mathfrak{R}$ in $\mathfrak{B}(G)$ relative to the uniform norm. Since, as noted above, $\mathfrak{B}(G)$ is a Banach algebra (see Hewitt and Stromberg [1] Chapter 2, § 7, say) and $\mathfrak{R}$ is a subalgebra of $\mathfrak{D}$, it follows that $\mathfrak{D}^*$ is also a Banach algebra. In fact, $\mathfrak{D}^*$ is a commutative Banach algebra with identity which is closed under the involution $f\to\bar{f}$. Therefore the Gelfand

Representation Theorem (see for example Loomis [1] § 26) is applicable and implies that $\mathfrak{D}^*$ is isomorphic and isometric to the algebra $C(X)$ of all continuous complex-valued functions on a certain compact Hausdorff space X — its maximal ideal space. The following theorem, which does not depend on the preceding comments or the Gelfand Representation Theorem, provides slightly sharper information as to how almost even functions can be represented by continuous functions on an explicit compact space.

1.12. Theorem. *There exists a commutative compact topological semigroup G^* with identity, and a monomorphism $\tau: G \to G^*$ which sends G onto a dense subset of G^*, with the property that the induced homomorphism $\tau^*: C(G^*) \to \mathfrak{B}(G)$ maps $C(G^*)$ isomorphically and isometrically onto $\mathfrak{D}^*$.*

Proof. Over here, a 'topological' semigroup will be understood to be a semigroup H with a Hausdorff topology such that the multiplication map $H \times H \to H$ is continuous (i.e., multiplication is jointly continuous). If Y is any topological space, $C(Y)$ will denote the algebra of all continuous complex-valued functions on Y. Lastly, as usual the homomorphism τ^* 'induced' by τ is defined by $f \to f\tau$.

In order to construct G^*, first consider any prime $p \in P$ and the set $G_{(p)}$ consisting of all powers p^r ($r=0, 1, 2, \ldots$) together with one extra element p^∞. Regard $G_{(p)}$ as the *one-point compactification* of the discrete topological space consisting of all powers p^r ($r=0, 1, 2, \ldots$), obtained by adjoining the extra point p^∞; for the definition of one-point compactification, see Dugundji [1] Chapter 11, § 8, say. Then $G_{(p)}$ becomes a commutative compact topological semigroup with identity if one uses the multiplication of prime-powers in G and in additions lets

$$(p^\infty)^2 = p^\infty,$$
$$p^\infty p^r = p^r p^\infty = p^\infty \quad \text{for any } r = 0, 1, 2, \ldots .$$

Now form the topological product

$$G^* = \prod_{p \in P} G_{(p)}.$$

Then it follows with the aid of Tychonoff's theorem that G^* is a compact topological semigroup. Further, the existence of unique factorization into prime powers for the elements of G gives rise to a natural mapping $\tau: G \to G^*$. This mapping is a semigroup monomorphism, and, using the definition of the product topology and of the topology in each $G_{(p)}$, it is easy to verify that the image of τ is dense in G^*.

Now consider the induced algebra homomorphism $\tau^*: C(G^*) \to \mathfrak{B}(G)$. Since $\tau(G)$ is dense in G^*, a continuous complex-valued function on the subspace $\tau(G)$ has at most one extension to a continuous function on G^*. Hence τ^* is an algebra monomorphism. Further, for $f \in C(G^*)$,

$$\|\tau^*(f)\|_u = \sup_{a \in G} |f(\tau(a))| \leqq \|f\|_u,$$

and, since $|f|$ is continuous on G^* and $\tau(G)$ is dense, strict inequality is impossible. Thus τ^* is a norm-preserving monomorphism.

Next consider the natural projection map $\pi_p: G^* \to G_{(p)}$ and, given the positive integer $n > 0$, define the sets

$$A = \{x \in G^*:\ \pi_p(x) = p^r \text{ for some } r < n-1\},$$
$$B = \{x \in G^*:\ \pi_p(x) = p^{n-1}\},$$
$$C = \{x \in G^*:\ \pi_p(x) = p^\infty \text{ or } p^r \text{ for some } r > n-1\}.$$

Then A, B and C are mutually disjoint open sets with union G^*. Therefore one may define a continuous function $c_{p^n}{}^*$ on G^* by the formula

$$c_{p^n}{}^*(x) = \begin{cases} 0 & \text{if } x \in A, \\ -|p|^{\delta(n-1)} & \text{if } x \in B, \\ |p|^{\delta n} - |p|^{\delta(n-1)} & \text{if } x \in C. \end{cases}$$

Then it follows from (1.5) that

$$c_{p^n}{}^*(\tau(a)) = c_{p^n}(a) \quad \text{for } a \in G.$$

Now define a continuous function $c_r{}^*$ for each $r \in G$ by letting $c_1{}^*$ be the identity element of $C(G^*)$, and

$$c_{s_1 s_2 \dots s_m}{}^* = c_{s_1}{}^* \cdot c_{s_2}{}^* \cdot \dots \cdot c_{s_m}{}^*$$

whenever $s_1, s_2, \dots, s_m$ are powers of distinct primes in G. In that case, it follows that $\tau^*(c_r{}^*) = c_r$ for each $r \in G$. Hence, if $\mathfrak{R}^*$ denotes the vector subspace of $C(G^*)$ spanned by all the functions $c_r{}^*$, one sees that τ^* maps $\mathfrak{R}^*$ onto $\mathfrak{R} \subseteqq \mathfrak{D}^*$. Therefore $\mathfrak{R}^*$ is a subalgebra of $C(G^*)$ which contains the identity element and is closed under complex conjugation.

In order to complete the proof, we note next that $\mathfrak{R}^*$ "separates" the points of G^*. For, if x, y are distinct points of G^*, then there is a prime

$p \in P$ such that $\pi_p(x) = p^m$, $\pi_p(y) = p^n$ with $m \neq n$ and either m or n is not ∞. Then, without loss of generality, it may be supposed that $m = \infty$ or $m > n$, so that the definition of $c_{p^n}{}^*$ shows that it takes different values on x and y. It follows that $\mathfrak{R}^*$ satisfies all the requirements of the stronger form of the Stone–Weierstrass theorem which applies to $C(X)$ when X is a compact Hausdorff space; see for example Hewitt and Stromberg [1] Chapter 2, § 7. Therefore $\mathfrak{R}^*$ is dense in $C(G^*)$ relative to the uniform topology. This implies that the norm-preserving monomorphism τ^* must send the closure $C(G^*)$ of $\mathfrak{R}^*$ onto the closure $\mathfrak{D}^*$ of $\mathfrak{R}$ in $\mathfrak{B}(G)$, and the theorem is proved. □

The property of G^* in Theorem 1.12 relative to the almost even functions on G is directly analogous to a well-known property of the Bohr compactification of a topological group, or the 'almost periodic compactification' of a topological semigroup, relative to the almost periodic functions on the group or semigroup; see Loomis [1] § 41, Rudin [1] Chapter 1 and Burckel [1] Chapter 1. Therefore it is reasonable to refer to G^* as the **almost-even compactification** of G.

It is interesting to note that the property of G^* in Theorem 1.12 characterizes it uniquely up to topological isomorphism:

1.13. Proposition. *Let G' denote a commutative compact topological semigroup with identity. Suppose that there exists a homomorphism $\varrho: G \to G'$ of G onto a dense subset of G', with the property that the induced homomorphism $\varrho^*: C(G') \to \mathfrak{B}(G)$ maps $C(G')$ isomorphically onto $\mathfrak{D}^*$ as an algebra. Then G' is topologically isomorphic to G^*.*

Proof. Under the above hypotheses, there exists an algebra isomorphism $\varrho^{*-1}\tau^*: C(G^*) \to C(G')$. The present proof, which uses weaker properties than those occurring in Theorem 1.12, will be based on a theorem of the theory of algebras of continuous functions on topological spaces. This theorem (see Semadeni [1] § 7.7) states that every identity-preserving algebra homomorphism $C(Y) \to C(X)$ with Y a compact Hausdorff space is induced by a unique continuous mapping $X \to Y$.

In the present case, it follows that there exists a (unique) homeomorphism $\alpha: G' \to G^*$ such that $\alpha^* = \varrho^{*-1}\tau^*$. Thus $\varrho^*\alpha^* = \tau^*$. Hence, for any $a \in G$ and $f \in C(G^*)$.

$$f(\tau a) = (\tau^* f)(a) = ((\alpha\varrho)^* f)(a) = f(\alpha\varrho a).$$

Since $C(G^*)$ "separates" the points of G^*, it follows that $\alpha\varrho = \tau$.

Now $\varrho: G \to G'$ must be a monomorphism. For, $\varrho(a) = \varrho(b)$ implies $f(\varrho a) = f(\varrho b)$ for every $f \in C(G')$, and this implies that $g(a) = g(b)$ for every $g \in \mathfrak{D}^*$. Since $\mathfrak{D}^*$ "separates" the points of G, it follows that $a = b$. Therefore ϱ^{-1} exists on $\varrho(G)$ and $\alpha = \tau\varrho^{-1}$ defines an isomorphism from $\varrho(G)$ to $\tau(G)$.

In order to show that α is an isomorphism from G' to G^*, suppose that there exist elements $x, y \in G'$ such that $\alpha(x)\alpha(y) \neq \alpha(xy)$. Then, since G^* is a Hausdorff space, there exist disjoint neighbourhoods U and V of $\alpha(x)\alpha(y)$ and $\alpha(xy)$, respectively. Since α and the multiplication operations in G^* and G' are continuous:

(i) there exist neighbourhoods U_1, U_2 of $\alpha(x)$, $\alpha(y)$ in G^*, respectively, such that $U_1 U_2 \subseteqq U$;

(ii) there exists a neighbourhood O of xy in G' such that $\alpha(O) \subseteqq V$;

(iii) there exist neighbourhoods O_1, O_2 of x, y in G', respectively, such that $O_1 O_2 \subseteqq O$ and hence $\alpha(O_1 O_2) \subseteqq V$;

(iv) there exist neighbourhoods O_1', O_2' of x, y in G', respectively, such that $\alpha(O_1') \subseteqq U_1$ and $\alpha(O_2') \subseteqq U_2$. Further, since $\varrho(G)$ is dense in G', there exist elements $a \in O_1 \cap O_1' \cap \varrho(G)$ and $b \in O_2 \cap O_2' \cap \varrho(G)$. Then $\alpha(a)\alpha(b) \in U_1 U_2 \subseteqq U$, while $\alpha(ab) \in \alpha(O) \subseteqq V$, which gives a contradiction. Therefore α must be a topological isomorphism. □

The other types of compactifications for topological groups and semigroups have similar uniqueness properties to the above one. It may be interesting to remark also in passing that it is not difficult to show that every almost even function on G is also almost-periodic in the sense of de Leeuw and Glicksberg discussed in Burckel [1]. Connections with ordinary periodicity in the case of $G_{\mathbf{Z}}$ will be mentioned again in § 5 below.

§ 2. Fourier theory of even functions

The preceding discussion suggests making a closer examination of the properties of even and almost even functions on G. In the even case, it is possible to develop a purely algebraic "Fourier theory" over an arbitrary arithmetical semigroup, and this will be discussed in the present section. A more analytical theory of almost even functions, based on Axiom A, will be considered in the next sections.

Consider then the algebra $\mathfrak{D}_k$ of all even functions on G (mod k), where k is an arbitrary element of G, supposed fixed throughout this section. Recall that $\mathfrak{D}_k = \mathfrak{R}_k$, the vector space spanned by all the Ramanujan sums c_r

with $r|k$, and recall that in this chapter φ will denote the function on G such that

$$\varphi(r) = \sum_{d|r} \mu(r/d)\,|d|^{\delta},$$

where $\delta > 0$ is still a fixed but unspecified constant.

2.1. Theorem. *The algebra $\mathfrak{D}_k$ forms a complex Hilbert space of finite dimension $d(k)$ under the inner product*

$$\langle f, g\rangle = \sum_{d|k} \varphi(d) f(k/d)\bar{g}(k/d).$$

The functions

$$c_r' = (\varphi(r)\,|k|^{\delta})^{-1/2} c_r, \quad \textit{for } r|k,$$

provide an orthonormal basis for $\mathfrak{D}_k$.

Proof. Since $\varphi(d) > 0$ for all $d \in G$, it is easy to verify that $\langle\ ,\ \rangle$ has the usual properties of a complex inner product. Therefore the theorem will follow as soon as it has been shown that the functions c_r', $r|k$, form an orthonormal set. This will require a number of auxiliary results.

2.2. Lemma. *If r and s divide k, then*

$$\sum_{d|k} c_r(k/d) c_d(k/s) = \begin{cases} |k|^{\delta} & \textit{if } s = r, \\ 0 & \textit{otherwise.} \end{cases}$$

Proof. First recall the relation

$$c_r(a) = \sum_{d|r} \mu(r/d)\eta_d(a)$$

noted in the proof of Proposition 1.6 above. By Möbius inversion, this implies

$$\sum_{d|r} c_d(a) = \eta_r(a) = \begin{cases} |r|^{\delta} & \textit{if } r|a, \\ 0 & \textit{otherwise.} \end{cases} \tag{2.3}$$

Given divisors r, s of k, then consider the sum

$$\begin{aligned} S &= \sum_{d|k} c_r(k/d) c_d(k/s) = \sum_{d|k} c_d(k/s) \sum_{d'|(r,\, k/d)} \mu(r/d')\,|d'|^{\delta} \\ &= \sum_{de=k} c_d(k/s) \sum_{\substack{d'f=r \\ d'g=e}} \mu(r/d')\,|d'|^{\delta} = \sum_{\substack{d'f=r \\ dd'g=k}} c_d(k/s)\mu(r/d')\,|d'|^{\delta} \\ &= \sum_{d'|(r,\, k)} \mu(r/d')\,|d'|^{\delta} \sum_{d|(k/d')} c_d(k/s) \\ &= \sum_{d'|r} \mu(r/d')\,|d'|^{\delta}\eta_{k/d'}(k/s), \end{aligned}$$

by (2.3), since $r=(r,k)$. If $d'|r$ and $s\nmid r$, then $s\nmid d'$, and so $\eta_{k/d'}(k/s)=0$. Therefore $S=0$ if $s\nmid r$.

However, if $s|r$, then $\eta_{k/d'}(k/s)=|k/d'|^\delta$ or 0 according as $s|d'$ or not, and so in that case

$$S = \sum_{\substack{d'|r \\ s|d'}} \mu(r/d')|k|^\delta = |k|^\delta \sum_{\substack{d'e=r \\ sf=d'}} \mu(r/d')$$

$$= |k|^\delta \sum_{sfe=r} \mu(r/sf) = |k|^\delta \sum_{f|(r/s)} \mu(r/sf)$$

$$= \begin{cases} |k|^\delta & \text{if } r = s, \\ 0 & \text{if } r \neq s, \end{cases}$$

by Corollary 2.5.3. This proves the lemma. □

Now call a function $f\in\mathfrak{D}$ **separable** if and only if $f(a)=f(a_*)$ for all $a\in G$, where a_* denotes the core of a (see Chapter 2, § 6). For example, both the unitary-divisor function d_* and the function $\bar{\varphi}_\delta(a)=|a|^{-\delta}\varphi(a)$ are separable.

2.4. Lemma. *A multiplicative function f on G is separable if and only if $(\mu*f)(a)=0$ whenever a is not square-free. If f is separable, then*

(i) $f(a)f(b) = f(ab)f((a, b))$,

(ii) $f(a) = f((a, b)) \sum \{(\mu*f)(d)\colon d|a,\ (d, b) = 1\}$

for all $a,b\in G$.

Proof. Let $F=\mu*f$. Then $f=\zeta_G*F$, so that $f(a)=\sum_{d|a} F(d)$. If $F(d)=0$ whenever d is not square-free, then $f(a)=\sum_{d|a_*} F(d)=f(a_*)$, and here f need not necessarily be multiplicative.

For the rest of the discussion, suppose that f is multiplicative and separable. Then, for any prime-power p^m, $m>1$,

$$F(p^m) = \sum_{d|p^m} \mu(d)f(p^m/d) = f(p^m)-f(p^{m-1}) = 0.$$

Since F is also multiplicative, it follows now that $F(a)=0$ whenever a is not square-free.

Next, for any prime-powers p^m, p^n with $m,n\geqq 1$,

$$f(p^m)f(p^n) = f(p)f(p) = f(p^{m+n})f((p^m, p^n)).$$

Since f is multiplicative, this implies the formula (i). For (ii), note firstly that

$$\sum_{\substack{d|a \\ (d,b)=1}} F(d) = \sum_{\substack{d|a_* \\ (d,b)=1}} F(d) = \sum_{d|(a_*/(a,b)_*)} F(d) = f(a_*/(a,b)_*).$$

Since $(a, b)_*$ and $a_*/(a, b)_*$ are coprime, it follows that

$$f(a) = f(a_*) = f((a, b)_*)f(a_*/(a, b)_*) = f((a, b)) \sum_{\substack{d|a \\ (d,b)=1}} F(d). \qquad \square$$

The following lemma provides a "closed" evaluation of the Ramanujan sums.

2.5. Lemma (Hölder's relation). *For any* $r,a \in G$,

$$c_r(a) = \frac{\varphi(r)\mu(c)}{\varphi(c)},$$

where $c = r/(r, a)$.

Proof. Given $r,a \in G$, let $b=(r, a)$ and $c=r/b$. Then

$$c_r(a) = \sum_{d|b} \mu(r/d)|d|^\delta = \mu(c) \sum_{\substack{d|b \\ (c, b/d)=1}} \mu(b/d)|d|^\delta,$$

since μ is multiplicative and $\mu(s)=0$ if s is not square-free. Thus

$$c_r(a) = \mu(c)|b|^\delta \sum_{\substack{s|b \\ (s, c)=1}} \mu(s)|s|^{-\delta}.$$

Now let

$$\Phi = \mu * \bar{\varphi}_\delta,$$

where

$$\bar{\varphi}_\delta(a) = |a|^{-\delta}\varphi(a).$$

Then

$$\begin{aligned} \Phi(s) &= \sum_{d|s} \mu(d)\bar{\varphi}_\delta(s/d) = \sum_{d|s} \mu(d) \sum_{e|(s/d)} \mu(e)|e|^{-\delta} \\ &= \sum_{def=s} \mu(d)\mu(e)|e|^{-\delta} = \sum_{e|s} \mu(e)|e|^{-\delta} \sum_{d|(s/e)} \mu(d) \\ &= \mu(s)|s|^{-\delta}, \end{aligned}$$

by Corollary 2.5.3. This conclusion and part (ii) of Lemma 2.4 then imply that

$$\begin{aligned} c_r(a) &= \mu(c)|b|^\delta \sum_{\substack{s|b \\ (s, c)=1}} \Phi(s) \\ &= \mu(c)|b|^\delta \frac{\bar{\varphi}_\delta(b)}{\bar{\varphi}_\delta((b, c))}, \end{aligned}$$

since $\bar{\varphi}_\delta$ is multiplicative, separable and non-zero everywhere. Therefore part (i) of Lemma 2.4 implies that

$$c_r(a) = \mu(c)|b|^\delta \frac{\bar{\varphi}_\delta(r)}{\bar{\varphi}_\delta(c)} = \frac{\mu(c)\varphi(r)}{\varphi(c)}. \qquad \square$$

2.6. Corollary. *If r and s divide k, then*

$$\varphi(r)c_s(k/r) = \varphi(s)c_r(k/s).$$

Proof. By Lemma 2.5,

$$\varphi(r)c_s(k/r) = \frac{\varphi(r)\varphi(s)\mu(s/(s, k/r))}{\varphi(s/(s, k/r))},$$

while

$$\varphi(s)c_r(k/s) = \frac{\varphi(s)\varphi(r)\mu(r/(r, k/s))}{\varphi(r/(r, k/s))}.$$

This proves the corollary, since

$$s/(s, k/r) = r/(r, k/s).$$ □

In order to show that the functions c_r', $r|k$, form an orthonormal set in $\mathfrak{D}_k$, now note that Corollary 2.6 implies that

$$\langle c_r, c_s\rangle = \sum_{d|k} \varphi(d)c_r(k/d)c_s(k/d) = \sum_{d|k} c_r(k/d)\varphi(s)c_d(k/s)$$

$$= \begin{cases} \varphi(s)|k|^\delta & \text{if } s = r, \\ 0 & \text{otherwise,} \end{cases}$$

by Lemma 2.2. This implies the required conclusion. □

It is a direct consequence of Theorem 2.1 that every function f (mod k) has a unique expansion of the form

$$f = \sum_{r|k}{}' \alpha_r c_r',$$

where $\alpha_r = \langle f, c_r'\rangle$ is the rth Fourier coefficient of f relative to the orthonormal basis $\{c_r' : r|k\}$. Therefore f has a unique **Ramanujan expansion**

$$f = \sum_{r|k} F(r)c_r,$$

where

$$F(r) = [\varphi(r)|k|^\delta]^{-1/2}\langle f, c_r'\rangle$$

$$= |k|^{-\delta/2}\boldsymbol{F}(r),$$

say. The function $\boldsymbol{F}(r)$ of $r|k$ may be extended in a unique way to an even function $\boldsymbol{F}$ (mod k), which will be called the **Fourier transform** of $f \in \mathfrak{D}_k$.

The following "Fourier inversion" theorem is valid.

2.7. Theorem. *Given $f \in \mathfrak{D}_k$, define the 'k-associated' function f^0 of f by*

$$f^0(a) = f(k/(k, a)) \qquad (a \in G).$$

Then $f^0 \in \mathfrak{D}_k$ *and* $f^{00} = f$. *In addition, if* $\boldsymbol{F}$ *denotes the Fourier transform of* f, *then* f^0 *is the Fourier transform of* $\boldsymbol{F}^0$.

Proof. The first assertions about f^0 are easy to verify. Then consider

$$\begin{aligned}\sum_{d|k} f(k/d)c_d(k/(k,a)) &= |k|^{-\delta/2} \sum_{d|k} c_d(k/(k,a)) \sum_{r|k} \boldsymbol{F}(r)c_r(k/d)\\ &= |k|^{-\delta/2} \sum_{r/k} \boldsymbol{F}(r) \sum_{d|k} c_r(k/d)c_d(k/(k,a))\\ &= |k|^{\delta/2}\boldsymbol{F}((k,a)) = |k|^{\delta/2}\boldsymbol{F}(a),\end{aligned}$$

by Lemma 2.2 and the fact that $\boldsymbol{F}$ is even (mod k). Therefore

$$\boldsymbol{F}(a) = |k|^{-\delta/2} \sum_{d|k} f(k/d)c_d(k/(k,a)).$$

A similar formula is then true for the Fourier transform Φ of $\boldsymbol{F}^0$, and this gives

$$\Phi(a) = |k|^{-\delta/2} \sum_{d|k} \boldsymbol{F}^0(k/d)c_d(k/(k,a)).$$

However, by definition,

$$\begin{aligned}f^0(a) = f(k/(k,a)) &= |k|^{-\delta/2} \sum_{d|k} \boldsymbol{F}(d)c_d(k/(k,a))\\ &= |k|^{-\delta/2} \sum_{d|k} \boldsymbol{F}^0(k/d)c_d(k/(k,a))\\ &= \Phi(a),\end{aligned}$$

by the preceding formula. Therefore f^0 is the Fourier transform of $\boldsymbol{F}^0$. □

It is interesting to note that the Parseval identity for the Hilbert space $\mathfrak{D}_k$ has the form

$$\begin{aligned}\langle f_1, f_2\rangle = \sum_{d|k} \varphi(d)f_1(k/d)\bar{f}_2(k/d) &= \sum_{d|k} \langle f_1, c_d'\rangle \overline{\langle f_2, c_d'\rangle}\\ &= \sum_{d|k} \varphi(d)\boldsymbol{F}_1(d)\bar{\boldsymbol{F}}_2(d)\\ &= \langle \boldsymbol{F}_1^0, \boldsymbol{F}_2^0\rangle,\end{aligned}$$

where $\boldsymbol{F}_1, \boldsymbol{F}_2$ denote the Fourier transforms of f_1 and f_2, respectively. It follows from this that the rule $f \to \boldsymbol{F}^0$ defines a unitary isometry of the Hilbert space $\mathfrak{D}_k$ ("Plancherel's theorem").

In order to give one fairly natural illustration of the Fourier transform and Ramanujan expansion of an explicit function in $\mathfrak{D}_k$, consider the function

$$\bar{\sigma}_w(a) = |a|^{-w}\sigma_w(a),$$

where

$$\sigma_w(a) = \sum_{d|a} |d|^w \qquad [a \in G,\ w \text{ real}].$$

Then note that, given any function $g \in \mathfrak{D}$, one obtains a function $g^\triangle \in \mathfrak{D}_k$ by defining

$$g^\triangle(a) = g((k, a));$$

then $g^{\triangle\triangle} = g^\triangle$.

2.8. Proposition. *Let* Σ_w *denote the Fourier transform of the function* $\bar{\sigma}_w{}^\triangle$. *Then*

$$\Sigma_w = |k|^{-\delta/2-w} \sigma_{w+\delta}{}^0$$

and $\bar{\sigma}_w{}^\triangle$ *has the Ramanujan expansion*

$$\bar{\sigma}_w{}^\triangle(a) = |(k, a)|^{-w} \sigma_w((k, a)) = |k|^{-\delta-w} \sum_{d|k} \sigma_{w+\delta}(k/d) c_d(a).$$

Proof. Consider

$$\sum_{d|k} \sigma_{w+\delta}(k/d) c_d(a) = \sum_{d|k} \sigma_{w+\delta}(k/d) \sum_{e|(d,a)} \mu(d/e)|e|^\delta$$

$$= \sum_{\substack{ef|k \\ e|a}} \sigma_{w+\delta}(k/ef) \mu(f) |e|^\delta$$

$$= \sum_{\substack{ef|k \\ e|a}} \mu(f) |e|^\delta \sum_{g|(k/ef)} |g|^{w+\delta}$$

$$= \sum_{\substack{eg|k \\ e|a}} |e|^\delta |g|^{w+\delta} \sum_{f|(k/eg)} \mu(f)$$

$$= \sum_{\substack{eg|k \\ e|a}} |e|^\delta |g|^{w+\delta},$$

by Corollary 2.5.3. Therefore

$$\sum_{d|k} \sigma_{w+\delta}(k/d) c_d(a) = |k|^{w+\delta} \sum_{\substack{e|k \\ e|a}} |e|^{-w} = |k|^{w+\delta} \sigma_{-w}((k, a))$$

$$= |k|^{w+\delta} |(k, a)|^{-w} \sigma_w((k, a)),$$

since

$$\sigma_{-w}(r) = \sum_{d|r} |r/d|^{-w} = |r|^{-w} \sum_{d|r} |d|^w = |r|^{-w} \sigma_w(r).$$

This proves the proposition. □

§ 3. Fourier theory of almost even functions

In this and the next section *it will again be assumed that G satisfies Axiom* A. In this case it is possible to develop a Fourier theory of almost even functions in which summation over the discrete space X_k as in § 2 is replaced by integration over the compactification G^*.

In order to establish the existence of a suitable integral over G^*, first consider:

3.1. Theorem. *Every almost even function f possesses a finite asymptotic mean-value $m(f)$ over G.*

Proof. The following lemma is needed both here and later.

3.2. Lemma. *For any fixed $r,k \in G$,*

$$\sum_{\substack{|a| \leq x \\ k|a}} c_r(a) = \begin{cases} |k|^{-\delta} A\varphi(r)x^{\delta} + O(|k|^{-\eta}x^{\eta}) & \textit{if } r|k, \\ O(|k|^{-\eta}x^{\eta}) & \textit{otherwise.} \end{cases}$$

In particular, c_r has the asymptotic mean-value 1 *or* 0 *according as* $r=1$ *or not.*

Proof. We have

$$\sum_{\substack{|a| \leq x \\ k|a}} c_r(a) = \sum_{|bk| \leq x} c_r(bk) = \sum_{\substack{|bk| \leq x \\ d|(r,bk)}} \mu(r/d)\,|d|^{\delta}.$$

Therefore, if $r|k$, then, with the aid of Axiom A,

$$\sum_{\substack{|a| \leq x \\ k|a}} c_r(a) = \sum_{\substack{|b| \leq x/|k| \\ d|r}} \mu(r/d)\,|d|^{\delta} = \varphi(r) N_G(x/|k|)$$

$$= |k|^{-\delta} A\varphi(r)x^{\delta} + O(|k|^{-\eta}x^{\eta}).$$

The case when $r \nmid k$ is more involved. Firstly, the initial equation gives

$$\sum_{\substack{|a| \leq x \\ k|a}} c_r(a) = \sum_{|bk| \leq x} \sum_{\substack{d|r,\, d'|b \\ k=tk',\, d=td' \\ (k',d')=1}} \mu(r/d)\,|d|^{\delta}$$

$$= \sum_{\substack{|bk| \leq x \\ k=tk'}} \sum_{\substack{td'|r,\, b=b'd' \\ (k',d')=1}} \mu(r/td')\,|td'|^{\delta}$$

$$= \sum_{\substack{k=tk' \\ t|r}} |t|^{\delta} \sum_{\substack{d'|(r/t) \\ (k',d')=1}} \mu(r/td')\,|d'|^{\delta} \sum_{|b'd'k| \leq x} 1.$$

But, for fixed d', k,

$$\sum_{|b'd'k|\leq x} 1 = N_G(x/kd')$$
$$= |kd'|^{-\delta} Ax^{\delta} + O(|kd'|^{-\eta} x^{\eta}),$$

and so

$$\sum_{\substack{|a|\leq x \\ k|a}} c_r(a) = \Sigma_1 + \Sigma_2,$$

where

$$\Sigma_1 = |k|^{-\delta} A x^{\delta} \sum_{\substack{k=tk' \\ t|r}} |t|^{\delta} \sum_{\substack{d'|(r/t) \\ (k',d')=1}} \mu(r/td'),$$

$$\Sigma_2 = O\Big\{|k|^{-\eta} x^{\eta} \sum_{\substack{k=tk' \\ t|r}} |t|^{\delta} \sum_{\substack{d'|(r/t) \\ (k',d')=1}} |d'|^{\delta-\eta}\Big\}$$
$$= O\Big\{|k|^{-\eta} x^{\eta} \sum_{t|r} |t|^{\delta} \sigma_{\delta-\eta}(r/t)\Big\}$$
$$= O(|k|^{-\eta} x^{\eta});$$

here the implied constant depends on r but not on k.

In order to deal with Σ_1, now consider:

3.3. Lemma. *For any given $r, a \in G$,*

$$\sum_{\substack{d|r \\ (d,a)=1}} \mu(r/d) = \begin{cases} \mu(r) & \textit{if } r|a, \\ 0 & \textit{otherwise.} \end{cases}$$

Proof. The cases in which $r|a$ or $(r, a)=1$ are easy to deal with if one recalls Corollary 2.5.3 again. Hence it may be assumed that $1 \neq (r, a) \neq r$. Then let $r=tc$ where $(r, a)|t$ and $(t, c)=1$, so that $d|c$ if and only if $d|r$ and $(d, a)=1$. In that case,

$$\sum_{\substack{d|r \\ (d,a)=1}} \mu(r/d) = \mu(t) \sum_{d|c} \mu(c/d) = \begin{cases} \mu(t) & \text{if } c=1, \\ 0 & \text{otherwise.} \end{cases}$$

Now, if $c=1$, then $r=t$. Also, the above assumption on (r, a) implies that there is a prime power $p^n \neq 1$ such that $p|(r, a)$, $p^n|r$ but $p^n \nmid (r, a)$; hence $n>1$ and r is not square-free. Therefore the last sum reduces to $\mu(r)=0$ when $c=1$. This proves the lemma. □

By means of Lemma 3.3, one now sees that $\Sigma_1=0$ when $r \nmid k$, since then $(r/t) \nmid k'$. This proves Lemma 3.2. □

It is an easy consequence of Lemma 3.2 that every even function g possesses a finite asymptotic mean-value $m(g)$, since g is a finite linear combination of Ramanujan sums. The similar conclusion for any almost even function f follows from the fact that f can be uniformly approximated by even functions. In order to show this in detail, suppose that (g_n) is a sequence of even functions that approximates f uniformly. Then, given any $\varepsilon>0$, there exists n_0 such that $\|f-g_n\|_u<\frac{1}{6}\varepsilon$ when $n\geqq n_0$. Hence, in the notation of Chapter 4, § 3, for any $x>0$,

$$|N(f, x)-N(g_n, x)| \leqq \sum_{|a|\leqq x} |f(a)-g_n(a)| < \tfrac{1}{6}\varepsilon N_G(x) \quad \text{for } n\geqq n_0.$$

Therefore, for any $x>0$,

$$\begin{aligned}|N(g_i, x)-N(g_j, x)| &\leqq |N(g_i, x)-N(f, x)|+|N(f, x)-N(g_j, x)| \\ &< \tfrac{1}{3}\varepsilon N_G(x) \quad \text{for } i,j \geqq n_0.\end{aligned}$$

Also, there exists $x_0=x_0(n, \varepsilon)$ such that

$$|m(g_n)-m(g_n, x)| < \tfrac{1}{3}\varepsilon \quad \text{for } x \geqq x_0.$$

Therefore, for $i,j\geqq n_0$ and any $x>0$,

$$\begin{aligned}|m(g_i)-m(g_j)| &\leqq |m(g_i)-m(g_i, x)|+|m(g_i, x)-m(g_j, x)| \\ &\quad +|m(g_j, x)-m(g_j)| \\ &< |m(g_i)-m(g_i, x)|+\tfrac{1}{3}\varepsilon+|m(g_j, x)-m(g_j)|.\end{aligned}$$

By choosing $x\geqq\max\,[x_0(i, \varepsilon), x_0(j, \varepsilon)]$, one therefore sees that

$$|m(g_i)-m(g_j)| < \varepsilon$$

whenever $i,j\geqq n_0$. Thus the asymptotic mean-values $m(g_n)$ form a Cauchy sequence, and hence have some limit m.

Then, for any x and n,

$$\begin{aligned}|m-m(f, x)| &\leqq |m-m(g_n)|+|m(g_n)-m(g_n, x)|+|m(g_n, x)-m(f, x)| \\ &< |m-m(g_n)|+|m(g_n)-m(g_n, x)|+\tfrac{1}{6}\varepsilon \quad \text{for } n \geqq n_0 \\ &< \tfrac{1}{2}\varepsilon+\tfrac{1}{3}\varepsilon+\tfrac{1}{6}\varepsilon = \varepsilon\end{aligned}$$

for $n\geqq$ some $n_1\geqq n_0$ and $x\geqq x_0(n, \varepsilon)$. Hence f has the asymptotic mean-value m. □

3.4. Corollary. *There exists a regular and complete probability measure ν on a σ-algebra of subsets of G^* containing all the open sets, such that, for every almost even function f,*

$$\int f^*\, d\nu = m(f),$$

where $f^*\in C(G^*)$ is the function corresponding to $f=\tau^*(f^*)$.

Proof. This is a direct consequence of Theorem 3.1 and the Riesz Representation Theorem (see Hewitt and Stromberg [1] 12.36, 12.55 and 11.21, say), since the rule $f \to m(f)$ and the isomorphism $\mathfrak{D}^* \cong C(G^*)$ induce a non-negative linear functional $m: C(G^*) \to \mathbf{C}$. The fact that ν is a probability measure, i.e. that $\nu(G^*)=1$, follows from the equations

$$\nu(G^*) = \int c_1{}^* \, d\nu = m(c_1) = 1. \qquad \square$$

(In passing, it may be noted that the equation $\int f^* \, d\nu = m(f)$ implies that the elements of $\tau(G)$ are uniformly distributed in G^* in the sense of the 'abstract theory of uniform distribution' as discussed in Helmberg [1].)

Corresponding to Theorem 2.1, we have:

3.5. Theorem. *The rule $f \to f^* = \tau^{*-1} f$ defines an embedding of the Banach algebra $\mathfrak{D}^*$ of all almost even functions on G into the Hilbert space $L^2(\nu)$ of all square-integrable functions on G^* relative to the measure ν. The (new) inner product on $\mathfrak{D}^*$ is given by*

$$\langle f, g \rangle = m(f \cdot \bar{g}) = \int f^* \bar{g}^* \, d\nu.$$

A complete orthonormal subset of $L^2(\nu)$ is provided by all the functions

$$C_r{}^* = (\varphi(r))^{-1/2} c_r{}^* \qquad (r \in G).$$

Proof. In view of the preceding results, the crux of the present proof is the statement that the functions $C_r{}^*$ $(r \in G)$ form a complete orthonormal set in $L^2(\nu)$. Firstly, the orthonormality of these functions is an immediate corollary of:

3.6. Lemma. *For any fixed $r, s \in G$,*

$$\sum_{|a| \leqq x} c_r(a) c_s(a) = \begin{cases} A\varphi(r) x^\delta + O(x^\eta) & \textit{if } r = s, \\ O(x^\eta) & \textit{otherwise.} \end{cases}$$

Proof. Consider

$$\begin{aligned} \sum_{|a| \leqq x} c_r(a) c_s(a) &= \sum_{|a| \leqq x} c_r(a) \sum_{d \mid (s,a)} \mu(s/d) |d|^\delta \\ &= \sum_{d \mid s} \mu(s/d) |d|^\delta \sum_{\substack{|a| \leqq x \\ d \mid a}} c_r(a) \\ &= \sum_{d \mid s} \mu(s/d) |d|^\delta \\ &\quad \times \begin{cases} |d|^{-\delta} A\varphi(r) x^\delta + O(|d|^{-\eta} x^\eta) & \text{if } r \mid d, \\ O(|d|^{-\eta} x^\eta) & \text{otherwise,} \end{cases} \end{aligned}$$

by Lemma 3.2. Therefore

$$\sum_{|a| \leqq x} c_r(a)c_s(a) = A\varphi(r)x^\delta \sum_{\substack{d|s \\ r|d}} \mu(s/d) + \mathrm{O}\big(x^\eta \sum_{d|s} |d|^{\delta-\eta}\big)$$

$$= A\varphi(r)x^\delta + \mathrm{O}(x^\eta),$$

if $r=s$. On the other hand, if $r \neq s$, then $r \nmid s$ or $s \nmid r$. If $r \nmid s$, then the above sum $\sum \mu(s/d)$ cannot occur, and the remaining term is $\mathrm{O}(x^\eta)$. Also, if $s \nmid r$, then a similar conclusion follows, because symmetry in r and s gives

$$\sum_{|a| \leqq x} c_r(a)c_s(a) = \sum_{|a| \leqq x} c_s(a)c_r(a)$$

$$= A\varphi(s)x^\delta \sum_{\substack{d|r \\ s|d}} \mu(r/d) + \mathrm{O}\big(x^\eta \sum_{d|r} |d|^{\delta-\eta}\big). \qquad \square$$

In order to show that the set $\{C_r^* : r \in G\}$ is complete, now let $\| \ \|_2$ denote the norm on $L^2(\nu)$ or $\mathfrak{D}^*$ derived from the present inner product. Then, for any $f \in \mathfrak{D}^*$,

$$\|f^*\|_2^2 = m(|f|^2) = \lim_{x \to \infty} (N_G(x))^{-1} \sum_{|a| \leqq x} |f(a)|^2$$

$$\leqq \|f\|_u^2 = \|f^*\|_u^2.$$

This implies that any sequence of functions in $C(G^*)$ that approximates f^* uniformly also approximates f^* in the L^2-norm. Since the algebra $\mathfrak{R}^*$ spanned by all the functions C_r^* $(r \in G)$ is uniformly dense in $C(G^*)$ by the proof of Theorem 1.12, and $C(G^*)$ is dense in $L^2(\nu)$ relative to the L^2-norm (see Hewitt and Stromberg [1] 13.21, say), it follows that $\mathfrak{R}^*$ is a dense subspace of the Hilbert space $L^2(\nu)$. Therefore a standard theorem on Hilbert spaces (see Hewitt and Stromberg [1] 16.26, say) implies that the set $\{C_r^* : r \in G\}$ is a complete orthonormal basis for $L^2(\nu)$. $\square$

A particular consequence of Theorem 3.5 is that every almost even function f on G has a unique L^2-convergent expansion of the form

$$f = \sum_{r \in G} \alpha_r C_r,$$

where $C_r = (\varphi(r))^{-1/2} c_r$ and $\alpha_r = \langle f, C_r \rangle = \langle f^*, C_r^* \rangle$ is the rth Fourier coefficient of f or f^* relative to the orthonormal basis $\{C_r : r \in G\}$. Thus f has a unique L^2-convergent **Ramanujan expansion**

$$f = \sum_{r \in G} F(r) c_r,$$

where

$$F(r) = (\varphi(r))^{-1}\langle f, c_r\rangle$$

may be called the rth **Ramanujan coefficient** of f. In addition, the Parseval identity for $L^2(\nu)$ implies the (ordinarily convergent) formula

$$\langle f_1, f_2\rangle = \sum_{r\in G} \varphi(r) F_1(r)\overline{F_2(r)},$$

where $F_i(r)$ is the rth Ramanujan coefficient of f_i.

For number-theoretical purposes, it would seem particularly interesting to obtain information about the pointwise convergence of the Ramanujan expansion $\sum_{r\in G} F(r)c_r$ to a given arithmetical function f, and some information in this direction is given in the next section.

Before turning to such questions, and considering the Ramanujan expansions of some specific arithmetical functions of interest, it may be interesting (even if not unexpected) to observe that the measure ν on G^* cannot be pulled back directly to a non-trivial measure on G:

3.7. Proposition. *The embedding $\tau: G \to G^*$ maps G onto a null set relative to ν, i.e., $\nu(\tau(G))=0$.*

Proof. Let the primes $p\in P$ be ordered in some sequence $p_1, p_2, \ldots$, and let $\pi_i: G^* \to G_{(p_i)}$ denote the natural projection map onto $G_{(p_i)}$. Define

$$Y_r = \{x\in G^*\colon \pi_r(x) = \pi_{r+1}(x) = \ldots = 1\},$$

$$Y_{r,s} = \{x\in G^*\colon \pi_r(x) = \pi_{r+1}(x) = \ldots = \pi_{r+s}(x) = 1\}.$$

Then $Y_r \subseteq Y_{r,s}$ and $\tau(G) \subseteq \bigcup_{r=1}^{\infty} Y_r$.

Since $Y_{r,s}$ is both open and closed in G^*, its characteristic function is continuous on G^*; hence this function is of the form q^* for some function $q=q_{r,s}\in\mathfrak{D}^*$ and

$$\nu(Y_{r,s}) = \int q^* \, d\nu = m(q).$$

Now let $c=p_r p_{r+1} \cdots p_{r+s} \in G$. Then τ maps an element $a\in G$ into $Y_{r,s}$ if and only if $(a, c)=1$. Thus

$$\tau(G) \cap Y_{r,s} = \tau(G\langle c\rangle),$$

where $G\langle c\rangle$ denotes the semigroup of all elements $a\in G$ that are coprime to c. It follows that q is the characteristic function of $G\langle c\rangle$ in G, and so

$$\nu(Y_{r,s}) = \lim_{x\to\infty} (N_G(x)^{-1} N_{G\langle c\rangle}(x) = |c|^{-\delta}\varphi(c),$$

by Proposition 4.1.3. Therefore

$$\nu(Y_{r,s}) = \prod_{i=r}^{r+s} (1-|p_i|^{-\delta})$$

and so, for every $s \geqq 1$,

$$\nu(Y_r) \leqq \nu(Y_{r,s}) = \nu(Y_{1,s}) \prod_{i=1}^{r} (1-|p_i|^{-\delta})^{-1}.$$

Also, for any $x > 0$,

$$\nu(Y_{1,s}) = \prod_{i=1}^{s+1} (1-|p_i|^{-\delta})$$

$$\leqq \prod_{\substack{p \in P \\ |p| \leqq x}} (1-|p|^{-\delta})$$

whenever $s \geqq s_0 = s_0(x)$, say. Since Lemma 6.3.1 implies that

$$\lim_{x\to\infty} \Big\{ \prod_{\substack{p \in P \\ |p| \leqq x}} (1-|p|^{-\delta})\Big\} = 0,$$

it follows now that Y_r is a null set. Therefore $\tau(G)$ is also a null set, since it is contained in $\bigcup_{r=1}^{\infty} Y_r$. □

§ 4. A wider type of almost evenness, and point-wise convergence of Ramanujan expansions

In the previous sections, analogy with the theory of almost periodic functions suggested the investigation of arithmetical functions which are 'almost even' in the sense they can be "nicely" approximated by even functions. There, as in classical function theory, "nice" approximation was implicitly taken to mean approximation in terms of the uniform metric $\|f-g\|_u$. However, as in ordinary function theory (see for example Besicovitch [1] Chapter 2), it may be interesting to investigate functions with slightly less stringent approximation properties, such as arithmetical functions which can be approximated by even functions in terms of weaker metrics or pseudo-metrics. Over here, we shall confine attention to one weaker mode of approximation which is analogous to that of Besicovitch in the classical theory of almost-periodic functions.

In order to discuss this, we first define the **upper** mean-value $\overline{m}(g)$ of a non-negative real-valued arithmetical function g by

$$\overline{m}(g) = \limsup_{x\to\infty} m(g, x) = \limsup_{x\to\infty} (N_G(x))^{-1} \sum_{|a| \leqq x} g(a).$$

Then, for $\lambda \geqq 1$, let $\mathfrak{B}^\lambda = \mathfrak{B}^\lambda(G)$ denote the set of all arithmetical functions f on G such that $|f|^\lambda$ has a finite upper mean value. In that case, it follows from Minkowski's inequality

$$\left(N^{-1}\sum_{i=1}^{N}(a_i+b_i)^\lambda\right)^{1/\lambda} \leqq \left(N^{-1}\sum_{i=1}^{N}a_i^\lambda\right)^{1/\lambda} + \left(N^{-1}\sum_{i=1}^{N}b_i^\lambda\right)^{1/\lambda} \qquad (a_i, b_i \geqq 0)$$

that $\mathfrak{B}^\lambda$ is a vector subspace of $\mathfrak{D}$ on which the function

$$\|f\|_\lambda = (\overline{m}(|f|^\lambda))^{1/\lambda}$$

behaves as a seminorm (i.e., $\| \ \|_\lambda$ has the usual properties of a norm except that $\|f\|_\lambda = 0$ need not imply $f = 0$). Consequently, $\mathfrak{B}^\lambda$ forms a pseudo-metric space under the pseudo-metric $\|f-g\|_\lambda$.

It is easy to see that $\mathfrak{B}(G) \subseteqq \mathfrak{B}^\lambda(G)$, and the standard inequality

$$\left(N^{-1}\sum_{i=1}^{N}a_i^\lambda\right)^{1/\lambda} \leqq \left(N^{-1}\sum_{i=1}^{N}a_i^{\lambda'}\right)^{1/\lambda'} \qquad (\lambda \leqq \lambda',\ a_i \geqq 0)$$

(see Hardy, Littlewood and Pólya [1] § 2.9) shows that $\|f\|_\lambda \leqq \|f\|_{\lambda'}$ and

$$\mathfrak{B}^{\lambda'}(G) \subseteqq \mathfrak{B}^\lambda(G) \quad \text{for } \lambda' \geqq \lambda.$$

We now define an arithmetical function f on G to be **almost even** (B^λ) if and only if f belongs to the closure $B^\lambda(G)$ of the vector space $\mathfrak{R}$ in $\mathfrak{B}^\lambda(G)$ relative to the topology defined by $\| \ \|_\lambda$. If $\lambda = 1$, we shall write $B(G) = B^1(G)$ and call its elements almost even (B), but the notation $\mathfrak{B}^1(G)$ will not be changed; it is the case $\lambda = 1$ that will be our main concern in the present discussion.

4.1. Proposition. *If f is an almost even (B^λ) function on G, then so are $\bar{f}$, $|f|$ and $f \cdot g$, where g is any uniformly almost even function on G. Further, f possesses a finite asymptotic mean-value $m(f)$.*

Proof. If g is any even function, then so are $\bar{g}$ and $|g| = g \cdot \bar{g}$, because $\mathfrak{R}$ is an algebra closed under complex conjugation. Therefore, if (f_n) is a sequence of even functions f_n that approximate a given almost even (B^λ) function f in the $\mathfrak{B}^\lambda$-seminorm, then $(\bar{f}_n)$ is a sequence of even functions which similarly approximate $\bar{f}$, while $(|f_n|)$ is a sequence of even functions that approximate $|f|$ in the $\mathfrak{B}^\lambda$-seminorm. Suppose further that g is any uniformly almost even function, which is uniformly approximated by some sequence (g_n) of even functions g_n. Then, for any $\varepsilon>0$, Minkowski's inequality implies that

$$(m(|fg-f_n g_n|^\lambda, x))^{1/\lambda} \leqq (m(|fg-f_n g|^\lambda, x))^{1/\lambda}+(m(|f_n g-f_n g_n|^\lambda, x))^{1/\lambda}$$
$$< \|g\|_u(m(|f-f_n|^\lambda, x))^{1/\lambda}+\varepsilon(m(|f_n|^\lambda, x))^{1/\lambda}$$

for $n\geqq$ some $n_0=n_0(\varepsilon)$. Then a second application of Minkowski's inequality gives

$$(m(|fg-f_n g_n|^\lambda, x))^{1/\lambda} < \|g\|_u(m(|f-f_n|^\lambda, x))^{1/\lambda}+\varepsilon(m(|f|^\lambda, x))^{1/\lambda}$$
$$+\varepsilon(m(|f-f_n|^\lambda, x))^{1/\lambda}.$$

Therefore

$$\|fg-f_n g_n\|_\lambda < \|g\|_u\|f-f_n\|_\lambda+\varepsilon\|f\|_\lambda+\varepsilon\|f-f_n\|_\lambda,$$

and this leads to the conclusion that $f\cdot g$ is also almost even (B^λ).

So far, Axiom A was not needed; however, it will now be used in order to show that f has a finite asymptotic mean-value. For this purpose, note that if (f_n) approximates f in the $\mathfrak{B}^{\lambda'}$-seminorm then (f_n) also approximates f in the $\mathfrak{B}^\lambda$-seminorm for any $\lambda\leqq\lambda'$, by the inequality $\|f\|_\lambda\leqq\|f\|_{\lambda'}$ mentioned earlier. It follows from this that $B^{\lambda'}(G)\subseteqq B^\lambda(G)$ for $\lambda'\geqq\lambda$, and also, in particular, that (f_n) approximates f in the $\mathfrak{B}^1$-seminorm. Now suppose further that each f_n has a finite asymptotic mean-value $m(f_n)$, as will happen if f_n is even.

Then, given any $\varepsilon>0$, there exists $x_0=x_0(n, \varepsilon)$ such that $|m(f_n)-$ $-m(f_n, x)|<\frac{1}{3}\varepsilon$ for $x\geqq x_0$. Also, there exists $n_0=n_0(\varepsilon)$ such that $\overline{m}(|f-f_n|)<$ $<\frac{1}{6}\varepsilon$ for $n\geqq n_0$. Hence, for $n\geqq n_0$, there exists $x_1=x_1(n, \varepsilon)$ such that

$$|m(f, x)-m(f_n, x)| \leqq m(|f-f_n|, x) < \tfrac{1}{6}\varepsilon \quad \text{for } x\geqq x_1.$$

Therefore, there exists $x_2=x_2(r, n, \varepsilon)$ such that

$$|m(f_r, x)-m(f_n, x)| \leqq m(|f_r-f_n|, x) \leqq m(|f_r-f|, x)+m(|f-f_n|, x)$$
$$< \tfrac{1}{6}\varepsilon+\tfrac{1}{6}\varepsilon = \tfrac{1}{3}\varepsilon \quad \text{for } r, n \geqq n_0 \text{ and } x\geqq x_2.$$

Then, as in the proof of Theorem 3.1, one sees that the mean-values $m(f_n)$ form a Cauchy sequence, with a limit m say, and this fact again enables one to deduce that f has the asymptotic mean-value m. □

4.2. Corollary. *If f is almost even (B), then all the Ramanujan coefficients*

$$F(r) = (\varphi(r))^{-1}\langle f, c_r\rangle = (\varphi(r))^{-1}m(f. c_r)$$

exist. Further, $\|f\|_1=m(|f|)$, and, if g is any almost even (B) function such that $\|f-g\|_1=0$, then g has the same Ramanujan coefficients as f. □

If all the Ramanujan coefficients of a given arithmetical function f are known to exist, it becomes of interest to investigate the convergence of the associated Ramanujan series $\sum_{r\in G} F(r)c_r$. The following theorem, though not intended as the most delicate possible theorem of its kind, provides number-theoretical criteria for both the almost evenness (B) of a given arithmetical function f and the point-wise convergence to f of the associated Ramanujan series, which are easy to apply in quite a number of natural situations.

4.3. Theorem. *Let f denote any arithmetical function on G such that $f(z)=$ $=\zeta_G(z)g(z)$, where $\tilde{g}(z)$ is absolutely convergent for all complex z with* Re $z>\alpha$, *for some $\alpha<\delta$. Then, f is almost even (B), and its rth Ramanujan coefficient is*

$$F(r) = (\varphi(r))^{-1}\langle f, c_r\rangle = \sum_{\substack{b\in G\\ r|b}} g(b)\,|b|^{-\delta}.$$

Further (in terms of ordinary convergence of complex numbers)

$$f(a) = \sum_{a\in G} F(r)c_r(a),$$

the right-hand series being absolutely convergent, for each $a\in G$.

Proof. Consider the functions η_r such that

$$\sum_{d|r} c_d(a) = \eta_r(a)$$

$$= \begin{cases} |r|^{\delta} & \text{if } r|a,\\ 0 & \text{otherwise.}\end{cases}$$

Then, since $f(z) = \zeta_G(z)g(z)$,

$$f(a) = \sum_{r|a} g(r) = \sum_{r\in G}' g(r)\,|r|^{-\delta}\eta_r(a).$$

For $n=1, 2, \ldots,$ define

$$f_n = \sum_{|r|\leqq n} g(r)\,|r|^{-\delta}\eta_r.$$

Then $f_n\in\mathfrak{R}$ in view of the above equation for η_r. Further,

$$\overline{m}(|f-f_n|) = \overline{m}\Big(\Big|\sum_{|r|>n} g(r)\,|r|^{-\delta}\eta_r\Big|\Big)$$

$$\leqq \sum_{|r|>n} |g(r)|\,|r|^{-\delta}\,\overline{m}(\eta_r) = \sum_{|r|>n} |g(r)|\,|r|^{-\delta},$$

since $\bar{m}(\eta_r)=m(\eta_r)=1$ by Lemma 3.2. Since $\tilde{g}(\delta)$ is absolutely convergent, it therefore follows that $\bar{m}(|f-f_n|)\to 0$ as $n\to\infty$. Hence $f\in B(G)$, and this fact does not depend on the convergence of $\tilde{g}(z)$ "past" δ. Next, consider

$$\sum_{|a|\leqq x} f(a)c_r(a) = \sum_{|a|\leqq x} c_r(a)\sum_{d|a} g(d) = \sum_{|bd|\leqq x} c_r(bd)g(d)$$
$$= \sum_{|d|\leqq x} g(d) \sum_{\substack{|a|\leqq x\\ d|a}} c_r(a)$$
$$= \sum_{|d|\leqq x} g(d)$$
$$\times \begin{cases} |d|^{-\delta}A\varphi(r)x^{\delta}+\mathrm{O}(|d|^{-\beta}x^{\beta}) & \text{if } r|d,\\ (|d|^{-\beta}x^{\beta}) & \text{otherwise,}\end{cases}$$

where, by Lemma 3.2, $\beta\geqq\eta$ may be chosen so that $\alpha<\beta<\delta$. Hence

$$\sum_{a|\leqq x} f(a)c_r(a) = A\varphi(r)x^{\delta}\sum_{\substack{|d|\leqq x\\ r|d}} g(d)|d|^{-\delta}+\mathrm{O}\big(x^{\beta}\sum_{|d|\leqq x}|g(d)||d|^{-\beta}\big)$$
$$= A\varphi(r)x^{\delta}\big\{\sum_{\substack{d\in G\\ r|d}} g(d)|d|^{-\delta}+\mathrm{o}(1)\big\}+\mathrm{O}(x^{\beta}).$$

This proves the stated formula for the Ramanujan coefficient $F(r)$ of f. (Regarding $F(1)=m(f)=\tilde{g}(\delta)$, compare Theorem 4.3.1.)

In order to discuss the pointwise convergence of $\sum_{r\in G}F(r)c_r$, note that it now follows that, for any $a\in G$,

$$\sum_{r\in G} F(r)c_r(a) = \sum_{r\in G} c_r(a)\sum_{\substack{b\in G\\ r|b}} g(b)|b|^{-\delta}.$$

Also, by (1.3),

$$\sum_{r\in G}|c_r(a)|\sum_{\substack{b\in G\\ r|b}}|g(b)||b|^{-\delta} \leqq \sigma(a)\sum_{r\in G}\sum_{\substack{b\in G\\ r|b}}|g(b)||b|^{-\delta}$$
$$= \sigma(a)\sum_{r,s\in G}|g(rs)||rs|^{-\delta}$$
$$= \sigma(a)\sum_{b\in G}|g(b)||b|^{-\delta}\sum_{rs=b}1.$$

Since $d(b) = \mathrm{O}(|b|^{\varepsilon})$ for every $\varepsilon>0$, by Proposition 5.2.3 it follows that

$$\sum_{r\in G}|c_r(a)|\sum_{\substack{b\in G\\ r|b}}|g(b)||b|^{-\delta} = \mathrm{O}\big(\sum_{b\in G}|g(b)||b|^{\varepsilon-\delta}\big) = \mathrm{O}(1),$$

if $\varepsilon>0$ is chosen so that $\delta-\varepsilon>\alpha$. Therefore the above double series for $\sum_{r\in G} F(r)c_r(a)$ is absolutely convergent, and so by interchanging the orders of summation one obtains

$$\sum_{r\in G} F(r)c_r(a) = \sum_{r\in G} c_r(a) \sum_{s\in G} g(rs)|rs|^{-\delta} = \sum_{s\in G}\sum_{r\in G} c_r(a)g(rs)|rs|^{-\delta}$$

$$= \sum_{b\in G} g(b)\,|b|^{-\delta} \sum_{rs=b} c_r(a) = \sum_{b\in G} g(b)|b|^{-\delta}\eta_b(a)$$

$$= f(a),$$

by the equation for $f(a)$ noted at the beginning of the present proof. Further, the left-hand series is absolutely convergent, and so the theorem is proved. (Note that the series cannot be uniformly convergent if f is not a bounded function.) □

The hypothesis on f in Theorem 4.3 is of a kind that was seen in Chapters 4 and 5 to be valid for various natural arithmetical functions of interest. Hence one may now note a number of applications of Theorem 4.3 to particular arithmetical functions. In these cases, and in general, it is interesting to observe how the equation

$$f(a) = \sum_{r\in G} F(r)c_r(a)$$

provides a rather graphic indication of how the value $f(a)$ differs from its asymptotic mean-value $m(f)=F(1)$.

4.4. Proposition. *For* $w>0$, *the functions*

$$\bar{\sigma}_w(a) = |a|^{-w}\sigma_w(a),$$

$$\bar{\varphi}_w(a) = |a|^{-w}\varphi_w(a) \qquad (a\in G)$$

are both almost even (B) *and have the pointwise absolutely convergent Ramanujan expansions*

$$\bar{\sigma}_w(a) = \zeta_G(\delta+w) \sum_{r\in G} |r|^{-\delta-w} c_r(a),$$

$$\bar{\varphi}_w(a) = (\zeta_G(\delta+w))^{-1} \sum_{r\in G} (\varphi_{\delta+w}(r))^{-1}\mu(r)c_r(a).$$

Proof. In the proof of Proposition 4.5.7, which gave the asymptotic mean-values of $\bar{\sigma}_w$ and $\bar{\varphi}_w$, it was noted that

$$\bar{\sigma}_w(z) = \zeta_G(z)\zeta_G(z+w), \qquad \bar{\varphi}_w(z) = \zeta_G(z)\mu(z+w),$$

and that $\zeta_G(z+w)$ and $\tilde{\mu}(z+w)$ are both absolutely convergent for $\operatorname{Re} z > \delta - w$. (Note that the functions $\bar{\sigma}_w$ and $\bar{\varphi}_w$ are real-valued, and that the use of the bar in this notation should not be confused with complex conjugation as earlier.) It follows that Theorem 4.3 applies to the present functions, and it remains only to calculate their Ramanujan coefficients.

Firstly, by Theorem 4.3, the rth Ramanujan coefficient of $\bar{\sigma}_w$ is

$$\sum_{\substack{b\in G\\ r|b}} |b|^{-w}|b|^{-\delta} = \sum_{s\in G} |rs|^{-\delta-w} = |r|^{-\delta-w} \sum_{s\in G} |s|^{-\delta-w}$$

$$= |r|^{-\delta-w}\zeta_G(\delta+w).$$

Also, the rth Ramanujan coefficient of $\bar{\varphi}_w$ is

$$\sum_{\substack{b\in G\\ r|b}} \mu(b)|b|^{-w}|b|^{-\delta} = \sum_{s\in G} \mu(rs)\,|rs|^{-\delta-w} = \sum_{\substack{s\in G\\ (r,s)=1}} \mu(rs)|rs|^{-\delta-w},$$

since $\mu(b)=0$ if b is not square-free. Hence this coefficient is

$$\mu(r)|r|^{-\delta-w} \sum_{\substack{s\in G\\ (r,s)=1}} \mu(s)|s|^{-\delta-w} = \mu(r)|r|^{-\delta-w}\tilde{\mu}_r(\delta+w),$$

where μ_r denotes the Möbius function of the arithmetical semigroup $G\langle r\rangle$ of all elements $s\in G$ that are coprime to r. Then

$$\mu_r(z) = \prod_{\substack{p\in P\\ p\nmid r}} (1-p^{-z}) = \mu(z) \prod_{\substack{p\in P\\ p|r}} (1-p^{-z})^{-1},$$

and since Proposition 4.1.3 states that $G\langle r\rangle$ also satisfies Axiom A (with unchanged exponents) it follows that not only does $\tilde{\mu}_r(\delta+w)$ converge but

$$\tilde{\mu}_r(\delta+w) = \tilde{\mu}(\delta+w) \prod_{\substack{p\in P\\ p|r}} (1-|p|^{-\delta-w})^{-1} = (\bar{\varphi}_{\delta+w}(r))^{-1}\tilde{\mu}(\delta+w).$$

The proposition therefore follows. □

4.5. Corollary. *The functions $\bar{\sigma}(n)=\sigma(n)/n$ and $\bar{\varphi}(n)=\varphi(n)/n$ of $n\in G_{\mathrm{Z}}$ are both almost even (B) and have the point-wise absolutely convergent Ramanujan expansions*

$$\bar{\sigma}(n) = \tfrac{1}{6}\pi^2 \sum_{r=1}^{\infty} \frac{c_r(n)}{r^2}, \qquad \bar{\varphi}(n) = 6\pi^{-2} \sum_{r=1}^{\infty} \frac{\mu(r)}{\varphi_2(r)} c_r(n),$$

where

$$\varphi_2(r) = r^2 \prod_{\substack{p\text{ prime}\\ p|r}} (1-p^{-2}).$$ □

The proofs of Propositions 4.4.3 to 4.4.5 show that the functions mentioned in the next proposition all satisfy the hypotheses of Theorem 4.3. Hence this proposition follows from Theorem 4.3.

4.6. Proposition. *For every* $k=1, 2, \ldots,$ *the point-wise powers* β^k, $(d_*/d)^k$, $(d/d_*)^k$, *and the function* q_k, *are all almost even* (B) *and possess point-wise absolutely convergent Ramanujan expansions which converge to their respective values on G.* □

Further results about the spaces $B^\lambda(G)$ appear in the author's paper [15]. For example, it is shown there that they are complete and isometrically isomorphic to the Lebesgue spaces $L^\lambda(\nu)$ over the almost even compactification G^*. In particular, $B^2(G)\cong L^2(\nu)$ is actually the Hilbert-space completion of $\mathfrak{D}^*$.

§ 5. Arithmetical functions over G_Z

Before considering some further applications of Theorem 4.3, to specific functions over G_Z, it may be interesting to interpret some of the previous concepts in terms of slightly more familiar ones. For example, Proposition 5.1 below relates the Ramanujan sums with ordinary trigonometrical sums, and as an immediate corollary of this and Corollary 4.5 one obtains the original Ramanujan expansion for $\sigma(n)/n$ which was quoted in the introduction to this chapter. It is then also possible to deduce results relating (almost) evenness of arithmetical functions on G_Z to classical notions of (almost) periodicity. It may be interesting to mention also that the elements of the almost even compactification $G_Z{}^*$ of G_Z appear classically in the theory of certain types of fields, where they are sometimes known as *Steinitz numbers;* see for example D. K. Harrison [1]. (The topological construction of $G_Z{}^*$, and also certain similar but slightly more general constructions over G_Z, were first introduced into the study of Ramanujan sums by Schwarz and Spilker [1].)

5.1. Proposition. *For any positive integers n and r,*

$$c_r(n) = \sum_{\substack{s\leqq r\\(r,s)=1}} \exp\left[-2\pi i sn/r\right]$$

$$= \sum_{\substack{s\leqq r\\(r,s)=1}} \cos\left(2\pi sn/r\right).$$

Proof. The second trigonometrical sum indicated is often taken to define $c_r(n)$ in elementary number theory. For the present proof, first note that

$$\sum_{s=1}^{r} \exp[-2\pi i\, sn/r] = \begin{cases} r & \text{if } r|n, \\ 0 & \text{otherwise}, \end{cases}$$
$$= \eta_r(n).$$

Therefore

$$c_r(n) = \sum_{d|r} \mu(r/d)\eta_d(n) = \sum_{d|r} \mu(r/d) \sum_{s=1}^{d} \exp[-2\pi i\, sn/d].$$

Now consider any complex-valued function F of real $x>0$, and observe that by reducing every fraction s/r to the form s'/d, where $d|r$ and $(s', d)=1$, one obtains the equation

$$\sum_{s=1}^{r} F(s/r) = \sum_{d|r} \sum_{\substack{s \leqq d \\ (s,d)=1}} F(s/d).$$

Therefore

$$f(r) = \sum_{d|r} g(d),$$

where

$$f(r) = \sum_{s=1}^{r} F(s/r),$$

$$g(d) = \sum_{\substack{s \leqq d \\ (s,d)=1}} F(s/d).$$

Thus Möbius inversion gives

$$g(r) = \sum_{d|r} \mu(r/d) f(d).$$

By substituting $F(s/r) = \exp[-2\pi i\, sn/r]$, it follows then that

$$c_r(n) = \sum_{d|r} \mu(r/d) f(d) = g(r)$$
$$= \sum_{\substack{s \leqq r \\ (r,s)=1}} \exp[-2\pi i\, sn/r].$$

Lastly,

$$\sum_{\substack{s \leqq r \\ (r,s)=1}} \cos(2\pi\, sn/r) = \mathrm{Re}\Big\{ \sum_{\substack{s \leqq r \\ (r,s)=1}} \exp[-2\pi i\, sn/r]\Big\}$$
$$= \mathrm{Re}\,\{c_r(n)\} = c_r(n). \qquad \square$$

The above proposition shows in more classical terms why it is the case that every even function over G_Z is also periodic. It also yields the following corollaries, after reference to Proposition 1.10 and the definitions of $\mathfrak{D}^*$ and $B^\lambda(G)$.

5.2. Corollary. *The set of all trigonometrical polynomials of the form*

$$T(n) = \sum_{s=1}^{m} \alpha_s \exp[-2\pi i \varrho_s n] \qquad (\alpha_s \text{ complex}, \varrho_s \text{ rational})$$

is dense in the topological algebra $\mathfrak{D}(G_Z)$. □

5.3. Corollary. (i) *Every function in* $\mathfrak{D}^*(G_Z)$ *is uniformly almost periodic i.e., is a uniform limit of trigonometrical polynomials (of the type* $T(n)$ *above).*

(ii) *Every function in* $B^\lambda(G_Z)$ *is almost periodic* (B^λ), *i.e., is the* $\mathfrak{B}^\lambda$*-limit of a sequence of trigonometrical polynomials (of the type* $T(n)$ *above).* □

By the proofs of Theorems 5.1.9 and 5.1.13, it follows that the functions over G_Z occurring in the next proposition satisfy the hypotheses of Theorem 4.3. Therefore that theorem and Corollary 5.3 yield the stated conclusion:

5.4. Proposition. *Let* $a(n)$ *and* $S(n)$ *denote the total numbers of isomorphism classes of finite abelian groups, or semisimple finite rings, of cardinal* n, *respectively. Then, for each* $k=1, 2, \ldots$, *the pointwise powers* $[a(n)]^k$ *and* $[S(n)]^k$ *are almost-periodic* (B) *functions of* $n \in G_Z$, *and they possess pointwise absolutely convergent Ramanujan expansions which converge to their respective values.* □

The last conclusion for $[a(n)]^k$ is actually a special case of a proposition concerning the function $[a(I)]^k$ of $I \in G_K$, where K is an arbitrary algebraic number field (see Theorem 5.1.9) and its proof). Similarly, by appealing to results in the paper of the author and Ridley [1], it may be deduced that the above statement about $[S(n)]^k$ is a special case of one about a corresponding function over G_K. (In connection with even functions over G_K, it may be mentioned that the Ramanujan sum $c_I(J)$, where $I,J \in G_K$, can also be represented as a trigonometrical sum whenever J is a principal ideal; for details, see Rieger [8].)

We end this chapter with some comments on further results about Ramanujan expansions. Firstly, for more delicate number-theoretical theorems about the convergence of such expansions for special classes of arithmetical functions over G_Z, particular reference may be made to the papers of Schwarz [6–8]. Secondly, there are some fascinating examples of pointwise convergent Ramanujan expansions for particular functions of interest, which do not fit within the previous framework.

For example, in his original paper, Ramanujan [2] established the following formulae for the divisor function $d(n)$ and the function $r(n)$ (see Proposition 4.1.2) of $n \in G_{\mathbf{Z}}$:

$$d(n) = -\sum_{k=1}^{\infty} \frac{\log k}{k} c_k(n); \tag{5.5}$$

$$r(n) = \pi \sum_{k=1}^{\infty} \frac{(-1)^{k-1}}{2k-1} c_{2k-1}(n). \tag{5.6}$$

The proofs of these formulae are *ad hoc* ones depending on results of roughly the same order of difficulty as the Prime Number Theorem. The fact that the formulae do not fit within the previous scheme may be seen in the first case from the fact that $d(n)$ does not have a finite asymptotic mean value, and so is certainly not almost periodic (B) (compare Proposition 4.3.2). In the second case, although Proposition 4.1.2 shows that $r(n)$ does have the asymptotic mean value π, so that (5.6) has the same general appearance as the formulae considered earlier, it turns out that $r(n)$ is nevertheless not almost periodic (B); see Kac [2].

Perhaps, as in the general theory of trigonometrical series, there exist "pathological" functions and expansions in the present arithmetical setting which cannot readily be covered by a systematic theory. In any case, it remains to be seen whether or not there may be some deeper conceptual explanation for such formulae as (5.5) and (5.6).

Selected bibliography for Chapter 7

Section 1: E. Cohen [4, 8], Delsarte [1], Horadam [8, 14], Knopfmacher [15], Schwarz and Spilker [1].

Section 2: Anderson and Apostol [1], E. Cohen [4, 6, 8, 9, 12], Horadam [8, 14], McCarthy [1].

Section 3: E. Cohen [5], Knopfmacher [15], Schwarz and Spilker [1, 2].

Section 4: E. Cohen [5, 15], Delsarte [1], Horadam [15], Knopfmacher [15], Rieger [8], Schwarz and Spilker [2], Wintner [2].

Section 5: Carmichael [1], E. Cohen [10, 15, 16], Delsarte [1], Erdös and Wintner [1], Hardy [1], Hartman and Wintner [1, 2], Kac [2], Kac, van Kampen and Wintner [1], Novoselov [1], Ramanujan [2], Schwarz [5, 8], Schwarz and Spilker [2], van Kampen [1], van Kampen and Wintner [1], Wintner [2–4].

PART III

ANALYTICAL PROPERTIES OF OTHER ARITHMETICAL SYSTEMS

Part III of this book is concerned with asymptotic arithmetical properties of two types of arithmetical systems that differ significantly from those treated in Part II. In the first place, Chapter 8 contains a discussion of asymptotic questions regarding a broad class of natural arithmetical semigroups which seem to be best viewed as *additive* arithmetical semigroups satisfying a certain axiom of a rather different kind from Axiom A. Then Chapter 9 treats *arithmetical formations*, which are systems whose theory allows both a generalization and a refinement of the theory of arithmetical semigroups. An arithmetical formation is, roughly, an arithmetical semigroup G together with an equivalence relation on G which partitions it into classes that are in many ways analogous to arithmetical progressions of positive integers. Discussion of one wide variety of natural arithmetical formations leads to the consideration of a suitable asymptotic axiom on formations, which generalizes Axiom A for an ordinary arithmetical semigroup. The theory of formations that satisfy the axiom in question consequently provides a refinement of the asymptotic arithmetical theory of Part II.

Amongst other matters, both Chapter 8 and Chapter 9 contain proofs of "abstract prime number theorems" that are appropriate to the types of system discussed in each case. Once again, the applications of such theorems to specific arithmetical categories, and also the proofs that the relevant axioms are valid for those categories, yield a variety of asymptotic enumeration theorems that seem to have an intrinsic interest independent of, or in addition to, any abstract number-theoretical considerations.

CHAPTER 8

ADDITIVE ARITHMETICAL SEMIGROUPS

This chapter is concerned mainly with asymptotic properties of a fairly wide class of natural arithmetical semigroups which seem to be most appropriately regarded as additive arithmetical semigroups. The semigroups of this class all have asymptotic properties of a kind that may be described axiomatically by a certain simple *Axiom* C concerning an abstract additive arithmetical semigroup G. An "additive abstract prime number theorem" is proved subject to Axiom C. When this theorem is applied to specific additive arithmetical categories of interest, it yields a variety of asymptotic enumeration theorems concerning the isomorphism classes of objects in such categories.

It is interesting to note that the abstract additive prime number theorem subject to Axiom C is closely related to classical theorems of additive analytic number theory due to Hardy and Ramanujan, and concerning arithmetical functions of the same general type as the classical partition function $p(n)$. In fact, the theorem itself may to a large extent be regarded as a reformulation of a certain well-known 'Tauberian' theorem of Hardy and Ramanujan.

The final section of the chapter contains some results about asymptotic average values and densities of certain specific arithmetical functions and sets, within the present abstract setting. In some cases, the conclusions are strikingly different from the corresponding ones that occur subject to Axiom A.

§ 1. Axiom C

In considering asymptotic properties of an arithmetical semigroup G, one might in theory begin with either $N_G(x)$ or $\pi_G(x)$ and investigate how assumptions about the asymptotic behaviour of one of these functions as $x \to \infty$ influences that of the other. In the general setting, any theorem of this kind could perhaps be called an "abstract prime number theorem", and the same term could be applied to theorems involving assumptions and conclusions about related functions such as $G(q)$ and $P(q)$, or $N_G^{\#}(x)$ and $\pi_G^{\#}(x)$ if G is regarded as an additive arithmetical semigroup. (Some references to abstract investigations of this kind, of a type fairly close to Axiom A and the asymptotic law $\pi_G(x) \sim x^\delta/\delta \log x$, are given in the article of Bateman and Diamond [1]. For a slightly different type of abstract theory, see also Rémond [1].)

For a purely abstract investigation of the kind just referred to, there appear to be no special grounds for preferring one starting point or asymptotic law to another. However, if, as in this book, one regards the natural examples as the primary objects of investigation, and views abstract axiomatic discussions of arithmetical semigroups largely as convenient methods for conducting unified investigations of classes of natural systems with similar properties, then certain particularly significant patterns do emerge. In the first place, semigroups satisfying Axiom A come to the fore, and in all natural instances of this type considered so far it appears to be easier (though by no means always trivial) to begin with a consideration of the functions $N_G(x)$. In the second place, a natural class of semigroups presents itself, whose members appear to be most conveniently viewed as additive arithmetical semigroups.* Various concrete examples of such semigroups were discussed in Chapters 1 and 3, and it will be seen shortly that for many of these it turns out to be very easy to determine the asymptotic behaviour of $\pi_G^{\#}(x)$ as $x \to \infty$. By contrast, it will be seen later that the asymptotic investigation of $N_G^{\#}(x)$ and of other arithmetical properties of such an additive arithmetical semigroup G is usually much more difficult. (There are some apparent exceptions to the above statement about semigroups satisfying Axiom A which arise if one first assumes the Prime Number Theorem for $G_{\mathbf{Z}}$ or the Prime Ideal Theorem for G_K, where K is an algebraic number field. It is then possible to deduce the corresponding "prime number" theorems for various specific arithmetical categories like $\mathscr{A}$, $\mathfrak{S}$ and $\mathfrak{F}$ by means of a certain elementary "transfer principle"; further, given error-term estimates in the assumed theorems about $G_{\mathbf{Z}}$ and G_K, this procedure also yields corresponding estimates for the categories in question. (See the author's paper [8] for further details.)

For an additive arithmetical semigroup G, one type of asymptotic law for $\pi_G^{\#}(x)$ which occurs in a variety of instances is specified by:

* Two subclasses of this class, with asymptotic properties distinet from those treated in this chapter, are studied in Knopfmacher [13, 16].

Axiom C. *There exist positive constants C and* κ, *and a real constant* ν, *such that*

$$\pi_G{}^{\#}(x) \sim Cx^{\kappa}(\log x)^{\nu} \quad \textit{as } x \to \infty.$$

(Although it is not necessary for the purposes of this book, there seems to be some desirability in reserving for "Axiom B" a certain statement of the general type contained in Theorem 5.2.4 and Corollary 5.2.5; this follows the procedure of the author's paper [6].)

Now consider some specific cases:

1.1. Example: Finite abelian p-groups and "arithmetically equivalent" objects. Let $\mathscr{A}(p)$ denote the category of all finite abelian p-groups, where p is any fixed rational prime. The discussion given in Chapters 1 and 3 shows that, for any other rational prime p', the categories $\mathscr{A}(p)$ and $\mathscr{A}(p')$ are 'arithmetically equivalent' in the sense that their corresponding additive arithmetical semigroups are isomorphic under degree-preserving semigroup isomorphisms. Further, these categories $\mathscr{A}(p)$ are also arithmetically equivalent in this sense to certain other arithmetical categories, such as the category $\mathfrak{S}_c(p)$ of all commutative semisimple finite p-rings, where p is a fixed rational prime, the category $\mathfrak{S}_c(F_q)$ of all commutative semisimple finite-dimensional algebras over the Galois field $F_q = \mathrm{GF}(q)$, and the category $\mathfrak{P}$ of all pseudo-metrizable finite topological spaces. If $\mathfrak{C}$ denotes any one of these arithmetical categories, its generating function is

$$Z_{\mathfrak{C}}(y) = \sum_{n=0}^{\infty} \boldsymbol{p}(n) y^n = \prod_{r=1}^{\infty} (1-y^r)^{-1},$$

where $\boldsymbol{p}(n)$ is the partition function (see Chapter 3), and here

$$\pi_{\mathfrak{C}}{}^{\#}(x) = \sum_{r \leqq x} 1 = x + \mathrm{O}(1) \quad \sim x \quad \text{as } x \to \infty.$$

This is perhaps the simplest non-trivial case of Axiom C in a natural setting.

1.2. Example: Semisimple finite p-rings and finite-dimensional algebras over GF(q). Let $\mathscr{D}$ denote either the category $\mathfrak{S}(p)$ of all semisimple finite p-rings, where p is a fixed rational prime, or the category $\mathfrak{S}(F_q)$ of all semisimple finite-dimensional algebras over the Galois field $F_q = \mathrm{GF}(q)$; by Chapter 3, $\mathfrak{S}(p)$ and $\mathfrak{S}(F_q)$ are arithmetically equivalent (in the new sense) for all primes p and prime powers q. Then the discussion given in Chapter 3 shows that $\mathscr{D}$ has the generating function

$$Z_{\mathscr{D}}(y) = \sum_{n=0}^{\infty} s(n) y^n = \prod_{\substack{r \geqq 1 \\ m \geqq 1}} (1-y^{-rm^2})^{-1},$$

where $s(n)$ is a generalized partition function. Also

$$\pi_{\mathscr{D}}{}^{\#}(x) = \sum_{rm^2 \leqq m} 1 = \sum_{m \leqq x^{1/2}} [m^{-2}x] = \sum_{m \leqq x^{1/2}} \{m^{-2}x + O(1)\}$$

$$= x\{\zeta(2) + O(x^{-1/2})\} + O(x^{1/2}) = \tfrac{1}{6}\pi^2 x + O(x^{1/2}),$$

by Proposition 4.2.6. Hence $\mathscr{D}$ satisfies Axiom C.

1.3. Example: Further semisimple finite-dimensional algebras. Let $\mathfrak{S}(F)$ denote the category of all semisimple finite-dimensional algebras over a given field F.

Case (i): F is an algebraically closed field. Here, the discussion of Chapter 3 shows that

$$\pi_{\mathfrak{S}(F)}{}^{\#}(x) = \sum_{m^2 \leqq x} 1 = x^{1/2} + O(1).$$

Case (ii): F is a real closed field R. By Chapter 3,

$$\pi_{\mathfrak{S}(R)}{}^{\#}(x) = \sum_{m^2 \leqq x} 1 + \sum_{2m^2 \leqq x} 1 + \sum_{4m^2 \leqq x} 1$$

$$= [x^{1/2}] + [(\tfrac{1}{2}x)^{1/2}] + [\tfrac{1}{2}x^{1/2}]$$

$$= \tfrac{1}{2}(3 + \sqrt{2})x^{1/2} + O(1).$$

1.4. Example: Compact Lie groups and semisimple Lie algebras. Let $\mathscr{E}$ denote either the category $\mathfrak{G}_c$ of all compact simply-connected Lie groups, or the category $\mathfrak{L}_s(F_0)$ of all semisimple finite-dimensional Lie algebras over a given algebraically closed field F_0 of characteristic zero. The description of the primes in the categories given in Chapter 1 shows that

$$\pi_{\mathscr{E}}{}^{\#}(x) = \sum_{m^2 + 2m \leqq x} 1 + 2 \sum_{2m^2 + m \leqq x} 1 + \sum_{2m^2 - m \leqq x} 1 + O(1).$$

A convenient way of determining the asymptotic behaviour of any sum of the type appearing here is to "complete the square": Given real constants $a > 0$ and b, there exist real constants c, d such that

$$\sum_{am^2 + bm \leqq x} 1 = \sum_{a(m+c)^2 \leqq x + d} 1 = [((x+d)/a)^{1/2} - c]$$

$$\sim (x/a)^{1/2} \quad \text{as } x \to \infty.$$

In particular, it follows that

$$\pi_{\mathscr{G}}^{\#}(x) \sim \left(1+\tfrac{3}{2}\sqrt{2}\right)x^{1/2} \quad \text{as } x\to\infty.$$

1.5. Example: Symmetric Riemannian manifolds. Let $\mathfrak{R}_{\mathrm{sc}}$ denote the category of all compact simply-connected globally symmetric Riemannian manifolds. The description of the primes of $\mathfrak{R}_{\mathrm{sc}}$ given in Chapter 1 shows that $\pi_{\mathfrak{R}_{\mathrm{sc}}}^{\#}(x)$ is the sum of an O(1) term, a finite number of "quadratic" sums of the type considered immediately above, and

$$\sum_{\substack{2pq\leqq x\\ p\geqq 1,\ q\geqq p}} 1+\sum_{\substack{pq\leqq x\\ p\geqq 1,\ q\geqq p}} 1+\sum_{\substack{4pq\leqq x\\ p\geqq 1,\ q\geqq p}} 1.$$

(In the last sums, one may ignore restrictions such as $p+q>4$ because asymptotically this only introduces an error of order O(1).) Now for $y>0$

$$\sum_{\substack{pq\leqq y\\ p\geqq 1,\ q\geqq p}} 1 = \sum_{\substack{p\geqq 1\\ p\leqq q\leqq y/p}} 1 = \sum_{p\leqq\sqrt{y}}\ \sum_{p\leqq q\leqq y/p} 1$$

$$= \sum_{p\leqq\sqrt{y}} \{[y/p]-p+1\}$$

$$= \sum_{p\leqq\sqrt{y}} \{y/p+\mathrm{O}(1)\}-\tfrac{1}{2}w^2,$$

where $w=[\sqrt{y}]$. Therefore Proposition 4.2.8 implies that the sum reduces to

$$y\sum_{p\leqq\sqrt{y}} p^{-1}+\mathrm{O}(\sqrt{y})+\mathrm{O}(y) = \tfrac{1}{2}y\log y+\mathrm{O}(y) \quad \text{as } y\to\infty.$$

Letting y take the values $\frac{1}{2}x$, x and $\frac{1}{4}x$ in succession, it is now easy to deduce that

$$\pi_{\mathfrak{R}_{\mathrm{sc}}}^{\#}(x) = \tfrac{7}{8}x\log x+\mathrm{O}(x) \quad \text{as } x\to\infty.$$

1.6. Example: Finite modules over a ring of algebraic integers. Let $\mathfrak{F}$ denote the category of all modules of finite cardinal over the ring D of all algebraic integers in a given number field K, and let $\mathfrak{F}(p)$ denote the subcategory of all modules of cardinal a power of p, where p is a fixed rational prime. Then $\mathfrak{F}(p)=\mathscr{A}(p)$ if $D=\mathbf{Z}$, and in general it was seen in Chapter 3 that $\mathfrak{F}(p)$ forms an additive arithmetical category relative to the degree mapping $\partial(M)=\log_p \operatorname{card}(M)$.

If the ideal $(p)=P_1^{e_1}P_2^{e_2}\dots P_m^{e_m}$, where the P_i are distinct prime ideals in D, $e_i>0$ and $|P_i|=p^{\alpha_i}$, define the constant

$$C_{D,p} = \alpha_1^{-1}+\dots+\alpha_m^{-1}.$$

Then the discussion of $\mathfrak{F}(p)$ in Chapter 3 shows that as $x\to\infty$

$$\pi_{\mathfrak{F}(p)}{}^{\#}(x) = \sum_{i=1}^{m} \sum_{\alpha_i r \leqq x} 1$$
$$= \sum_{i=1}^{m} [x/\alpha_i] = C_{D,p}x + \mathrm{O}(1).$$

The above examples provide a selection of additive arithmetical categories which satisfy Axiom C. Lastly, we mention a less important but illustrative example, which is of a different type.

1.7. Example: Cyclotomic integers. Let G_Φ denote the multiplicative semigroup of all 'cyclotomic integers' as defined in Chapter 2, § 7. Then G_Φ is the semigroup of all polynomials with rational integer coefficients which are either 1 or admit (unique) factorization into powers of the various cyclotomic polynomials Φ_n $(n=1, 2, \ldots)$. Also, deg $\Phi_n = \varphi(n)$, and since Proposition 5.2.3 shows that $\varphi(n)\to\infty$ as $n\to\infty$, it follows that the degree mapping $\partial(f)=\deg f$ makes G_Φ into an additive arithmetical semigroup such that

$$\pi_{G_\Phi}{}^{\#}(x) = \sum_{\varphi(m)\leqq x} 1.$$

This last sum has been studied for its own sake, and it is known (see Dressler [1]) that it is asymptotic to $(\zeta(2)\zeta(3)/\zeta(6))x$ as $x\to\infty$. Hence the semigroup G_Φ satisfies Axiom C.

§ 2. Analytical properties of the zeta function

Let G denote an additive arithmetical semigroup. In certain cases, particularly when G satisfies Axiom C, special interest attaches to properties of G that are related to its degree mapping. Nevertheless, even when the main interest lies in that direction, it can sometimes still be useful to regard G simultaneously as an ordinary arithmetical semigroup. In this chapter, this will be done by associating with G and ∂ the particular norm $|\ |$ such that $|a|=\mathrm{e}^{\partial(a)}$ $(a\in G)$, where e denotes the base for natural logarithms. In that case, the zeta (enumerating) function of G and its Euler product formula may be written as

$$\zeta_G(z) = \sum_{u\in\partial(G)} G^{\#}(u)\mathrm{e}^{-uz} = \prod_{p\in P} (1-\mathrm{e}^{-\partial(p)z})^{-1}$$
$$= \prod_{u\in\partial(G)_0} (1-\mathrm{e}^{-uz})^{-P^{\#}(u)},$$

where $\partial(G)_0 = \partial(G)\setminus\{0\}$.

As in Chapter 3, one may also associate with G the formal generating function

$$Z_G(y) = \sum_{u\in\partial(G)} G^{\#}(u)\, y^u = \prod_{u\in\partial(G)_0} (1-y^u)^{-P^{\#}(u)},$$

where

$$y^u = e^{-uz} = (e^u)^{-z} \qquad [u\in\partial(G)];$$

this term will be used here even if the degree mapping ∂ is not integer-valued, although, as was noted in Chapter 3, there is a considerable simplification in the integer-valued case. Regardless of what real values ∂ may take, when G satisfies axioms such as Axiom C, it becomes useful to study the analytic properties of $\zeta_G(z)$ and $Z_G(y)$ when z and y are replaced by complex numbers.

2.1. Proposition. *Let G denote an additive arithmetical semigroup such that, for any $\varepsilon>0$,*

$$\pi_G{}^{\#}(x) = O(e^{\varepsilon x}) \quad \textit{as } x\to\infty.$$

Then the series $\zeta_G(z)$ and $Z_G(y)$ define analytic functions satisfying the respective Euler product formulae analytically, for all complex numbers z and y such that $\operatorname{Re} z>0$ *and* $|y|<1$, *respectively. In particular, this is true when G satisfies Axiom* C.

Proof. Consider

$$P_G(z) = \sum_{p\in P} |p|^{-z} = \sum_{p\in P} e^{-\partial(p)z}.$$

Then, for any complex z, partial summation gives

$$\sum_{\partial(p)\leqq x} e^{-\partial(p)z} = \pi_G{}^{\#}(x)e^{-xz} + z\int_{u_0}^{x} \pi_G{}^{\#}(t)e^{-tz}\, dt,$$

where $u_0=\min\{\partial(p): p\in P\}$. If $\pi_G{}^{\#}(x)=O(e^{\varepsilon x})$ as $x\to\infty$, and $\operatorname{Re} z>\varepsilon$, then the integral converges, and it follows from Lemma 4.2.4 that the series $P_G(z)$ converges. Hence, for $\sigma_0>\varepsilon$, Proposition 4.2.1 implies that $P_G(z)$, $\zeta_G(z)$ and the Euler product are absolutely and uniformly convergent (and the last two have the same value) whenever $\operatorname{Re} z\geqq\sigma_0$. In particular, $P_G(z)$ and $\zeta_G(z)$ define analytic functions of z for all z with $\operatorname{Re} z>\varepsilon$, since $\sigma_0>\varepsilon$ is arbitrary.

These statements now carry over to corresponding statements about $P_G{}^{\#}(y)=\sum_{p\in P} y^{\partial(p)}$ and $Z_G(y)$ as functions of complex y with $|y|<e^{-\varepsilon}$. Finally, if the assumption on $\pi_G{}^{\#}(x)$ is true for every $\varepsilon>0$, it follows that the previous statements hold whenever $\operatorname{Re} z>0$ and $|y|<1$. Since the assumption on $\pi_G{}^\#(x)$ certainly holds when G satisfies Axiom C, the proposition follows. (Note that, if desired, absolute convergence allows one to collect terms in the various series and products as in Corollary 4.2.2, within the above ranges for z and y.) □

Although the generating function is often more appropriate to the discussion of an additive arithmetical semigroup G than the corresponding zeta function, it is of course only a formal variant of the zeta function. Also, in the following pages we shall be particularly concerned with $Z_G(y)$ and $\zeta_G(s)$ as functions of *real* y and s with $0<y<1$ and $s>0$, and here especially there is little advantage in distinguishing between $Z_G(y)=\zeta_G(\log y^{-1})$ and $\zeta_G(s)=Z_G(e^{-s})$. Since these functions may to a large extent be viewed as tools for the investigation of intrinsic arithmetical questions about G, in this chapter it will suffice to study the zeta function.

2.2. Proposition. *Let G denote an additive arithmetical semigroup such that, for any* $\varepsilon>0$,

$$\pi_G{}^\#(x) = O(e^{\varepsilon x}) \quad \textit{as } x\to\infty.$$

Then, for any $\varepsilon>0$,

$$N_G{}^\#(x) = O(e^{\varepsilon x}) \quad \textit{as } x\to\infty,$$

and, for real $s>0$,

$$\zeta_G(s) = s\int_0^\infty N_G{}^\#(t)e^{-st}\,dt$$

$$= \exp\left\{s\int_0^\infty (e^{st}-1)^{-1}\pi_G{}^\#(t)\,dt\right\}.$$

In particular, these statements hold when G satisfies Axiom C.

Proof. For any fixed $\varepsilon>0$,

$$N_G(x)e^{-\varepsilon x} = \sum_{u\leqq x} G^\#(u)e^{-\varepsilon x} \leqq \sum_{u\leqq x} G^\#(u)e^{-\varepsilon u} \leqq \zeta_G(\varepsilon).$$

Therefore

$$N_G{}^\#(x) \leqq \zeta_G(\varepsilon)e^{\varepsilon x},$$

and the statement about $N_G{}^\#(x)$ follows. By using this and the convergence of $\zeta_G(z)$, one may now deduce from Lemma 4.2.4 that

$$\zeta_G(z) = z\int_0^\infty N_G{}^\#(t)e^{-tz}\,dt$$

whenever $\operatorname{Re} z>0$.

Now let $0<u_0<u_1<u_2<\ldots$ denote the ascending sequence of distinct values taken by ∂ on G. Then, for $s>0$, the Euler product formula implies that

$$\begin{aligned}\log \zeta_G(s) &= \sum_{r=0}^{\infty} P^{\#}(u_r) \log (1-e^{-su_r})^{-1} \\ &= P^{\#}(u_0) \log (1-e^{-su_0})^{-1} \\ &\quad + \sum_{r=1}^{\infty} [\pi_G{}^{\#}(u_r) - \pi_G{}^{\#}(u_{r-1})] \log (1-e^{-su_r})^{-1} \\ &= \sum_{r=0}^{\infty} \pi_G{}^{\#}(u_r) \{\log (1-e^{-su_r})^{-1} - \log (1-e^{-su_{r+1}})^{-1}\} \\ &= \sum_{r=0}^{\infty} \pi_G{}^{\#}(u_r) \int_{u_r}^{u_{r+1}} (e^{st}-1)^{-1} s \, dt \\ &= s \int_{u_0}^{\infty} (e^{st}-1)^{-1} \pi_G{}^{\#}(t) \, dt,\end{aligned}$$

where the rearrangement of terms is permissible since, for $0<\varepsilon<s$,

$$\pi_G{}^{\#}(u_n) \log (1-e^{-su_n})^{-1} = O(e^{\varepsilon u_n} e^{-su_n}) = o(1)$$

as $u_n \to \infty$, i.e., as $n \to \infty$. (Since it is always assumed that P is non-empty, $u_n \to \infty$ as $n \to \infty$.) This proves the proposition. □

In the following discussion, some of the constants that arise involve values of the Riemann zeta function $\zeta(s)$ for $s>1$, and of the Gamma function

$$\Gamma(s) = \int_0^{\infty} e^{-t} t^{s-1} \, dt$$

for $s>0$. For later applications to specific arithmetical semigroups, it will be useful to recall the following standard formulae in particular:

$$\begin{aligned}\zeta(2) &= \tfrac{1}{6}\pi^2, \\ \Gamma(\tfrac{1}{2}) &= \sqrt{\pi}, \\ \Gamma(s+1) &= s\Gamma(s), \\ \Gamma(n+1) &= n! \quad \text{for } n = 0, 1, 2, \ldots .\end{aligned}$$

In the first place, special values of $\zeta(s)$ and $\Gamma(s)$ will occur in the derivation of an abstract prime number theorem, valid for any additive arithmetical semigroup G satisfying Axiom C. In accordance with previous comments, such a theorem will here be interpreted as one which provides asymptotic information about $N_G{}^{\#}(x)$ as $x\to\infty$, or about $G^{\#}(u)$ as $u\to\infty$. The first main theorem which we shall prove states that

$$N_G{}^{\#}(x) = \exp\{[c_G+o(1)]x^{\kappa/(\kappa+1)}(\log x)^{\nu/(\kappa+1)}\} \quad \text{as } x\to\infty,$$

where

$$c_G = \kappa^{-1}(\kappa+1)^{(\kappa-\nu+1)/(\kappa+1)}[\kappa C\Gamma(\kappa+1)\zeta(\kappa+1)]^{1/(\kappa+1)}$$

and C, κ and ν are the constants specified by Axiom C.

In order to achieve this aim, use will be made of methods of the *analytic theory of partitions*. These methods and in fact the entire theory in question were initiated by **Hardy** and **Ramanujan** [1] in 1917, with the object of obtaining asymptotic information about purely arithmetical functions of the same general type as the classical partition function $p(n)$. Questions of the latter kind were investigated by Hardy and Ramanujan and later authors as problems of intrinsic interest. Although the present main emphasis is very different, the following discussion could also be used as a brief introduction to this branch of classical additive analytic number theory.

If questions of emphasis or eventual applications of a not necessarily "purely" arithmetical kind are ignored, the formal equivalence between *generalized partition theory* in the classical sense and certain aspects of the theory of additive arithmetical semigroups may be seen as follows: Let $0<u_0<u_1<u_2<\ldots$ denote an arbitrary ascending sequence of positive real numbers u_n such that $u_n\to\infty$ as $n\to\infty$. If $u>0$, let $p(u)$ denote the total number of ways of "partitioning" u into a sum of the terms u_n, with repetitions allowed. In other words, let $p(u)$ denote the total number of ways of writing u in the form $\sum_i n_i u_i$, where the n_i are non-negative integers. Then $p(u)=0$ unless u lies in the (countable) semigroup of real numbers S which is generated by $\{0, u_0, u_1, u_2, \ldots\}$ relative to addition. Further, it is clear that one has the formal identity

$$\sum_{u\in S} p(u)e^{-us} = \prod_{r=0}^{\infty}(1+e^{-u_r s}+e^{-2u_r s}+\ldots)$$

$$= \prod_{r=0}^{\infty}(1-e^{-u_r s})^{-1}.$$

More generally, given such a sequence of numbers u_n and a non-negative integer $w(u_n)$ for each $n=0, 1, 2, \ldots$, it is customary to define the corresponding *weighted* partition function $p_w(u)$ by the equation

$$\sum_u p_w(u)e^{-us} = \prod_{r=0}^{\infty}(1-e^{-u_r s})^{-w(u_r)}.$$

Usually, convergence conditions are imposed in order to make sense of such equations, but these can also be avoided with the aid of a suitable specialization of the formal algebraic theory of series given in Part I.

In fact, one way of doing this is to now formally introduce an additive arithmetical semigroup G as follows: For each n such that $w(u_n)>0$, consider the ordered pairs $(n, 1), \ldots, (n, w(u_n))$, and let P denote the set of all ordered pairs so constructed. Let G denote the free commutative semigroup with identity generated by the elements of P. (Roughly, this may be described as the set of all formal "commutative words" $x_1x_2\ldots x_n$ of elements $x_i\in P$, with the empty product acting as identity element and multiplication taken as formal juxtaposition of words subject to the commutative law.) Then define $\partial(n, r)=u_n$ and extend ∂ in the obvious way to a completely additive function on G. In that case, G becomes an additive arithmetical semigroup with

$$P^{\#}(u_n) = w(u_n), \qquad G^{\#}(u) = p_w(u),$$

and the above equation becomes the Euler product formula for $\zeta_G(s)$.

Conversely, given any additive arithmetical semigroup G, it is clear that $G^{\#}(u)$ can be interpreted as a weighted partition function in the above sense. (The above description of partition functions does not cover all the variations of this concept considered in additive number theory, but it will suffice for present purposes.)

In order to prove the abstract prime number theorem quoted above, some preliminary results will be needed which also have applications to other arithmetical questions (as well as to generalized partition theory, of course). The results considered are not intended to represent the most general or sharpest ones of their kind. Subject to some restrictions or specializations that seem convenient here, the proofs follow the basic pattern adopted by Kohlbecker [1], who also derives some more general partition theorems which will be referred to again later.

2.3. Proposition. *Let $W(x)\geqq 0$ be a step function of $x\geqq 0$ such that*

$$W(x) \sim Rx^{\alpha}(\log x)^{\beta} \quad \textit{as } x\to\infty,$$

where $R>0$, $\alpha>0$ and β are constants. If

$$f(s) = \exp\left\{s\int_0^{\infty} (e^{st}-1)^{-1}W(t)\,dt\right\}$$

for $s>0$, then, as $s\to 0+$,

$$\log f(s) \sim R\Gamma(\alpha+1)\zeta(\alpha+1)s^{-\alpha}(\log(1/s))^{\beta}.$$

Proof. By definition,

$$\Gamma(\alpha+1)\zeta(\alpha+1) = \sum_{n=1}^{\infty} n^{-\alpha-1}\int_0^\infty e^{-t}t^\alpha\,dt$$

$$= \sum_{n=1}^{\infty}\int_0^\infty e^{-nu}u^\alpha\,du$$

$$= \int_0^\infty (e^{-u}+e^{-2u}+\ldots)u^\alpha\,du$$

$$= \int_0^\infty (e^u-1)^{-1}u^\alpha\,du,$$

since $e^{-nu}u^\alpha>0$ for $u>0$. Now, for $u\geqq 1$, let

$$F(u) = \int_u^\infty (e^t-1)^{-1}t^\alpha\,dt$$

$$< 2\int_u^{2u} e^{-t}t^\alpha\,dt+2\int_{2u}^\infty e^{-t/2}(e^{-t/2}t^\alpha)\,dt$$

$$\leqq 2(2u)^\alpha\int_u^{2u} e^{-t}\,dt+2e^{-u}\int_{2u}^\infty e^{-t/2}t^\alpha\,dt$$

$$\leqq K_1u^\alpha e^{-u}$$

for some constant K_1, since the last integral is less than $2^{\alpha+1}\Gamma(\alpha+1)$.

Given any positive $\varepsilon<\min(1, \frac{1}{3}K_1)$, then choose positive η and δ such that $5\eta^\alpha/\alpha<\varepsilon$ and

$$0 \leqq \Gamma(\alpha+1)\zeta(\alpha+1)-\int_\eta^\delta (e^t-1)^{-1}t^\alpha\,dt < \varepsilon.$$

Then, from the definition of $f(s)$,

$$\log f(s) = \int_0^\infty (e^{-u}-1)^{-1}W(u/s)\,du$$

$$= \left(\int_0^\eta+\int_\eta^\delta+\int_\delta^\infty\right)(e^u-1)^{-1}W(u/s)\,du$$

$$= I_1+I_2+I_3,$$

say.

In order to discuss I_1, consider the asymptotic assumption on $W(x)$ and accordingly choose x_0 such that

$$W(x) \leqq 2Rx^\alpha(\log x)^\beta$$

whenever $x \geqq x_0$. Then, for sufficiently small s, Lemma 5.2.6 implies

$$\begin{aligned} I_1 &\leqq \int_0^\eta u^{-1} W(u/s)\,\mathrm{d}u \\ &\leqq \int_0^{x_0} t^{-1} W(t)\,\mathrm{d}t + 2R \int_{x_0}^{\eta/s} t^{\alpha-1} (\log t)^\beta\,\mathrm{d}t \\ &\leqq K_2 + (3R/\alpha)\,(\eta/s)^\alpha \big(\log(\eta/s)\big)^\beta \\ &\leqq K_2 + (4R/\alpha)\,\eta^\alpha s^{-\alpha} \big(\log(1/s)\big)^\beta \\ &< \varepsilon R s^{-\alpha} \big(\log(1/s)\big)^\beta \end{aligned}$$

(s being sufficiently small).

Next, let s be so small that for $u \geqq \eta$

$$(1-\varepsilon)\,R(u/s)^\alpha \big(\log(u/s)\big)^\beta \leqq W(u/s) \leqq (1+\varepsilon)\,R(u/s)^\alpha (\log u/s)^\beta.$$

Then

$$\begin{aligned} (1-\varepsilon)\,R\big(\log(1/s)\big)^\beta \int_\eta^\delta \frac{(u/s)^\alpha \big(\log(u/s)\big)^\beta}{(\mathrm{e}^u-1)\big(\log(1/s)\big)^\beta}\,\mathrm{d}u &\leqq I_2 \\ \leqq (1+\varepsilon)\,R\big(\log(1/s)\big)^\beta \int_\eta^\delta \frac{(u/s)^\alpha \big(\log(u/s)\big)^\beta}{(\mathrm{e}^u-1)\big(\log(1/s)\big)^\beta}\,\mathrm{d}u. \end{aligned}$$

Since $\big(\log(u/s)\big)^\beta / \big(\log(1/s)\big)^\beta \to 1$ uniformly for $\eta \leqq u \leqq \delta$ as $s \to 0+$, when s is sufficiently small,

$$\begin{aligned} (1-2\varepsilon)\,Rs^{-\alpha}\big(\log 1/s\big)^\beta \int_\eta^\delta (\mathrm{e}^u-1)^{-1} u^\alpha\,\mathrm{d}u &\leqq I_2 \\ \leqq (1+2\varepsilon)\,Rs^{-\alpha}\big(\log(1/s)\big)^\beta \int_\eta^\delta (\mathrm{e}^u-1)^{-1} u^\alpha\,\mathrm{d}u; \end{aligned}$$

therefore

$$\begin{aligned} (1-2\varepsilon)\,Rs^{-\alpha}\big(\log(1/s)\big)^\beta [\Gamma(\alpha+1)\zeta(\alpha+1)-\varepsilon] &\leqq I_2 \\ \leqq (1+2\varepsilon)\,Rs^{-\alpha}\big(\log(1/s)\big)^\beta [\Gamma(\alpha+1)\zeta(\alpha+1)+\varepsilon]. \end{aligned}$$

Lastly, for sufficiently small s,

$$I_3 \leqq 2R \int_\delta^\infty (\mathrm{e}^u-1)^{-1} (u/s)^\alpha \big(\log(u/s)\big)^\beta\,\mathrm{d}u = 2Rs^{-\alpha} X,$$

where, recalling the function $F(u)$ defined earlier,

$$X = -F(u)\big(\log u/s\big)^\beta\big|_\delta^\infty + \beta \int_\delta^\infty u^{-1} F(u)\big(\log(u/s)\big)^{\beta-1}\,\mathrm{d}u.$$

Now

$$F(\delta) \leqq \Gamma(\alpha+1)\zeta(\alpha+1) - \int_\eta^\delta (\mathrm{e}^u-1)^{-1} u\,\mathrm{d}u < \varepsilon,$$

$$F(u) \leqq K_1 u^\alpha \mathrm{e}^{-u} < K_1(\mathrm{e}^u-1)^{-1} u^\alpha \quad \text{for } u \geqq 1.$$

Hence

$$\begin{aligned} X &\leqq F(\delta)\big(\log(\delta/s)\big)^\beta + \big(\delta \log(\delta/s)\big)^{-1} |\beta|\, K_1 X \\ &< 2\varepsilon\big(\log(1/s)\big)^\beta + \varepsilon K_1 X \end{aligned}$$

for s sufficiently small. Since, $\varepsilon<\frac{1}{3}K_1$, this implies that

$$X < 2(1-\varepsilon K_1)^{-1}(\log(1/s))^\beta < 3\varepsilon(\log(1/s))^\beta,$$

and therefore

$$I_3 < 6\varepsilon R s^{-\alpha}(\log(1/s))^\beta.$$

Since ε is arbitrary, the estimates for I_1, I_2 and I_3 therefore give the required conclusion. □

2.4. Corollary. *Let G denote an additive arithmetical semigroup satisfying Axiom* C *as stated earlier. Then, as* $s\to 0+$,

$$\log \zeta_G(s) \sim C\Gamma(\kappa+1)\zeta(\kappa+1)s^{-\kappa}(\log(1/s))^\nu. \qquad \square$$

As one would hope, the estimates for particular zeta functions that were obtained by *ad hoc* methods during the proofs of Lemma 5.1.11, Corollary 5.1.12 and Lemma 5.1.14 also follow as special cases of the present corollary.

§ 3. The additive abstract prime number theorem

In order to deduce the abstract prime number theorem subject to Axiom C stated in § 2, now consider the following 'Tauberian' theorem; the required conclusion about $N_G{}^{\#}(x)$ will then follow immediately with the aid of Corollary 2.4 and Proposition 2.2.

3.1. Theorem. *Let* $M(x)$ *be a non-decreasing step function of* $x\geqq 0$ *such that*

$$f(s) = s\int_0^\infty M(t)e^{-st}\,dt$$

exists for $s>0$. *Suppose that, as* $s\to 0+$,

$$\log f(s) \sim B\Gamma(\alpha+1)\zeta(\alpha+1)s^{-\alpha}(\log(1/s))^\beta,$$

where $B>0$, $\alpha>0$ *and* β *are constants. Then, as* $x\to\infty$,

$$M(x) = \exp\{[b+o(1)]x^{\alpha/(\alpha+1)}(\log x)^{\beta/(\alpha+1)}\},$$

where

$$b = \alpha^{-1}(\alpha+1)^{(\alpha-\beta+1)/(\alpha+1)}[\alpha B\Gamma(\alpha+1)\zeta(\alpha+1)]^{1/(\alpha+1)}.$$

Proof. There is no loss of generality in assuming that $M(x)\geqq 0$ for $x\geqq 0$, since addition of a suitable constant would make this true and only alter $f(s)$ by the same constant. Then, for positive u and s,

$$M(u)e^{-su} = s\int_u^\infty M(u)e^{-st}\,dt \leqq s\int_u^\infty M(t)e^{-st}\,dt \leqq f(s).$$

Therefore, for any fixed positive $\varepsilon<1$, when s is sufficiently small,

$$M(u) < \exp\{(1+\varepsilon)Ds^{-\alpha}(\log(1/s))^{\beta}+su\},$$

where $D=B\Gamma(\alpha+1)\zeta(\alpha+1)$.

In order to derive the required conclusion about $M(u)$, our next step will be to replace s by a function s_u of u as specified in the following lemma.

3.2. Lemma. *There exists a function s_u of u for all sufficiently large u, such that s_u is a differentiable function of u which tends to zero monotonically as $u\to\infty$, and*

(i) $u = \alpha D s_u^{-\alpha-1}(\log(1/s_u))^{\beta}$;

(ii) $\mathrm{d}s_u/\mathrm{d}u \sim -s_u/\{(\alpha+1)u\}$, $\mathrm{d}(us_u)/\mathrm{d}u \sim \alpha s_u/(\alpha+1)$ *as* $u\to\infty$;

(iii) $g(u)=u^{1/(\alpha+1)}s_u$ *is a 'slowly oscillating' function of u in the sense that* $g(cu)/g(u)\to 1$ *as* $u\to\infty$, *for every fixed* $c>0$;

(iv) *if s_u is defined for $u\geqq u_0$, then*

$$\int_{u_0}^{u} s_t\,\mathrm{d}t \sim ((\alpha+1)/\alpha)\,us_u \quad \text{as } u\to\infty.$$

Proof. The function

$$F(s) = \alpha D s^{-\alpha-1}(\log(1/s))^{\beta}\to\infty \quad \text{as } s\to 0+,$$

and is a differentiable function of $s>0$ such that

$$\mathrm{d}F(s)/\mathrm{d}s = -\alpha D s^{-\alpha-2}(\log(1/s))^{\beta}[\alpha+1+\beta(\log(1/s))^{-1}] < 0$$

for s sufficiently small. It therefore follows from the inverse function theorem that there exists a differentiable function s_u of u such that (i) is true for $u\geqq u_0$, say. Since

$$\mathrm{d}F(s)/\mathrm{d}s \sim -((\alpha+1)/s)\,F(s) \quad \text{as } s\to 0+,$$

the inverse function theorem then implies (ii) as well as the fact that s_u tends to zero monotonically as $u\to\infty$.

In order to prove (iii), note by (ii) that, for any given $\varepsilon>0$ and sufficiently large u,

$$-\frac{(1+\varepsilon)s_u}{(\alpha+1)u} \leqq \frac{\mathrm{d}s_u}{\mathrm{d}u} \leqq -\frac{(1-\varepsilon)s_u}{(\alpha+1)u}.$$

Therefore, for any fixed $c\geqq 1$ and sufficiently large u,

$$-\frac{1+\varepsilon}{\alpha+1}\int_u^{cu} t^{-1}\,\mathrm{d}t \leqq \int_u^{cu} s_t^{-1}\frac{\mathrm{d}s_t}{\mathrm{d}t}\,\mathrm{d}t \leqq -\frac{1-\varepsilon}{\alpha+1}\int_u^{cu} t^{-1}\,\mathrm{d}t,$$

and so

$$(1-\varepsilon)\log c^{1/(\alpha+1)} \leqq \log(s_u/s_{cu}) \leqq (1+\varepsilon)\log c^{1/(\alpha+1)}.$$

This implies that $g(u)/g(cu) \to 1$ as $u \to \infty$, if $g(u) = u^{1/(\alpha+1)} s_u$, and the same conclusion follows when $c < 1$ by now reversing the limits of integration.

Lastly

$$\int_{u_0}^{u} s_t \, dt = us_u - u_0 s_{u_0} - \int_{t=u_0}^{t=u} t \, ds_t,$$

$$\int_{t=u_0}^{t=u} t \, ds_t = D\int_{t=u_0}^{t=u} s_t^{-\alpha-1}(\log(1/s_t))^\beta \, ds_t$$

$$= -D\int_{t=u_0}^{t=u} v^{\alpha-1}(\log v)^\beta \, dv,$$

where $v = 1/s_t$. Conclusion (iv) then follows with the aid of Lemma 5.2.6. □

Now consider the inequality for $M(u)$ obtained earlier when s is sufficiently small, and for sufficiently large u replace s by s_u so as to obtain

$$M(u) < \exp\{(1+\varepsilon)((\alpha+1)/\alpha)\, us_u\}.$$

Then, by part (i) of Lemma 3.2,

$$us_u = u^{\alpha/(\alpha+1)}\left[\alpha D(\log(1/s_u))^\beta\right]^{1/(\alpha+1)},$$

and

$$\log u \sim (\alpha+1)\log(1/s_u) \quad \text{as } u \to \infty.$$

Hence

$$((\alpha+1)/\alpha)us_u \sim bu^{\alpha/(\alpha+1)}(\log u)^{\beta/(\alpha+1)} \quad \text{as } u \to \infty,$$

where b is the constant defined in the statement of Theorem 3.1. This shows that the required result will follow after it has been proved that

$$M(u) > \exp\{(1-\varepsilon)((\alpha+1)/\alpha)us_u\}$$

for sufficiently large u.

In order to do this, consider a positive $\delta < 1$ and a fixed

$$c \geqq \max\{1, (8\alpha+8)^{(\alpha+1)/\alpha}, (8/\alpha+8)^{\alpha+1}\}.$$

Write $f(s_u)$ as

$$f(s_u) = s_u\left(\int_0^{u/c} + \int_{u/c}^{(1-\delta)u} + \int_{(1-\delta)u}^{(1+\delta)u} + \int_{(1+\delta)u}^{cu} + \int_{cu}^{\infty}\right) M(t)e^{-s_u t}\, dt$$

$$= J_1 + J_2 + J_3 + J_4 + J_5,$$

say.

Now

$$M(u) < \exp\{2((\alpha+1)/\alpha)us_u\},$$

and (by part (ii) of Lemma 3.2) us_u is an increasing function of u, for u sufficiently large. Hence

$$\begin{aligned} J_1 &\leqq s_u \int_0^{u/c} M(u/c)\mathrm{e}^{-s_u t}\,\mathrm{d}t \\ &\leqq s_u \exp\{2((\alpha+1)u/\alpha c)s_{u/c}\}\int_0^\infty \mathrm{e}^{-s_u t}\,\mathrm{d}t \\ &\leqq \exp\{4((\alpha+1)/\alpha)c^{-\alpha/(\alpha+1)}us_u\} \end{aligned}$$

for u sufficiently large, since $s_{u/c} \sim c^{1/(\alpha+1)}s_u$ by part (iii) of Lemma 3.2. Therefore

$$J_1 \leqq \exp\{\tfrac{1}{2}\alpha^{-1}us_u\}$$

for large u, because $c \geqq (8\alpha+8)^{(\alpha+1)/\alpha}$.

Next, s_u is a decreasing function of u for large u, and so for such u we have $s_x \leqq s_{cu}$ whenever $x \geqq cu$. Also, by Lemma 3.2 (iii),

$$s_{cu} \sim c^{-1/(\alpha+1)}s_u,$$

and so

$$s_{cu}/s_u \leqq 2c^{-1/(\alpha+1)}$$

when u is sufficiently large. Since $c \geqq (8/\alpha+8)^{\alpha+1}$, it then follows that

$$s_{cu} \leqq (\alpha/(4\alpha+4))s_u$$

and hence

$$\begin{aligned} M(x) &< \exp\{2((\alpha+1)/\alpha)xs_x\} \\ &\leqq \exp\{2((\alpha+1)/\alpha)xs_{cu}\} \leqq \exp(\tfrac{1}{2}xs_u), \end{aligned}$$

for $x \geqq cu$ and u sufficiently large. Therefore

$$J_5 \leqq s_u \int_{cu}^\infty \mathrm{e}^{ts_u/2}\mathrm{e}^{-s_u t}\,\mathrm{d}t \leqq s_u \int_0^\infty \mathrm{e}^{-s_u t/2}\,\mathrm{d}t = 2.$$

If s_u is defined for $u \geqq u_0$, now define

$$\theta(x) = \int_{u_0}^x s_t\,\mathrm{d}t.$$

Then Lemma 3.2 implies that

$$\theta(x) \sim ((\alpha+1)/\alpha)xs_x, \qquad \theta''(x) = \mathrm{d}s_x/\mathrm{d}x \sim -s_x/((\alpha+1)x) \qquad \text{as } x \to \infty.$$

Given a positive number $\eta < 1$, now consider u so large that, for $x \geqq u/c$,

$$\theta(x) \leqq \frac{(1+2\eta)(\alpha+1)}{(1+\eta)\alpha}xs_x, \qquad M(x) \leqq \exp\{(1+\eta)\theta(x)\}.$$

Then

$$M(x)\mathrm{e}^{-s_u x} \leqq \mathrm{e}^{\varphi(x)},$$

where

$$\varphi(x) = (1+\eta)\theta(x) - s_u x,$$

for $x \geqq u/c$.

In order to obtain an upper bound for the function $\varphi(x)$, now note that $\varphi'(u) = \eta s_u$ and

$$\varphi(u) = (1+\eta)\theta(u) - s_u u \leqq [(1+2\eta)((\alpha+1)/\alpha) - 1] u s_u.$$

Then consider u so large that in addition

$$\theta''(x) \leqq \frac{-s_x}{(1+\eta)(\alpha+1)x} \quad \text{for } x \geqq u/c;$$

then, for $u/c \leqq x \leqq cu$ and u large enough for s_x to be decreasing,

$$\begin{aligned}\varphi''(x) &= (1+\eta)\theta''(x) \\ &\leqq \frac{-s_x}{(\alpha+1)x} \leqq \frac{-s_{cu}}{(\alpha+1)cu}.\end{aligned}$$

Since

$$s_{cu} \sim c^{-1/(\alpha+1)} s_u,$$

one may then take u so large that now $s_{cu} \geqq c^{-1} s_u$. This gives

$$\varphi''(x) \leqq \frac{-s_u}{(\alpha+1)c^2 u};$$

therefore, for $u/c \leqq x \leqq cu$, the Taylor expansion about $x = u$ yields

$$\varphi(x) \leqq [(1+2\eta)((\alpha+1)/\alpha) - 1] u s_u + cu\eta s_u - \frac{s_u(x-u)^2}{2(\alpha+1)c^2 u}.$$

If now $|x-u| \geqq \delta u$ and

$$\eta = \frac{\delta^2}{6c^2(\alpha+1)(2+2/\alpha+c)},$$

the inequality for $\varphi(x)$ reduces to

$$\varphi(x) \leqq \left[\frac{1}{\alpha} - \frac{\delta^2}{3c^2(\alpha+1)}\right] u s_u.$$

Consequently, when u is sufficiently large,

$$\begin{aligned}J_2 + J_4 &\leqq s_u \left(\int_{u/c}^{(1-\delta)u} + \int_{(1+\delta)u}^{cu}\right) e^{\varphi(x)}\, dx \\ &\leqq cu s_u \exp\left\{\left[\frac{1}{\alpha} - \frac{\delta^2}{3c^2(\alpha+1)}\right] u s_u\right\}.\end{aligned}$$

Letting $\Delta = \delta^2/\{8c^2(\alpha+1)\}$, the above estimates show that for u large

$$\begin{aligned} J_1+J_2+J_4+J_5 &\leqq 3\exp\{\tfrac{1}{2}us_u/\alpha\} + cus_u \exp\left\{\left[\frac{1}{\alpha} - \frac{\delta^2}{3c^2(\alpha+1)}\right]us_u\right\} \\ &\leqq (3+c)us_u \exp\left\{\left[\frac{1}{\alpha} - \frac{\delta^2}{3c^2(\alpha+1)}\right]us_u\right\} \\ &< \exp\left\{\left[\frac{1}{\alpha} - \frac{\delta^2}{4c^2(\alpha+1)}\right]us_u\right\} \\ &< \exp\{(1/\alpha - 2\Delta)us_u\}. \end{aligned}$$

In order to deal with J_3, now note that the hypothesis about $\log f(s)$ implies that for u sufficiently large

$$\log f(s_u) \geqq (1-\alpha\Delta)Ds_u^{-\alpha}(\log(1/s_u))^\beta \geqq ((1-\alpha\Delta)/\alpha)us_u.$$

Hence, for large u,

$$\begin{aligned} J_3 &= f(s_u) - (J_1+J_2+J_4+J_5) \\ &\geqq \exp\{(1/\alpha-\Delta)us_u\} - \exp\{(1/\alpha-2\Delta)us_u\} \\ &\geqq \exp\{(1/\alpha-2\Delta)us_u\}. \end{aligned}$$

Also,

$$\begin{aligned} J_3 &= s_u\int_{(1-\delta)u}^{(1+\delta)u} M(t)\mathrm{e}^{-s_ut}\,\mathrm{d}t \\ &\leqq M((1+\delta)u)\,s_u\int_{(1-\delta)u}^{(1+\delta)u} \mathrm{e}^{-s_ut}\,\mathrm{d}t \\ &< M((1+\delta)u)\exp\{-(1-\delta)us_u\}. \end{aligned}$$

It follows that, for large u,

$$\begin{aligned} M((1+\delta)u) &> J_3\exp\{(1-\delta)us_u\} \geqq \exp\{(1/\alpha-2\Delta+1-\delta)us_u\} \\ &\geqq \exp\{(1-2\Delta-\delta)((\alpha+1)/\alpha)us_u\} \geqq \exp((1-2\delta)((\alpha+1)/\alpha)us_u\}. \end{aligned}$$

Since $s_u \geqq s_{(1+\delta)u}$, this implies that

$$M((1+\delta)u) \geqq \exp\{(1-3\delta)(1+\delta)((\alpha+1)/\alpha)us_u\}$$

for large u. Writing $y=(1+\delta)u$ and $\delta=\frac{1}{3}\varepsilon$, it follows finally that for sufficiently large y

$$M(y) > \exp\{(1-\varepsilon)((\alpha+1)/\alpha)ys_y\}.$$

This proves Theorem 3.1. □

Now consider a weighted partition function $p_w(u)$ such that

$$\sum_u p_w(u)\mathrm{e}^{-us} = \prod_{r=0}^{\infty}(1-\mathrm{e}^{-u_rs})^{-w(u_r)},$$

where $0<u_0<u_1<u_2<\ldots$ and each $w(u_n)$ is a non-negative integer. In certain cases, such as when $p_w(u)$ arises from the study of a given additive arithmetical semigroup G satisfying Axiom C, Theorem 3.1 provides some information about the asymptotic behaviour of $M_w(x)=\sum_{u\leqq x} p_w(u)$ as $x\to\infty$. However, it would also be interesting to determine how $p_w(u)$ behaves as $u\to\infty$. One situation in which this is possible within the context of Axiom C or Theorem 3.1 is *when the* u_n *are integers and* $p_w(u)$ *is a non-decreasing function of* u *as* u *increases* (for all u past a certain point). For, in that case the above formula for $p_w(u)$ may be written in the form

$$\sum_{n=0}^{\infty} p_w(n)e^{-ns} = \prod_{r=1}^{\infty} (1-e^{-rs})^{-w(r)},$$

where $w(r)=0$ if r was not included in the previous sequence (u_n). If $p_w(n)$ is non-decreasing from some $n=n_0$ onwards, one can now deduce that

$$\begin{aligned} M_w(n) &= O(1)+\sum_{n_0\leqq r\leqq n} p_w(r) \\ &\leqq O(1)+np_w(n) \leqq O(1)+nM_w(n). \end{aligned}$$

Therefore an asymptotic formula for $\log M_w(x)$ of the type occurring in Theorem 3.1 here allows one to deduce that

$$\log p_w(n) \sim \log M_w(n) \quad \text{as } n\to\infty.$$

In particular, $p_w(n)$ *will be non-decreasing for* $n\geqq 1$ *and the above conclusion will be applicable, whenever* $w(1)\geqq 1$.

Another situation in which one may deduce that $\log p_w(n)\sim\log M_w(n)$ subject to a formula for $\log M_w(x)$ of the type occurring in Theorem 3.1 is *when the* u_n *are all integers again and there exist positive integers* $b_1, \ldots, b_k$ *such that* $w(b_i)\geqq 1$ *and* $b_1, \ldots, b_k$ *have g.c.d.* 1. For, in that case, a well-known algebraic result implies that there exist rational integers $z_1, \ldots, z_k$ such that

$$b_1z_1+b_kz_k = 1.$$

Let $y_i=\max\{0, z_i\}$ and $x_i=y_i-z_i$ for $i=1, \ldots, k$. Then, if

$$x_0 = b_1x_1+\ldots+b_kx_k,$$

we have

$$x_0+1 = b_1y_1+\ldots+b_ky_k.$$

Now consider any integer $m\geqq x_0^2$. If $x_0\neq 0$, then $m=qx_0+r$, where $0\leqq r<x_0$ and $q\geqq x_0$; hence

$$m = (t+r)x_0+r = tx_0+r(x_0+1),$$

where $t=q-r>0$. Therefore, if $x_0 \neq 0$, then every integer $m \geqq x_0^2$ can be partitioned in the form $m=s_1 b_1+\ldots+s_k b_k$, where the s_i are non-negative integers. On the other hand, if $x_0=0$, then 1 can be partitioned in this way and so every positive integer can be partitioned in such a way; in fact, $b_i=1$ for some i in that case.

Now let $n_0=\max\{x_0^2, 1\}$. Then, if $n \geqq m+n_0 \geqq n_0$, it follows that $p_{\mathrm{w}}(n) \geqq$ $\geqq p_{\mathrm{w}}(m)$, and so

$$M_{\mathrm{w}}(n-n_0) = \sum_{0 \leqq m \leqq n-n_0} p_{\mathrm{w}}(m) \leqq (n-n_0+1) p_{\mathrm{w}}(n) \leqq (n-n_0+1) M_{\mathrm{w}}(n).$$

Therefore the asymptotic conclusion about $\log p_{\mathrm{w}}(n)$ follows in this case also.

For convenience, we now formulate both the theorem stated earlier and the present refinements (when applied to an additive arithmetical semigroup) under the general heading of:

3.3. Additive Abstract Prime Number Theorem. *Let G denote an additive arithmetical semigroup satisfying Axiom* C *with*

$$\pi_G{}^{\#}(x) \sim C x^{\kappa} (\log x)^{\nu} \quad \text{as } x \to \infty.$$

Then, as $x \to \infty$,

$$N_G{}^{\#}(x) = \exp\{[c_G+\mathrm{o}(1)] x^{\kappa/(\kappa+1)} (\log x)^{\nu/(\kappa+1)}\},$$

where

$$c_G = \kappa^{-1} (\kappa+1)^{(\kappa-\nu+1)/(\kappa+1)} [\kappa C \Gamma(\kappa+1) \zeta(\kappa+1)]^{1/(\kappa+1)}.$$

If in addition the degree mapping ∂ on G is integer-valued, and

(i) *$G^{\#}(n)$ is a non-decreasing function of n for all $n \geqq$ some n_0, or*

(ii) *there exist positive integers $b_1, \ldots, b_k$ such that $P^{\#}(b_i) \geqq 1$ and $b_1, \ldots, b_k$ have g.c.d.* 1, *then, as $n \to \infty$,*

$$G^{\#}(n) = \exp\{[c_G+\mathrm{o}(1)] n^{\kappa/(\kappa+1)} (\log n)^{\nu/(\kappa+1)}\}. \qquad \square$$

By substituting the particular constants that occur in Axiom C for the special additive arithmetical semigroups discussed in § 1, one may use this theorem and Corollary 2.4 to verify the details of Table 3.1; in this table, G denotes the semigroup associated with the particular category indicated in each case, or else G_{Φ}, in the final case. It may be worth repeating that the results about $G^{\#}(n)$ for the various categories in question can be viewed as asymptotic enumeration theorems for the isomorphism classes of objects in those categories, and as such they also have an interest independent of the present number-theoretical background.

TABLE 3.1

G	$\pi_G{}^{\#}(x)$ as $x \to \infty$	$\log G^{\#}(n)$ as $n \to \infty$	$\log \zeta_G(s)$ as $s \to 0+$
$\mathscr{A}(p)$, $\mathfrak{S}_c(p)$, $\mathfrak{S}_c(F_q)$	x	$\pi \sqrt{\frac{2}{3}n}$	$\frac{1}{6}\pi^2/s$
$\mathfrak{S}(p)$, $\mathfrak{S}(F_q)$	$\frac{1}{6}\pi^2 x$	$\frac{1}{3}\pi^2 \sqrt{n}$	$\frac{1}{36}\pi^4/s$
$\mathfrak{S}(F)$ with F algebraically closed	$\sqrt{x}$	$3\left[\frac{1}{16}\pi\zeta\left(\frac{3}{2}\right)^2 n\right]^{1/3}$	$\frac{1}{2}\zeta\left(\frac{3}{2}\right)\sqrt{(\pi/s)}$
$\mathfrak{S}(R)$	$\frac{1}{2}(3+\sqrt{2})\sqrt{x}$	$\frac{3}{4}\left[(3+\sqrt{2})^2\pi\zeta\left(\frac{3}{2}\right)^2 n\right]^{1/3}$	$\frac{1}{4}(3+\sqrt{2})\zeta\left(\frac{3}{2}\right)\sqrt{(\pi/s)}$
$\mathfrak{G}_c$, $\mathfrak{L}_s(F_0)$	$(1+\frac{3}{2}\sqrt{2})\sqrt{x}$	$\frac{3}{2}\left[\frac{1}{2}(1+\frac{3}{2}\sqrt{2})^2\pi\zeta\left(\frac{3}{2}\right)^2 n\right]^{1/3}$	$\frac{1}{2}(1+\frac{3}{2}\sqrt{2})\zeta\left(\frac{3}{2}\right)\sqrt{(\pi/s)}$
$\mathfrak{R}_{sc}$	$\frac{7}{8}x\log x$	$\frac{1}{2}\pi\sqrt{(\frac{7}{6}n\log n)}$	$\frac{7}{48}(\pi^2/s)\log(1/s)$
$\mathfrak{F}(p)$, $(\alpha_1, \ldots, \alpha_m)=1$	$C_{D,p}x$	$\pi\sqrt{(\frac{2}{3}nC_{D,p})}$	$\frac{1}{6}(\pi^2/s)C_{D,p}$
G_Φ	$\{\zeta(2)\zeta(3)/\zeta(6)\}x$	$\frac{1}{3}\pi^2\sqrt{(\{\zeta(3)/\zeta(6)\}n)}$	$\frac{1}{36}(\pi^4/s)\zeta(3)/\zeta(6)$

§ 4. Further additive prime number theorems

Although Theorems 3.1 and 3.3 are quite widely applicable and provide fairly precise information, there remain questions that may be asked in connection with them. In the first place, these theorems only provide asymptotic estimates of a logarithmic type, and so one may ask to what extent it may be possible to obtain more precise asymptotic results. Questions of this kind regarding particular partition functions or types of partition functions have received considerable attention, and a variety of theorems have been proved which could easily be re-formulated (at least partly) as sharper types of additive abstract prime number theorems. In some special cases, the sharper results are extremely detailed and precise. For example, Hardy and Ramanujan [2] and Rademacher [1] established the existence of an explicit infinite series which converges to the classical partition function $\mathbf{p}(n)$. (It is an interesting fact that this theoretical work is also of definite practical value for the purpose of computing $\mathbf{p}(n)$ for explicit values of n.)

There are other analytic partition theorems which are not as precise as the last-mentioned one, but have the advantage of being more widely applicable. As may be expected, results which sharpen the conclusions of Theorems 3.1 and 3.3, even under stronger hypotheses, tend to have more delicate proofs. Over here, we shall content ourselves with quoting without proof a reformulation for additive arithmetical semigroups of one interesting case of a partition theorem of Ingham [2], as extended by Auluck and Haselgrove [1].

4.1. Theorem. *Let G denote an additive arithmetical semigroup which satisfies Axiom* C *in the special form*

$$\pi_G{}^\#(x) \sim Cx^\kappa \quad \text{as } x \to \infty,$$

where, for some positive integer q and constants A_i,

$$\int_0^x (1-t/x)^{q-1} t^{-1} R(t)\, dt = A_1 \log x + A_2 + o(1) \quad \text{as } x \to \infty$$

if $R(x) = \pi_G{}^\#(x) - Cx^\kappa$. *Suppose that the degree map* ∂ *on* G *is integer-valued and that*

(i) $G^\#(n)$ *is an increasing function of* n, *or*

(ii) *there exist positive integers* $b_1, \ldots, b_k$ *such that* $P^\#(b_i) \geqq 1$ *and* $b_1, \ldots, b_k$ *have g.c.d.* 1.

Then, as $n \to \infty$,

$$G^\#(n) \sim ((1-\alpha)/2\pi)^{1/2} e^b M^{-(a-1/2)\alpha} n^{(a-1/2)(1-\alpha)-1/2} \exp[M^\alpha n^\alpha/\alpha],$$

where

$$\alpha = \kappa/(\kappa+1), \qquad M = \{\kappa C\Gamma(\kappa+1)\zeta(\kappa+1)\}^{1/\kappa},$$

$$\frac{1}{(q-1)!}\left(\frac{d}{dx}\right)^{q-1}(A_1 x^{q-1}\log x + A_2 x^{q-1}) = a\log x + b. \qquad \square$$

Although the proof of this theorem will be omitted, it may be interesting to note some of its applications to questions of asymptotic enumeration:

4.2. Corollary. *Consider the category* $\mathfrak{F}(p)$ *over a given number domain* D, *where* p *is a fixed rational prime. Let* $a_D(p^n)$ *denote the total number of isomorphism classes of modules of cardinal* p^n *in* $\mathfrak{F}(p)$, *and in the notation of* § 1 *let*

$$C_{D,p} = \alpha_1^{-1} + \ldots + \alpha_m^{-1}.$$

Then, if $\alpha_1, \ldots, \alpha_m$ *have g.c.d.* 1,

$$a_D(p^n) \sim A n^{-(m+3)/4} \exp\left[\pi \sqrt{(\tfrac{2}{3} n C_{D,p})}\right] \quad \textit{as } n \to \infty,$$

where

$$A = (\alpha_1 \alpha_2 \ldots \alpha_m)^{1/2} 2^{-1-m/2} (\tfrac{1}{6} C_{D,p})^{(m+1)/4}.$$

In particular, when $D = \mathbf{Z}$,

$$a(p^n) = p(n) \sim (\tfrac{1}{12}\sqrt{3})\, n^{-1} \exp\left[\pi \sqrt{\tfrac{2}{3} n}\right] \quad \textit{as } n \to \infty.$$

Proof. Let

$$R(x) = \pi_{\mathfrak{F}(p)}{}^{\#}(x) - C_{D,p}x.$$

Then

$$\int_0^x t^{-1} R(t)\, dt = \sum_{i=1}^{m} \int_0^x t^{-1}\{[t/\alpha_i] - t/\alpha_i\}\, dt$$

$$= \sum_{i=1}^{m} \int_0^{x/\alpha_i} u^{-1}([u] - u)\, du.$$

Now, for $y > 0$,

$$\int_0^y u^{-1}([u]-u)\, du = \sum_{r=1}^{[y]} \int_{r-1}^{r} u^{-1}[u]\, du + \int_{[y]}^{y} u^{-1}[u]\, du - y$$

$$= [y]\log y - \sum_{r=1}^{[y]} \log r - y$$

$$= -\tfrac{1}{2}\log y - \tfrac{1}{2}\log 2\pi + o(1) \quad \text{as } y \to \infty,$$

by Stirling's formula for $\log n!$, $n = [y]$. It therefore follows that

$$\int_0^x t^{-1} R(t)\, dt = -\sum_{i=1}^{m} \{\tfrac{1}{2}\log(x/\alpha_i) + \tfrac{1}{2}\log 2\pi\} + o(1)$$

$$= a\log x + b + o(1), \quad \text{as } x \to \infty,$$

where

$$a = -\tfrac{1}{2}m, \qquad b = \log\{(\alpha_1\alpha_2 \ldots \alpha_m)^{1/2}(2\pi)^{-m/2}\}.$$

The corollary then follows by substituting values in the formulae of the theorem. □

4.3. Corollary. *Let $s_A(n)$ denote the total number of isomorphism classes of n-dimensional semisimple algebras over an algebraically closed field A, and let $s_R(n)$ denote the corresponding number for semisimple algebras over a real closed field R. Then, as $n \to \infty$,*

$$s_A(n) \sim An^{-7/6}\exp[Bn^{1/3}], \qquad s_R(n) \sim Cn^{-11/6}\exp[Dn^{1/3}],$$

where

$$A = (\tfrac{1}{3}\sqrt{3})(4\pi)^{-7/6}[\zeta(\tfrac{3}{2})]^{2/3}, \qquad B = 3\{\tfrac{1}{16}\pi\zeta(\tfrac{3}{2})^2\}^{1/3},$$

$$C = (\tfrac{1}{3}\sqrt{3})\{\tfrac{1}{8}(3+\sqrt{2})\pi^{-1}\zeta(\tfrac{3}{2})\}^{3/4}, \qquad D = \tfrac{3}{4}\{(3+\sqrt{2})^2\pi\zeta(\tfrac{3}{2})^2\}^{1/3}.$$

Proof. Let

$$r(x) = \pi_{\mathfrak{S}(A)}{}^{\#}(x) - \sqrt{x}, \qquad r^*(x) = \pi_{\mathfrak{S}(R)}{}^{\#}(x) - \tfrac{1}{2}(3+\sqrt{2})\sqrt{x}.$$

Then

$$\int_0^x t^{-1}r(t)\,dt = \int_0^x t^{-1}([\sqrt{t}] - \sqrt{t})\,dt = 2\int_0^{\sqrt{x}} u^{-1}([u]-u)\,du$$

$$= -\tfrac{1}{2}\log x - \log 2\pi + o(1) \quad \text{as } x \to \infty,$$

by a formula established above. Also

$$\int_0^x t^{-1}r^*(t)\,dt = \int_1 + \int_2 + \int_4,$$

where

$$\int_n = \int_0^x \{[\sqrt{(t/n)}] - \sqrt{(t/n)}\}t^{-1}\,dt.$$

Then

$$\int_n = 2\int_0^{\sqrt{(x/n)}} u^{-1}([u]-u)\,du$$

$$= -\tfrac{1}{2}\log x + \tfrac{1}{2}\log(\tfrac{1}{4}\pi^{-1}n) + o(1).$$

Hence

$$\int_0^x t^{-1}r^*(t)\,dt = -\tfrac{3}{2}\log x - \tfrac{1}{2}\log\pi^3 + o(1) \quad \text{as } x \to \infty.$$

The rest of the proof is now straightforward. □

Another question one may ask about Theorems 3.1 and 3.3 lies in the opposite direction and concerns the extent to which the stated hypotheses may be weakened. Firstly, the restriction to step functions in Proposition 2.3 and Theorem 3.1 may be removed if certain integrability assumptions are introduced instead; see Kohlbecker [1]. However, a possibly more interesting question concerns how far one may weaken the assumptions corresponding to Axiom C and yet still be able to obtain asymptotic information of a reasonably precise kind. The above paper of Kohlbecker and subsequent papers of Parameswaran [1] and Schwarz [2] contain a number of general partition theorems which could be re-interpreted partly as the abstract additive prime number theorems subject to less restricted axioms than Axiom C. For example, some of the results of Kohlbecker and Parameswaran would allow one to replace Axiom C by the weaker assumption

$$\pi_G{}^{\#}(x) \sim x^{\kappa} L(x) \quad \text{as } x \to \infty,$$

where $\kappa \geqq 0$ and where $L(x)$ is any slowly oscillating function of x, and yet still obtain a precise asymptotic conclusion about

$$\log N_G{}^{\#}(x) \quad \text{as } x \to \infty.$$

Although it is interesting to keep the above facts in mind, as providing first steps towards more general theories of abstract additive arithmetical semigroups, so far the most interesting natural semigroups that have fallen within this framework seem to be those that are already covered by Axiom C. Within the area covered by Axiom C, the most urgent questions are perhaps those concerned with obtaining at least partly sharper conclusions than those available for the general case at the time of writing of this book. For example, the next section contains some results about asymptotic average values and densities subject to Axiom C. It will be seen that the order of precision of the results obtained (as well as the possibility of obtaining corresponding theorems in some cases not treated below), appears to be limited by the types of analytical tools that are available at present. (This comment is in no way meant to disparage the striking analytical results that have been obtained; its purpose is merely to indicate the theoretical possibility of further developments.)

One further question worth commenting on concerns the establishment of converses to Theorems 3.1 and 3.3 and also Proposition 2.3, such as deductions about $\pi_G{}^{\#}(x)$ from given information about $N_G{}^{\#}(x)$, say. Theorems of this kind, which may be applied to $\pi_G{}^{\#}(x)$ and $N_G{}^{\#}(x)$ in particular, are contained in the above-mentioned paper of Kohlbecker (in a more general setting again), and similar results are obtained in Parameswaran's paper, for example. For purposes of the next section, we shall now state an 'Abelian' theorem corresponding to Theorem 3.1; the proof of this theorem will be omitted since it is basically of the same general type as, but slightly shorter than, the proof of Theorem 3.1 (compare, in particular, Theorem 4 of Kohlbecker [1]).

4.4. Theorem. *Let $M(x)$ be a step function of $x \geqq 0$ such that*

$$f(s) = s\int_0^\infty M(t)\mathrm{e}^{-st}\,\mathrm{d}t$$

exists for $s > 0$. Suppose that

$$\log M(x) \sim bx^{\alpha/(\alpha+1)}(\log x)^{\beta/(\alpha+1)} \quad as\ x \to \infty,$$

where $b > 0$, $\alpha > 0$ and β are constants. Then, as $s \to 0+$,

$$\log f(s) \sim B\Gamma(\alpha+1)\zeta(\alpha+1)s^{-\alpha}(\log(1/s))^{\beta},$$

where $B > 0$ is the constant such that

$$b = \alpha^{-1}(\alpha+1)^{(\alpha-\beta+1)/(\alpha+1)}[\alpha B\Gamma(\alpha+1)\zeta(\alpha+1)]^{1/(\alpha+1)}. \qquad \square$$

§ 5. Asymptotic average values and densities

Throughout this section, *G denotes an additive arithmetical semigroup satisfying Axiom* C.

The purpose of the following discussion is to obtain some results about asymptotic average values and densities of certain specific arithmetical functions and sets. In doing this, we shall orient the discussion in terms of the given degree mapping ∂ on G, and hence examine slightly different types of functions from those defined at the beginning of Chapter 4, § 3. Thus, if f denotes a given arithmetical function on G, we shall investigate $N^{\#}(f, x) = \sum_{\partial(a) \leqq x} f(a)$ or $N^{\#}(f, x)/N_G^{\#}(x)$ instead of $N(f, x)$ or $N(f, x)/N_G(x)$. In particular, when f is the characteristic function of some subset E of G, we shall now consider the "density" function

$$d^{\#}(E, x) = N^{\#}(f, x)/N_G^{\#}(x)$$

instead of the function $d(E, x)$ defined in Chapter 4. (The present discussion will be too brief to justify the formal introduction of a new terminology relating to these and similar concepts.)

5.1. Proposition. *As $x \to \infty$,*

$$N^{\#}(d_k, x) = \exp\{[k^{1/(\kappa+1)} + \mathrm{o}(1)]c_G x^{\kappa/(\kappa+1)}(\log x)^{\nu/(\kappa+1)}\}.$$

In particular,

$$N^{\#}(d, x) = \exp\{[2^{1/(\kappa+1)} + \mathrm{o}(1)]c_G x^{\kappa/(\kappa+1)}(\log x)^{\nu/(\kappa+1)}\}.$$

By contrast,

$$N^{\#}(d^2, x) = \exp\{[(4-2^{-\kappa})^{1/(\kappa+1)} + \mathrm{o}(1)]c_G x^{\kappa/(\kappa+1)}(\log x)^{\nu/(\kappa+1)}\},$$

where d^2 denotes the point-wise square of the divisor function d. Thus the approximate average value of d^2 is of larger order than the square of the approximate average value of d.

Proof. Since $d_k(z)=[\zeta_G(z)]^k$, Corollary 2.4 implies that, as $s\to 0+$,

$$\log \tilde{d}_k(s) = k \log \zeta_G(s) \sim kC\Gamma(\kappa+1)\zeta(\kappa+1)s^{-\kappa}(\log(1/s))^\nu.$$

Now a simple variation of the first part of the proof of Proposition 2.2 leads to:

5.2. Lemma. *Let $f\in \mathrm{Dir}\,(G)$ be a function such that $\tilde{f}(z)$ converges for* $\mathrm{Re}\, z>0$, *and, for any $\varepsilon>0$,*

$$N^\#(f,x) = \mathrm{O}(\mathrm{e}^{\varepsilon x}) \quad \text{as } x\to\infty.$$

Then

$$\tilde{f}(z) = z\int_0^\infty N^\#(f,t)\mathrm{e}^{-zt}\,\mathrm{d}t \quad \text{for } \mathrm{Re}\, z>0. \qquad \square$$

By Proposition 2.2, this lemma applies to d_k since $N^\#(d_k,x)\leqq[N_G{}^\#(x)]^k$. Hence Theorem 3.1 implies that

$$\log N^\#(d_k,x) \sim bx^{\kappa/(\kappa+1)}(\log x)^{\nu/(\kappa+1)} \quad \text{as } x\to\infty,$$

where $b=k^{1/(\kappa+1)}c_G$.

Next,

$$d^2(z) = [\zeta_G(z)]^4/\zeta_G(2z),$$

and so, as $s\to 0+$,

$$\log \tilde{d}^2(s) = 4\log\zeta_G(s)-\log\zeta_G(2s)$$
$$\sim (4-2^{-\kappa})C\Gamma(\kappa+1)\zeta(\kappa+1)s^{-\kappa}(\log(1/s))^\nu.$$

Therefore Proposition 5.2.3 and Proposition 2.2 imply that, for any $\varepsilon>0$,

$$N^\#(d^2,x) = \sum_{\partial(a)\leqq x} d(a)\mathrm{O}(\mathrm{e}^{\varepsilon\partial(a)}) = \mathrm{O}(\mathrm{e}^{\varepsilon x}N^\#(d,x)) = \mathrm{O}(\mathrm{e}^{2\varepsilon x}).$$

Hence Lemma 5.2 and Theorem 3.1 yield the stated conclusion about d^2. $\square$

By closely similar methods, one may deduce:

5.3. Proposition. *As $x\to\infty$,*

$$N^\#(d_*,x) = \exp\{[(2-2^{-\kappa})^{1/(\kappa+1)}+\mathrm{o}(1)]c_G x^{\kappa/(\kappa+1)}(\log x)^{\nu/(\kappa+1)}\},$$
$$N^\#(\beta,x) = \exp\{[(1+2^{-\kappa}+3^{-\kappa}-6^{-\kappa})^{1/(\kappa+1)}+\mathrm{o}(1)]c_G x^{\kappa/(\kappa+1)}(\log x)^{\nu/(\kappa+1)}\}.$$

$\square$

Lastly, we consider two propositions whose conclusions are strikingly different from the corresponding ones subject to Axiom A.

5.4. Proposition. *Let k be any fixed positive integer. Then 'almost every' element of G is divisible by a non-trivial k^{th} power. Equivalently, the subset G_k of all k-free elements of G has asymptotic density zero.*

Proof. For the characteristic function q_k of G_k, we have

$$q_k(z) = \zeta_G(z)/\zeta_G(kz).$$

Therefore, as $s \to 0+$,

$$\log \tilde{q}_k(s) = \log \zeta_G(s) - \log \zeta_G(ks) \sim (1-k^{-\kappa})C's^{-\kappa}(\log(1/s))^{\nu},$$

where $C' = C\Gamma(\kappa+1)\zeta(\kappa+1)$. As before, this leads to

$$\log N^{\#}(q_k, x) \sim (1-k^{-\kappa})c_G x^{\kappa/(\kappa+1)}(\log x)^{\nu/(\kappa+1)} \quad \text{as } x \to \infty.$$

It follows that

$$\log\{N^{\#}(q_k, x)/N_G{}^{\#}(x)\} \to -\infty \quad \text{as } x \to \infty,$$

and hence G_k has asymptotic density zero in G. □

5.5. Proposition. *Suppose that, in addition to G satisfying Axiom* C, *∂ is integer-valued and*

$$\log G^{\#}(n) \sim \log N_G{}^{\#}(n) \quad \textit{as } n \to \infty.$$

Then 'almost every' two elements of G have a non-trivial common divisor. Equivalently, the set of all ordered pairs of coprime elements of G has asymptotic density zero amongst all pairs of elements of G.

Proof. The Euler function φ defined in Chapter 2, § 6 can be expressed in the form

$$\varphi(a) = \sum \{1 : \partial(b) \leq \partial(a), \ (b, a) = 1\}.$$

Also, Theorem 2.6.1 implies that

$$\varphi(z) = \mu(z)\Big(\sum_{a \in G} N_G(|a|)a^{-z}\Big) = \mu(z)\Big(\sum_{a \in G} N_G{}^{\#}(\partial(a))a^{-z}\Big)$$
$$= \mu(z)h(z),$$

say. Therefore

$$\tilde{\varphi}(z) = \tilde{\mu}(z)\tilde{h}(z),$$

where

$$\tilde{h}(z) = \sum_{a \in G} N_G{}^{\#}(\partial(a))e^{-\partial(a)z} = \sum_{n=0}^{\infty} G^{\#}(n)N_G{}^{\#}(n)e^{-nz}.$$

Then

$$G^{\#}([x])N_G{}^{\#}([x]) \leqq \sum_{n \leqq x} G^{\#}(n)N_G{}^{\#}(n) \leqq [N_G{}^{\#}(x)]^2,$$

and hence Theorem 3.3 and the assumption about $\log G^{\#}(n)$ imply that

$$\log\Big(\sum_{n \leqq x} G^{\#}(n)N_G{}^{\#}(n)\Big) \sim 2\log N_G{}^{\#}(x) \sim 2c_G x^{\kappa/(\kappa+1)}(\log x)^{\nu/(\kappa+1)}$$

as $x \to \infty$. Therefore Theorem 4.4 implies that, as $s \to 0+$,

$$\log \tilde{h}(s) \sim 2^{\kappa+1} C\Gamma(\kappa+1)\zeta(\kappa+1)s^{-\kappa}(\log(1/s))^{\nu}.$$

This implies that, as $s \to 0+$,

$$\log \tilde{\varphi}(s) \sim (2^{\kappa+1}-1)C\Gamma(\kappa+1)\zeta(\kappa+1)s^{-\kappa}(\log(1/s))^{\nu}.$$

Now

$$N^{\#}(\varphi, x) \leqq \sum_{n \leqq x} G^{\#}(n)^2 \leqq [N_G{}^{\#}(x)]^2,$$

and so Theorem 3.1 and Lemma 5.2 imply that, as $x \to \infty$,

$$\log N^{\#}(\varphi, x) \sim (2^{\kappa+1}-1)^{1/(\kappa+1)} c_G x^{\kappa/(\kappa+1)}(\log x)^{\nu/(\kappa+1)}.$$

Next, let $\Delta(x)$ denote the total number of all ordered pairs (a, b) of coprime elements $a,b \in G$ with $\partial(a) \leqq x$ and $\partial(b) \leqq x$. Then

$$\Delta(x) = 2N^{\#}(\varphi, x) - V(x),$$

where $V(x)$ is the number of pairs of coprime elements (a, b) with $\partial(a) = \partial(b) \leqq x$. It follows that

$$\log\big(\Delta(x)/N_G{}^{\#}(x)^2\big) \leqq \log 2N^{\#}(\varphi, x) - 2\log N_G{}^{\#}(x)$$

$$\to -\infty \quad \text{as } x \to \infty.$$

Thus the set of all ordered pairs of coprime elements of G has asymptotic density zero amongst all ordered pairs of elements of G. □

5.6. Corollary. *If ∂ is integer-valued and*

$$\log G^{\#}(n) \sim \log N_G{}^{\#}(n) \quad \textit{as } n \to \infty,$$

then, as $x \to \infty$.

$$N^{\#}(\varphi, x) = \exp\{[(2^{\kappa+1}-1)^{1/(\kappa+1)} + o(1)]c_G x^{\kappa/(\kappa+1)}(\log x)^{\nu/(\kappa+1)}\}. \quad □$$

Selected bibliography for Chapter 8

Section 1: Knopfmacher [3, 6, 7].

Sections 2 *and* 3: Brigham [1], Hardy and Ramanujan [1], Knopfmacher [3, 5—7, 12], Kohlbecker [1], Parameswaran [1], Schwarz [2, 3].

Section 4: Auluck and Haselgrove [1], Hardy and Ramanujan [2], Ingham [2], Knopfmacher [5—7, 12], Meinardus [1], Kohlbecker [1], Parameswaran [1], Rademacher [1, 2, 4], Schwarz [2, 3].

Section 5: Knopfmacher [6].

CHAPTER 9

ARITHMETICAL FORMATIONS

This chapter contains a discussion of some of the basic features of certain abstract systems, here called *(arithmetical) formations*, which include the previous arithmetical semigroups as special cases. Roughly put, a formation consists of an arithmetical semigroup G together with an equivalence relation on G which partitions it into classes possessing certain specific algebraic properties. Some initial examples of such formations arise by considering arithmetical progressions of positive integers and ideal classes in algebraic number fields. After a discussion of these and various other natural examples, attention is restricted to a widely applicable asymptotic 'equi-distribution' axiom (*Axiom* A*) concerning the classes of a given formation. In particular, this axiom implies Axiom A for the semigroup G containing all the classes, and it reduces to Axiom A in the special case in which there is only one class.

It follows that the theory of formations satisfying Axiom A* generalizes and refines that of arithmetical semigroups satisfying Axiom A. Therefore, after considering the validity of Axiom A* in various special cases, the remainder of the chapter is devoted to extending a selection of basic asymptotic theorems of Part II to the present context. In particular, an *Abstract Prime Number Theorem* is proved for formations subject to Axiom A* and one further axiom. Put roughly, this theorem states that the primes of G are asymptotically equi-distributed amongst the various classes of the formation structure on G, in addition to satisfying the asymptotic law implied by the validity of Axiom A for G. This and other conclusions provide interesting corollaries for various specific formations, subject to the (usually non-trivial) theorems which establish the validity of the axioms in particular situations. Applications of this kind are usually easy to read off, given the detailed information about a particular system, and so, despite their special interest in various specific cases, will normally not be stated explicitly.

It is, however, worth mentioning that one of the corollaries of the abstract prime number theorem for formations is the ordinary *Prime Number*

Theorem for Arithmetical Progressions. This theorem states that, for any pair of coprime positive integers m and r,

$$\pi_{m,r}(x) =_{def.} \sum \{1: \text{ rational primes } p,\ p \leqq x,\ p \equiv r (\operatorname{mod} m)\}$$

$$\sim \frac{x}{\varphi(m) \log x} \quad \text{as } x \to \infty.$$

§ 1. Natural examples

As usual, it is convenient to start with a definition of the basic concept to be considered.

Let G denote an arbitrary arithmetical semigroup on which there is given an equivalence relation $\sim$ with the property that

$$a \sim a' \quad \text{and} \quad b \sim b' \quad \text{implies that} \quad ab \sim a'b',$$

and such that the corresponding set $\Gamma = G/\sim$ of equivalence classes $[a]$ $(a \in G)$ forms a *finite abelian group* under the operation defined by

$$[a][b] = [ab].$$

In such a case, $(G, \sim)$ or (G, Γ) will be called an **(arithmetical) formation**, Γ will be called the **class group** of the formation, and, for reasons appearing below, the classes $[a] \in \Gamma$ may sometimes be referred to as *generalized arithmetical progressions* or *generalized ideal classes*; the order $h = \operatorname{card} \Gamma$ of Γ will be called the **class number** of (G, Γ).

If (G, Γ) is a given formation, it is clear that the natural mapping rule $a \to [a]$ defines an identity-preserving homomorphism β of G onto Γ. Conversely, given any identity-preserving homomorphism $\beta': G \to \Gamma'$ of an arithmetical semigroup G onto a finite abelian group Γ', one can define a formation structure on G by letting

$$a \sim b \quad \text{if and only if} \quad \beta'(a) = \beta'(b);$$

the resulting set Γ of equivalence classes will then form an abelian group that is naturally isomorphic to Γ'. Thus it would be equivalent, and for some purposes neater, to define a formation to be a pair (G, β') consisting of an arithmetical semigroup G and a homomorphism β' as above. However, since special interest attaches to the classes $[a]$ $(a \in G)$ and the actual quotient set $\Gamma = G/\sim$, the first definition will be preferred.

The simplest example of a formation, which may be called the **trivial** formation structure, is obtained by taking any arithmetical semigroup G and considering the trivial equivalence relation $\sim_0$ such that $a \sim_0 b$ for all $a,b \in G$. More interesting examples arise as follows by considering various types of integral domains:

1.1. Example: Arithmetical progressions. Let $G_{\mathbf{Z}}\langle m\rangle$ denote the arithmetical semigroup of all positive integers that are coprime to a given positive integer m, and consider the relation $\equiv \pmod{m}$ of congruence modulo m on $G_{\mathbf{Z}}\langle m\rangle$. Then the resulting set $\Gamma_{\mathbf{Z}}\langle m\rangle$ of congruence classes (arithmetical progressions mod m) forms a finite abelian group of order $\varphi(m)$ under the induced multiplication operation, where φ denotes the Euler function on $G_{\mathbf{Z}}$.

1.2. Example: Generalized progressions in Euclidean domains. Let D denote an arbitrary integral domain, and let G_D denote the corresponding semigroup of all associate classes $\bar{a}$ of non-zero elements $a \in D$. If I is any ideal in D, we define the relation of *congruence modulo I* on D by letting $a \equiv b \pmod{I}$ if and only if $a-b \in I$ (i.e., a and b have the same coset in D/I). This relation may be transferred to G_D by defining

$$\alpha \equiv \beta \pmod{I} \quad \text{if and only if} \quad a \equiv b \pmod{I} \quad \text{for some} \quad a \in \alpha$$
$$\text{and } b \in \beta \qquad (\alpha,\beta \in G_D).$$

It is not difficult to verify that this congruence relation on G_D is an equivalence relation such that

$$\alpha \equiv \alpha' \pmod{I} \text{ and } \beta \equiv \beta' \pmod{I} \text{ implies } \alpha\beta \equiv \alpha'\beta' \pmod{I}.$$

Thus the set $\Gamma^*\langle I\rangle$ of all congruence classes in G_D forms a commutative semigroup under the induced multiplication, with the class $[\bar{1}]$ containing $\bar{1}$ acting as an identity element.

Now let D be a unique factorization domain, i.e., any domain for which the Unique Factorization Theorem is valid. In that case, every element $\alpha \neq \bar{1}$ in G_D admits unique factorization into powers of the classes $\bar{p}$ of the prime (irreducible) elements $p \in D$. In such a situation, there is an obvious definition of 'coprime' elements in G_D and, given any element $\xi \in G_D$, the set $G_D\langle\xi\rangle$ of all elements coprime to ξ in G_D will form a commutative semigroup with identity $\bar{1}$ such that every element $\alpha \neq \bar{1}$ in $G_D\langle\xi\rangle$ admits unique factorization into powers of those prime classes $\bar{p}$ (p prime in D) that do not divide ξ.

Given $\xi \in G_D$, now let $I = I(\xi)$ denote the principal ideal in D generated by any (or all) elements $x \in \xi$, and for both D and G_D define *congruence modulo* ξ to be congruence modulo this ideal I. In particular, this relation will induce an equivalence relation $\equiv \pmod{\xi}$ on $G_D\langle\xi\rangle$, and we shall let $\Gamma_D\langle\xi\rangle$ denote the corresponding set of equivalence classes in G_D. The elements $[\alpha]$ of $\Gamma_D\langle\xi\rangle$ will be called **generalized progressions mod** ξ in D or G_D; under the induced multiplication of generalized progressions, $\Gamma_D\langle\xi\rangle$ forms a commutative semigroup with the previous class $[\bar{1}]$ as identity element.

1.3. Proposition. (i) *If D is a unique factorization domain, then $\Gamma_D\langle\xi\rangle$ is a 'cancellation' semigroup, i.e., $AB = AC$ in $\Gamma_D\langle\xi\rangle$ implies $B = C$.*

(ii) *If D is a principal ideal domain, then $\Gamma_D\langle\xi\rangle$ is an abelian group.*

(iii) *If D is a Euclidean domain such that G_D forms an arithmetical semigroup under the induced norm, then $\Gamma_D\langle\xi\rangle$ is a finite abelian group.*

Proof. Suppose that $AB = AC$ in $\Gamma_D\langle\xi\rangle$. Then $\alpha\beta \equiv \alpha'\gamma \pmod{\xi}$ for some $\alpha, \alpha' \in A$, $\beta \in B$ and $\gamma \in C$. Hence the definition of congreunce modulo ξ implies that $ab - a'c = kx$ and $a' = a_1 + k_1 x$ for some $a, a_1 \in \alpha$, $a' \in \alpha'$, $b \in \beta$, $c \in \gamma$, $x \in \xi$ and $k, k_1 \in D$. Therefore $a(b - uc) = k'x$ for some unit u in D and $k' \in D$. Since α and ξ are coprime, a and x have no common prime factors in D, and therefore $b - uc = k''x$ for some $k'' \in D$. This shows that $\beta \equiv \gamma \pmod{\xi}$, and thus $B = C$.

Next suppose that D is a principal ideal domain, and let $B \in \Gamma_D\langle\xi\rangle$. Then $B = [\beta]$, where β is coprime to ξ; hence, given $b \in \beta$ and $x \in \xi$, there exist elements $s, t \in D$ such that $sb + tx = 1$. If $s = 0$, this implies that x is a unit; hence $I = D$, and all elements of $G_D\langle\xi\rangle$ are congruent. Therefore $\Gamma_D\langle\xi\rangle$ is the trivial group with one element in this case. On the other hand, if $s \neq 0$, then $\sigma\beta \equiv \bar{1} \pmod{\xi}$, where $\sigma = \bar{s}$; then $[\sigma]$ is an inverse for $B = [\beta]$ in $\Gamma_D\langle\xi\rangle$. Therefore $\Gamma_D\langle\xi\rangle$ is a group.

Lastly, if D is a Euclidean domain and $x \in \xi$, then every element $a \in D$ can be expressed in the form $a = xq + r$, where $|r| < |x|$. Hence every element $\alpha \in G_D\langle\xi\rangle$ is congruent to an element ϱ with $|\varrho| < |x| = |\xi|$. If G_D forms an arithmetical semigroup, then there can be only finitely many such elements ϱ, and so $\Gamma_D\langle\xi\rangle$ is finite. □

1.4. Corollary. *If D is a unique factorization domain and $\xi \in G_D$ is an associate class such that $D/I(\xi)$ is a finite ring, then $\Gamma_D\langle\xi\rangle$ is a finite abelian group.*

Proof. In this case, $\Gamma_D\langle\xi\rangle$ will be a finite abelian semigroup with identity satisfying the cancellation law. Hence it is a group. □

In particular, part (iii) of Proposition 1.3 shows that every Euclidean domain D such that G_D is an arithmetical semigroup also provides a variety of non-trivial formation structures as well. For example, this applies to such domains as $\mathbf{Z}$, $\mathbf{Z}[\sqrt{-1}]$, $\mathbf{Z}[\sqrt{2}]$ and $\mathrm{GF}[q, t]$, the case of $\mathbf{Z}$ having already been treated explicitly in Example 1.1 of course. In terms of the correspondence between associate classes and principal ideals, some examples of these types also fall essentially within the scope of:

1.5. Example: Ideal class groups of algebraic number fields. Although once again detailed knowledge of the facts in question will not be presupposed, it seems desirable to at least mention some of the most interesting classical examples leading to the concept of a formation (which also have the asymptotic properties that provide some of the motivation for introducing and investigating consequences of Axiom A* later on).

In the first place, consider the arithmetical semigroup G_K of all non-zero integral ideals in a given algebraic number field K (i.e., non-zero ideals in the domain D of all algebraic integers of K). A classical method of measuring how far removed D is from being a principal ideal domain is provided by studying the equivalence relation $\sim$ on G_K such that

$$I \sim J \quad \text{if and only if} \quad (a)I = (b)J \text{ for some principal ideals}$$
$$(a) \text{ and } (b) \text{ in } G_K.$$

It is a fact that this definition provides an equivalence relation on G_K which has the property that the corresponding quotient set $\Gamma_K = G_K/\sim$ forms a finite abelian group under the induced operation of multiplication. This group is called the *ideal class group* of K, and its order h_K is called the *class number* of K. In particular, D is a principal ideal domain if and only if $h_K = 1$, and *here* this condition is also equivalent to that of D being a unique factorization domain. In general, (G_K, Γ_K) will be referred to here as the **absolute** *ideal class formation* over K. (For some elementary discussions of these concepts, see for example Pollard [1] and LeVeque [1].)

Now let $G_K\langle A\rangle$ denote the semigroup of all ideals in G_K that are co-prime to a given ideal $A \in G_K$. A number of equivalence relations, giving rise to formation structures on $G_K\langle A\rangle$ in the present sense, have received special attention. One of these (sometimes called the "widest" relation — a slightly inaccurate terminology), which is perhaps the simplest to describe, is that in which two ideals $I, J \in G_K\langle A\rangle$ are said to be equivalent if and only if $(a)I = (b)J$ for some principal ideals (a) and (b) such that $a \equiv b \equiv 1$

(mod A) in D. (A treatment of this and similar relations is given in detail by Landau [8]. Different approaches to these ideas, depending partly on concepts not discussed in this book, are given by Hasse [1] and Lang [2].)

1.6. Example: Arithmetical categories and further examples. The above discussion has indicated a variety of instances in which formations arise naturally as the result of studying some particular integral domain of interest. Such examples often lead to further interesting ones in the following manner.

Let $\Phi: G' \to G$ denote an identity-preserving semigroup homomorphism of an arithmetical semigroup G' into an arithmetical semigroup G, and let $\beta: G \to \Gamma$ be an identity-preserving homomorphism of G onto a finite abelian group Γ. Then, if

$$\Gamma' = \operatorname{Im} \beta\Phi,$$

$\beta' = \beta\Phi: G' \to \Gamma'$ defines a formation structure on G'. For, Γ' is a subsemigroup of Γ containing the identity element, and if $b = \beta\Phi(a) \in \Gamma'$ has order k in Γ, then $b^{k-1} = \beta\Phi(a^{k-1}) \in \Gamma'$; thus $b^{-1} \in \Gamma'$ and so Γ' is a group. In particular, if Φ maps G' onto G, then β' induces a formation structure on G' which partitions G' into classes that are in 1–1 correspondence with the classes defined by β, and have the property that *the corresponding class groups are isomorphic.*

An interesting special case to which these considerations may be applied arises when the norm mapping on a given arithmetical semigroup G is integer-valued. If m is a given positive integer, let $G\langle m\rangle$ denote the subsemigroup of all elements $a \in G$ such that $|a|$ is coprime to m. Then $G\langle m\rangle$ is also an arithmetical semigroup, and the norm mapping $|\ |: G\langle m\rangle \to G_{\mathbf{Z}}\langle m\rangle$ is an identity-preserving homomorphism which, in the above manner, now leads to the partitioning of $G\langle m\rangle$ into a group of *generalized arithmetical progressions* mod m. For example, we have seen that most natural arithmetical categories possess integer-valued norm functions. Therefore such categories give rise to a variety of natural formations in this way. In particular, later in this chapter we shall pay special attention to generalized arithmetical progressions in such categories as the category $\mathscr{A}$ of all finite abelian groups and the category $\mathfrak{S}$ of all semisimple finite rings. (Up to now, very little attention appears to have been paid to generalized arithmetical progressions in such semigroups as G_K above. For G_K, interest has usually centred about the types of ideal class groups described earlier, and it is the corresponding classes that have usually been taken to provide the "correct" analogues of arithmetical progressions for K.)

Another interesting application of the general principle outlined above is obtained by considering the category $\mathfrak{F}=\mathfrak{F}_D$ of all modules of finite cardinal over the domain D of all algebraic integers in a given algebraic number field K. Consider the homomorphism $\Phi:G_{\mathfrak{F}}\to G_K$ defined in Chapter 3, § 2. This homomorphism maps $G_{\mathfrak{F}}$ onto G_K, and hence in the first instance it induces a partitioning of $G_{\mathfrak{F}}$ into a group of exactly h_K *generalized ideal classes*. Secondly, by considering only those modules M in $\mathfrak{F}$ such that $\Phi(\overline{M})$ is coprime to a given ideal $A\in G_K$ and restricting Φ to the corresponding arithmetical sub-semigroup $G_{\mathfrak{F}}\langle A\rangle$ of $G_{\mathfrak{F}}$, one obtains generalized *relative* ideal classes in 1–1 correspondence with the classes in $G_K\langle A\rangle$ that arise from the various types of formation structures on $G_K\langle A\rangle$ referred to in Example 1.5. Similar comments apply to the category $\mathfrak{S}_D$ alluded to in Chapter 3, § 2.

Lastly it may be noted that, if $\beta:G\to\Gamma$ defines a given formation structure on an arithmetical semigroup, then, by composition, any group epimorphism $\theta:\Gamma\to\Gamma'$ that is not an isomorphism will induce a "wider" formation structure on G, the "widest" such structure being the trivial one in which all elements of G are equivalent (i.e., $\Gamma'=\{1\}$).

§ 2. Characters and formations

Let (G, Γ) denote an arbitrary arithmetical formation. In order to discuss questions about the behaviour of arithmetical functions or the densities of certain sets relative to a given class $H\in\Gamma$, it is often convenient to make use of the characters of Γ, i.e., the group homomorphisms of Γ into the multiplicative group $\mathbf{C}^{\times}$ of all non-zero complex numbers. Certain standard facts about the characters of finite abelian groups will be needed for this purpose. (The reader will find a treatment of all such facts quoted below in the book of M. Hall [1] for example; in referring to this book, the reader should bear in mind the fact that the additive group of all positive real numbers modulo one is isomorphic to the multiplicative group of all complex numbers of absolute value one.)

Now let X denote an arbitrary finite abelian group. Then the set X^* of all characters of X forms an abelian group under point-wise multiplication:

$$(\chi_1\chi_2)(x) = \chi_1(x)\chi_2(x) \quad \text{for } x\in X.$$

This group X^* is called the *character group* of X and it is isomorphic to X. Amongst other properties of X^*, the following are worth noting for later purposes:

2.1. *The values $\chi(x)$ of any character $\chi \in X^*$ all have absolute value one. The inverse χ^{-1} of a character $\chi \in X^*$ is the 'conjugate' character $\bar{\chi}$ such that $\bar{\chi}(x) = \overline{\chi(x)}$, the complex conjugate of $\chi(x)$ $[x \in X]$. The identity element of X^* is the character 1 taking value 1 at each $x \in X$.*

These statements may be treated as an exercise for the reader; they depend only on the definition of multiplication in X^* and the fact that every element $x \in X$ has finite order. For the next statements, reference may be made to the book of M. Hall [1], for example.

2.2. *Given any element $x \neq 1$ in X, there exists a character $\chi \in X^*$ with the property that $\chi(x) \neq 1$.* □

2.3. 'Orthogonality' relations. *Let $h = \text{card } X = \text{card } X^*$. Then*

$$\sum_{x \in X} \chi(x)\bar{\psi}(x) = \begin{cases} h & \text{for } \chi = \psi \text{ in } X^*, \\ 0 & \text{for } \chi \neq \psi \text{ in } X^*, \end{cases}$$

$$\sum_{\chi \in X^*} \chi(x)\bar{\chi}(y) = \begin{cases} h & \text{for } x = y \text{ in } X, \\ 0 & \text{for } x \neq y \text{ in } X. \end{cases}$$

In particular,

$$\sum_{x \in X} \chi(x) = \begin{cases} h & \text{for } \chi = 1, \\ 0 & \text{for } \chi \neq 1, \end{cases}$$

$$\sum_{\chi \in X^*} \chi(x) = \begin{cases} h & \text{for } x = 1, \\ 0 & \text{for } x \neq 1. \end{cases}$$ □

Now consider the character group Γ^*, where (G, Γ) is a given formation. Let $\beta: G \to \Gamma$ denote the homomorphism sending any element $a \in G$ to its class $[a] \in \Gamma$. Then, for any $\chi \in \Gamma^*$, $\chi\beta: G \to \mathbf{C}^\times$ is an identity-preserving homomorphism of G into $\mathbf{C}^\times$ which will be called a **character** of the formation (G, Γ). It is easy to verify that *the set $X = X(G, \Gamma)$ of all these characters of (G, Γ) forms an abelian group isomorphic to Γ^** under point-wise multiplication. This group X will be called the **character group** of (G, Γ).

For example, for the formation $(G_{\mathbf{Z}}\langle m\rangle, \Gamma_{\mathbf{Z}}\langle m\rangle)$ described in Example 1.1, there are exactly $\varphi(m)$ characters. These are known as the *(residue class) characters* mod m; in some discussions, they are regarded as functions on $G_{\mathbf{Z}}$ by introducing the convention that they take value 0 on all integers not coprime to m.

In the discussion below, it will be useful to know that the character group of a formation (G, Γ) characterizes (G, Γ) in the following sense: Let G be an arbitrary arithmetical semigroup, and let X denote any finite set of identity-preserving homomorphisms $\chi: G \to \mathbf{C}^{\times}$, which forms an abelian group under point-wise multiplication. Define an equivalence relation $\sim_X$ on G by letting

$$a \sim_X b \quad \text{if and only if} \quad \chi(a) = \chi(b) \quad \text{for every } \chi \in X.$$

Let Γ_X denote the set of all equivalence classes under this relation $\sim_X$.

2.4. Proposition. *The pair (G, Γ_X) forms an arithmetical formation with X as its character group. If X is the character group of a given formation structure (G, Γ) on G, then*

$$(G, \Gamma_X) = (G, \Gamma).$$

Proof. It is easy to see that Γ_X forms a commutative semigroup with identity under the induced multiplication $[a][b]=[ab]$ of equivalence classes. Also, given any $\chi \in X$, one may define an identity-preserving homomorphism $\chi^*: \Gamma_X \to \mathbf{C}^{\times}$ by letting $\chi^*([a]) = \chi(a)$; the set of all these homomorphisms χ^* then forms an abelian group isomorphic to X, relative to pointwise multiplication.

Now define a mapping $\theta: \Gamma_X \to X^*$ by letting $\theta([a])$ be the character of X such that

$$\theta([a])(\chi) = \chi^*([a])$$

for $[a] \in \Gamma_X$ and $\chi \in X$. After verifying that $\theta([a])$ is indeed a character of X, a further verification shows that the mapping θ is actually a 1–1 identity-preserving homomorphism. Since X^* is a group, this implies that Γ_X is a finite semigroup satisfying the cancellation law. Hence Γ_X is a group, and (G, Γ_X) is a formation. Further, a remark above about the homomorphism χ^* now shows that the rule $\chi \to \chi^*$ defines a group monomorphism of X into Γ_X^*, while consideration of the mapping θ shows that card $\Gamma_X \leqq$ $\leqq$ card $X^* =$ card X. Thus the rule $\chi \to \chi^*$ provides an isomorphism of X with Γ_X^*. Since every character of Γ_X is therefore of the form χ^* for $\chi \in X$, and $\chi^* \beta = \chi$, it follows that X is the character group of the formation (G, Γ_X).

Now suppose that X is the character group of a given formation structure (G, Γ) on G. Then, by its definition, X is the set of all homomorphisms of the form $\chi\beta$, where $\chi \in \Gamma^*$ and $\beta: G \to \Gamma$ is the natural epimorphism. Therefore

$$\begin{aligned} a \sim_X b &\Leftrightarrow (\chi\beta)(a) = (\chi\beta)(b) && \text{for every } \chi \in \Gamma^* \\ &\Leftrightarrow \chi([a][b]^{-1}) = 1 && \text{for every } \chi \in \Gamma^*, \text{ where } [c] = \beta(c), \\ &\Leftrightarrow [a] = [b] && \text{by 2.2 above,} \\ &\Leftrightarrow a \sim b && \text{relative to } (G, \Gamma). \end{aligned}$$

Therefore X induces the given relation $\sim$ on G, and so $\Gamma_X = \Gamma$. □

The above proposition shows that one could actually define an arithmetical formation to be a pair (G, X) consisting of an arithmetical semigroup G together with a finite set X of homomorphisms $\chi: G \to \mathbf{C}^{\times}$ which forms an abelian group relative to point-wise multiplication. For some purposes this slightly less intuitive formulation is particularly convenient. In fact, it is at least partly because such a formulation is possible that it becomes fairly easy to refine previous results concerning distributions of values of arithmetical functions and densities of subsets in arithmetical semigroups so as to obtain a corresponding **relative theory** of the classes of formations. (As was mentioned before, in the discussion below, attention will be restricted to refining the theory of arithmetical semigroups satisfying Axiom A. An example of a different kind of relative asymptotic theory may be found in the paper of Hayes [1], which deals with the special ring GF$[q, t]$.) For particular semigroups such as $G_{\mathbf{Z}}$, $G_{\mathscr{A}}$, G_K and so on, or more generally, it might perhaps also be interesting to investigate different "relative" theories, involving other modes of partitioning into "classes" besides that occurring in a formation. However, in general, this would seem to require additional, or other, techniques.

Now let X denote the character group of a given formation (G, Γ). In the discussion below, it will sometimes be convenient to identify the character $\chi \in X$ with the corresponding character χ^* of Γ defined in the proof of Proposition 2.4. Also, given a subset E of G and a function $f \in \text{Dir}(G)$, frequent use will now be made of the notations

$$E[x] = \{a \in E: |a| \leqq x\}, \qquad N_E(f, x) = \sum_{a \in E[x]} f(a) \qquad (x > 0).$$

The following lemma provides a basic tool for interchanging asymptotic conclusions over G with conclusions relative to the classes $H \in \Gamma$.

2.5. Lemma. *Let $f \in \mathrm{Dir}(G)$ and suppose that, for each $\chi \in X$,*

$$\sum_{a \in G[x]} \chi(a) f(a) = F_\chi(x) + O(R(x)) \quad \text{as } x \to \infty.$$

Then, for any class $H \in \Gamma$,

$$N_H(f, x) = h^{-1} \sum_{\chi \in X} \bar{\chi}(H) F_\chi(x) + O(R(x)),$$

where $h = \mathrm{card}\, X = \mathrm{card}\, \Gamma$ is the class number of the formation. Further,

$$N(f, x) = F(x) + O(R(x)),$$
$$\sum_{a \in G[x]} \chi(a) f(a) = O(R(x)) \quad \text{for each } \chi \neq 1 \in X$$

if and only if

$$N_H(f, x) = h^{-1} F(x) + O(R(x)) \quad \text{for each class } H \in \Gamma.$$

Proof. The proof depends on the following direct consequence of the orthogonality relations 2.3 above:

2.6. 'Orthogonality' lemma for formations. *Let $h = \mathrm{card}\, X = \mathrm{card}\, \Gamma$. Then*

$$\sum_{H \in \Gamma} \chi(H) \bar{\psi}(H) = \begin{cases} h & \text{if } \chi = \psi \text{ in } X, \\ 0 & \text{otherwise,} \end{cases}$$

$$\sum_{\chi \in X} \chi(a) \bar{\chi}(b) = \begin{cases} h & \text{if } a \sim b \text{ in } G \text{ (i.e., } [a] = [b] \text{ in } \Gamma), \\ 0 & \text{otherwise.} \end{cases}$$

In particular,

$$\sum_{H \in \Gamma} \chi(H) = \begin{cases} h & \text{if } \chi = 1, \\ 0 & \text{if } \chi \neq 1, \end{cases}$$

$$\sum_{\chi \in X} \chi(a) = \begin{cases} h & \text{if } a \sim 1 \text{ in } G, \\ 0 & \text{otherwise.} \end{cases}$$ □

Now suppose that, for $\chi \in X$,

$$\sum_{a \in G[x]} \chi(a) f(a) = F_\chi(x) + O(R(x)).$$

Then, for any class $H \in \Gamma$,

$$N_H(f, x) = \sum_{a \in G[x]} f(a) h^{-1} \sum_{\chi \in X} \chi(a) \bar{\chi}(H) = h^{-1} \sum_{\chi \in X} \bar{\chi}(H) \sum_{a \in G[x]} \chi(a) f(a)$$
$$= h^{-1} \sum_{\chi \in X} \bar{\chi}(H) F_\chi(x) + O(R(x)),$$

since X is finite.

In particular, if $F_1(x)=F(x)$ and $F_\chi(x)=0$ for $\chi\neq 1$, then

$$N_H(f, x) = h^{-1}F(x)+O(R(x)).$$

Conversely, if the latter relation is true for every class $H\in\Gamma$, then firstly

$$N(f, x) = \sum_{H\in\Gamma} N_H(f, x) = F(x)+O(R(x)).$$

Secondly, if $\chi\neq 1$, then

$$\begin{aligned}\sum_{a\in G[x]} \chi(a)f(a) &= \sum_{H\in\Gamma}\sum_{a\in H[x]} \chi(a)f(a) = \sum_{H\in\Gamma} \chi(H)N_H(f, x)\\ &= \sum_{H\in\Gamma} \chi(H)[h^{-1}F(x)+O(R(x))]\\ &= h^{-1}F(x)\sum_{H\in\Gamma} \chi(H)+O(R(x)) = O(R(x)),\end{aligned}$$

since $\sum_{H\in\Gamma}\chi(H)=0$. This proves Lemma 2.5. □

§ 3. The *L*-series of a formation

Let X denote the character group of some given formation (G, Γ). Given $\chi\in X$, consider the formal Dirichlet series

$$\zeta_G(z, \chi) = \chi(z) = \sum_{a\in G} \chi(a)a^{-z}.$$

Since χ is completely multiplicative, Corollary 2.4.2 implies that

$$\zeta_G(z, \chi) = \prod_{p\in P} (1-\chi(p)p^{-z})^{-1};$$

this formula will be called the *Euler product formula* for $\zeta_G(z, \chi)$. The formal series

$$L_G(z, \chi) = \tilde{\zeta}_G(z, \chi) = \sum_{a\in G} \chi(a)|a|^{-z}$$

will be called the **L-series** for χ over G; since $\tilde{}$ is a continuous algebra homomorphism, $L_G(z, \chi)$ in turn has the 'Euler product formula':

$$L_G(z, \chi) = \prod_{p\in P} (1-\chi(p)|p|^{-z})^{-1}.$$

For a given formation, the *L*-series play a rôle similar to that of the zeta function of a given arithmetical semigroup. (Here, the prefix *L* follows historical usage: For the formation $(G_{\mathbf{Z}}\langle m\rangle, \Gamma_{\mathbf{Z}}\langle m\rangle)$ of Example 1.1, these series become the ordinary *Dirichlet L-series* $L(z, \chi)$ mod m (which are sometimes written as series over all $n \in G_{\mathbf{Z}}$, with the convention that $\chi(n)=0$ for $(n, m)>1$); for formations over a number field K of the types described in Example 1.5, the series $L_G(z, \chi)$ become the various *Hecke–Landau L-series* for K.) A related concept is that of the **zeta function** of a **class** $H \in \Gamma$:

$$\zeta_H(z) = \sum_{a \in H} a^{-z};$$

as in the discussion of $\zeta_G(z)$, if confusion seems unlikely we shall later sometimes use the same name and notation for the corresponding series $\sum_{a \in H} |a|^{-z}$.

Strictly speaking, $\zeta_H(z)$ is simply the characteristic function of H. By reference to Lemma 2.6, an argument similar to that for Lemma 2.5 yields:

3.1. Proposition (Inversion formulae). *For any class* $H \in \Gamma$,

$$\zeta_H(z) = h^{-1} \sum_{\chi \in X} \bar{\chi}(H) \zeta_G(z, \chi), \qquad \tilde{\zeta}_H(z) = h^{-1} \sum_{\chi \in X} \bar{\chi}(H) L_G(z, \chi).$$

For any character $\chi \in X$,

$$\zeta_G(z, \chi) = \sum_{H \in \Gamma} \chi(H) \zeta_H(z), \qquad L_G(z, \chi) = \sum_{H \in \Gamma} \chi(H) \tilde{\zeta}_H(z). \qquad \square$$

We conclude this section with some examples of L-series for particular formations. For this purpose, first let $\Phi: G' \to G$ denote an identity-preserving semigroup epimorphism of an arithmetical semigroup G' onto the given arithmetical semigroup G. Then, as in § 1, the formation epimorphism $\beta: G \to \Gamma$ induces an epimorphism $\beta' = \beta\Phi: G' \to \Gamma$. This induces a partitioning of G' into a group Γ' of classes, which is isomorphic to Γ via the isomorphism Φ_* sending the class $[a] \in \Gamma'$ to the class $[\Phi(a)] \in \Gamma$. Then the rule $\chi \to \chi\Phi_*$ may be verified to define an isomorphism of Γ^* with Γ'^*. This fact allows one to check that the characters χ' in the character group X' of (G', Γ') are simply the various homomorphisms $\chi\beta\Phi: G' \to \mathbf{C}^{\times}$ for which $\chi \in \Gamma^*$; therefore the rule $\chi \to \chi\beta\Phi$ defines an isomorphism of Γ^* with X' — bear in mind the isomorphism of X' with Γ'^* defined in the proof of Proposition 2.4. In terms of the isomorphism of X with Γ^* defined in the proof of Proposition 2.4, it follows that the composition

$$\chi \to \chi^* \to \chi^* \beta\Phi = \chi\Phi \quad (\chi \in X)$$

defines an isomorphism of X with X'. In other words, *the natural rule* $\Phi^*: \chi \to \chi\Phi$ by which X and Φ induce a group of homomorphisms $G' \to \mathbf{C}^{\times}$ actually *defines an isomorphism of X with the character group X' of the induced formation structure* (G', Γ').

Now let G be an arithmetical semigroup with an integer-valued norm mapping such that, for some positive integer m, the mapping $|\ |: G\langle m\rangle \to G_{\mathbf{Z}}\langle m\rangle$ is an epimorphism. Then, by § 1, the formation structure $(G_{\mathbf{Z}}\langle m\rangle, \Gamma_{\mathbf{Z}}\langle m\rangle)$ defines a formation structure on $G\langle m\rangle$ which partitions it into a group of $\varphi(m)$ generalized arithmetical progressions mod m. In that case, the above discussion implies that the characters of this formation structure on $G\langle m\rangle$ are all of the form $\chi|\ |$ where χ is a residue class character mod m. Therefore

$$L_{G\langle m\rangle}(z, \chi|\ |) = \sum_{a\in G\langle m\rangle} \chi(|a|)|a|^{-z} = \sum_{(n,m)=1} \chi(n)G(n)n^{-z}.$$

3.2. Example: Finite abelian groups. As a particular case, let $\mathscr{A}\langle m\rangle$ denote the category of all finite abelian groups of order coprime to m. Then $\mathscr{A}\langle m\rangle$ is an arithmetical category with norm mapping induced by the orders of its objects. The corresponding L-series are then of the form

$$L_{\mathscr{A}\langle m\rangle}(z, \chi|\ |) = \sum_{(n,m)=1} \chi(n)a(n)n^{-z} = \prod_{r=1}^{\infty} L(rz, \chi^r),$$

where $L(z, \psi)$ denotes a Dirichlet L-series mod m, the last equation following from the equation

$$\zeta_{\mathscr{A}}(z) = \prod_{r=1}^{\infty} \zeta(rz)$$

and Proposition 2.6.3. (Here it is useful to follow the custom which extends χ to a completely multiplicative function on $G_{\mathbf{Z}}$ by letting $\chi(n)=0$ if $(n, m)>1$.)

3.3. Example: Semisimple finite rings. Another interesting case arises by considering the arithmetical category $\mathfrak{S}\langle m\rangle$ of all semisimple finite rings of cardinal coprime to m. Then in the same general way we obtain L-series of the form

$$L_{\mathfrak{S}\langle m\rangle}(z, \chi|\ |) = \sum_{(n,m)=1} \chi(n)S(n)n^{-z} = \prod_{r=1}^{\infty}\prod_{s=1}^{\infty} L(rs^2z, \chi^{rs^2}).$$

Analogous formulae, involving the Hecke–Landau L-series referred to above, are valid for the formation structures over the categories $\mathfrak{F}_D$ and $\mathfrak{S}_D$ mentioned in § 1; for details see the author [8].

Lastly, in connection with examples of zeta functions of classes, we remark that the zeta function of an ordinary arithmetical progression is essentially just a multiple of one of the well-known *Hurwitz* zeta functions $\zeta(z, w)$; these are usually defined for $0<w\leqq 1$ and $\operatorname{Re} z>1$ by the equation

$$\zeta(z, w) = \sum_{n=0}^{\infty} (n+w)^{-z};$$

compare LeVeque [1], say.

§ 4. Axiom A*

The classical examples involving positive integers or ideals in a number field, as well as certain arithmetical categories, provide a variety of instances of arithmetical formations obeying the following basic asymptotic 'equi-distribution' axiom concerning a given formation (G, Γ).

Axiom A*. *There exist positive constants A and δ, and a constant η with $0\leqq\eta<\delta$, such that, for any class $H\in\Gamma$,*

$$N_H(x) = h^{-1}Ax^{\delta}+\mathrm{O}(x^{\eta}) \quad \textit{as } x \to \infty.$$

Here $h=\operatorname{card}\Gamma$ denotes the class number of (G, Γ), and as before, for any subset E of G and $x>0$,

$$N_E(x) = \operatorname{card} E[x].$$

It follows immediately from this axiom that

$$N_G(x) = \sum_{H\in\Gamma} N_H(x) = Ax^{\delta}+\mathrm{O}(x^{\eta}),$$

i.e., G itself satisfies Axiom A. Hence the theory of formations satisfying Axiom A* is a generalization of the theory of arithmetical semigroups satisfying Axiom A. The main significance of this abstract generalization seems to lie, as in other cases, in the existence of natural examples to which the theory may be applied. In particular, most of the comments concerning Axiom A made at the beginning of Part II apply with very little change to Axiom A*.

Before considering some specific examples, we note the following immediate consequence of Lemma 2.5.

4.1. Proposition. *The arithmetical formation* (G, Γ) *satisfies Axiom* A* *as stated above if and only if G itself satisfies Axiom* A *with*

$$N_G(x) = Ax^\delta + O(x^\eta),$$

and in addition

$$N(\chi, x) = O(x^\eta) \quad \textit{for every } \chi \neq 1 \in X. \qquad \square$$

The simplest non-trivial instance of Axiom A* arises from

4.2. Example: Ordinary arithmetical progressions. Consider the formation $(G_Z\langle m\rangle, \Gamma_Z\langle m\rangle)$, where m is a given positive integer. Let $H_{m,r} \in \Gamma_Z\langle m\rangle$ denote the progression $\{r, m+r, 2m+r, \ldots\}$, where $1 \leqq r < m$ and $(r, m) = 1$. Then

$$N_{H_{m,r}}(x) = \sum_{nm+r \leqq x} 1 = [x/m - r/m]$$

$$= m^{-1}x + O(1) \quad \text{as } x \to \infty.$$

Thus this formation satisfies Axiom A* with $A = \varphi(m)/m$, $\delta = 1$ and $\eta = 0$.

4.3. Example: Ideal class formations over number fields. It was proved by Landau [8] that the various formations described (in out-line) in Example 1.5 have asymptotic equi-distribution properties amounting to the validity of Axiom A* in the present terminology. In particular, the absolute ideal class formation (G_K, Γ_K) over a given number field K has the property that, for any class $H \in \Gamma_K$,

$$N_H(x) = h_K{}^{-1} A_K x + O(x^{\eta_0}),$$

where h_K is the class number of K, and A_K and η_0 are the constants referred to in Chapter 4, § 1; the algebraic and asymptotic properties of this particular formation over K are also treated in detail in Landau's book [9].

The following lemma is useful in establishing the validity of Axiom A* for various other formations.

4.4. Lemma (Transfer Principle). *Let* (G, Γ) *and* (G', Γ') *denote arithmetical formations with character groups* X *and* X', *respectively. Let* $\Phi: G' \to G$ *be an identity-preserving semigroup homomorphism such that the homomorphism* $\chi \to \chi\Phi$ *maps* X *onto* X'. *Suppose that* (G, Γ) *satisfies Axiom* A* *as stated above. Also suppose that* $|G'| \subseteq |G|$ *and that there exists a function* $g \in \mathrm{Dir}\,(G)$ *with the properties that:*

(i) $L_{G'}(z, \chi\Phi) = L_G(z, \chi)\tilde{g}(z, \chi)$ *for* $\chi \in X$,

(ii) $N(|g|, x) = O(x^\nu)$ *as* $x \to \infty$, *where* $0 \leqq \nu < \delta$.

Then, for any class $H' \in \Gamma'$,

$$N_{H'}(x) = (A/h')\tilde{g}(\delta)x^{\delta} + \begin{cases} O(x^{\nu}) & \textit{if } \nu > \eta, \\ O(x^{\eta}\log x) & \textit{if } \nu = \eta, \\ O(x^{\eta}) & \textit{if } \nu < \eta, \end{cases}$$

where $h' = \operatorname{card} \Gamma'$ *is the class number of* (G', Γ'). *Hence* (G', Γ') *also satisfies Axiom* A* *if* $\tilde{g}(\delta) > 0$.

Proof. By condition (i) on the function g, if $\chi \in X$, then

$$N_{G'}(\chi\Phi, x) = \sum_{\substack{a, b \in G \\ |ab| \leq x}} \chi(a)\chi(b)g(b)$$

$$= \sum_{b \in G[x]} \chi(b)g(b)N_G(\chi, x/|b|).$$

Now, condition (ii) on g and partial summation (see also the proof of Lemma 4.5.2) gives

$$\sum_{b \in G[x]} |g(b)|\,|b|^{-\eta} = O(x^{\nu-\eta}) + \begin{cases} O(x^{\nu-\eta}) & \text{if } \nu > \eta, \\ O(\log x) & \text{if } \nu = \eta, \\ O(1) & \text{if } \nu < \eta. \end{cases}$$

Therefore Axiom A* on G and Proposition 4.1 imply that, for $\chi \neq 1$ in X,

$$N_{G'}(\chi\Phi, x) = O\Big(\sum_{b \in G[x]} |g(b)|(x/|b|)^{\eta}\Big) = \begin{cases} O(x^{\nu}) & \text{if } \nu > \eta, \\ O(x^{\eta}\log x) & \text{if } \nu = \eta, \\ O(x^{\eta}) & \text{if } \nu < \eta. \end{cases}$$

Also, Corollary 4.5.3* (see the end of Chapter 4, § 5) yields

$$N_{G'}(x) = A\tilde{g}(\delta)x^{\delta} + \begin{cases} O(x^{\nu}) & \text{if } \nu > \eta, \\ O(x^{\eta}\log x) & \text{if } \nu = \eta, \\ O(x^{\eta}) & \text{if } \nu < \eta. \end{cases}$$

Since every character in X' has the form $\chi\Phi$ ($\chi \in X$), the required conclusion therefore follows from Lemma 2.5 (because, if $1 \neq \chi \in X'$, then $\chi' = \chi\Phi$ for some $\chi \neq 1$ in X). □

Now consider the arithmetical category $\mathscr{A}\langle m\rangle$ of all finite abelian groups of order coprime to a given positive integer m, and the previously discussed formation structure on $G_{\mathscr{A}\langle m\rangle}$ which partitions it into $\varphi(m)$ generalized arithmetical progressions mod m.

4.5. Theorem. *Let $H_{m,r}(\mathscr{A})$ denote the generalized arithmetical progression* mod m *in $G_{\mathscr{A}\langle m\rangle}$ consisting of all isomorphism classes of abelian groups of order congruent to r* (mod m), *where* $(r, m)=1$. *Then, as $x\to\infty$,*

$$N_{H_{m,r}(\mathscr{A})}(x) = \sum_{\substack{n\leq x\\ n\equiv r\,(\mathrm{mod}\,m)}} a(n)$$

$$= \left[m^{-1}\prod_{k=2}^{\infty} L(k, 1)\right]x+\mathrm{O}(x^{1/2}).$$

In particular, $G_{\mathscr{A}\langle m\rangle}$ together with these generalized arithmetical progressions forms a formation satisfying Axiom A*.

Proof. Let χ denote a residue class character mod m (and note that in the above formula 1 denotes the identity character of $G_{\mathbf{Z}}\langle m\rangle$, *not* of $G_{\mathbf{Z}}$). Then

$$L_{\mathscr{A}\langle m\rangle}(z, \chi|\ |) = L(z, \chi)\tilde{g}(z, \chi),$$

where

$$\tilde{g}(z, \chi) = \prod_{k=2}^{\infty} L(kz, \chi^k).$$

Thus, if χ is extended to $G_{\mathbf{Z}}$ in the standard way mentioned before, $\tilde{g}(z, \chi)$ may be written in the form

$$\tilde{g}(z, \chi) = \sum_{n=1}^{\infty} \chi(n)g_0(n)^{-z},$$

where

$$\tilde{g}_0(z) = \sum_{n=1}^{\infty} g_0(n)n^{-z} = \prod_{k=2}^{\infty} \zeta(kz).$$

It therefore follows from Theorem 5.1.6 that

$$N(|g|, x) \leq N(g_0, x) = \mathrm{O}(x^{1/2}) \quad \text{as } x\to\infty,$$

and a slight variation of the proof of Lemma 5.1.4 shows in particular that

$$\tilde{g}(1) = \prod_{k=2}^{\infty} L(k, 1).$$

Since Example 4.2 shows that $(G_{\mathbf{Z}}\langle m\rangle, \Gamma_{\mathbf{Z}}\langle m\rangle)$ satisfies Axiom A* with $A=\varphi(m)/m$, $\delta=1$ and $\eta=0$, the present theorem now follows from Lemma 4.4. □

Next, consider the analogous formation structure over the arithmetical category $\mathfrak{S}\langle m\rangle$ whose objects are all semisimple finite rings of cardinal coprime to m.

4.6. Theorem. *Let $H_{m,r}(\mathfrak{S})$ denote the generalized arithmetical progression* mod m *in $G_{\mathfrak{S}\langle m\rangle}$ consisting of all isomorphism classes of rings in $\mathfrak{S}\langle m\rangle$ that have cardinal congruent to r* (mod m), *where* $(r, m)=1$. *Then as $x\to\infty$,*

$$N_{H_{m,r}(\mathfrak{S})}(x) = \sum_{\substack{n\leq x\\ n\equiv r(\operatorname{mod} m)}} S(n)$$

$$= \left[m^{-1} \prod_{ks^2>1} L(ks^2, 1)\right] x+O(x^{1/2}).$$

In particular, $G_{\mathfrak{S}\langle m\rangle}$ together with these generalized arithmetical progressions forms a formation satisfying Axiom A*.

Proof. By Example 3.3, if χ denotes a residue class character mod m, then

$$L_{\mathfrak{S}\langle m\rangle}(z, \chi|\;|) = L(z, \chi)\tilde{f}(z, \chi),$$

where

$$\tilde{f}(z, \chi) = \prod_{ks^2>1} L(ks^2 z, \chi^{ks^2}).$$

Then, if χ is extended to $G_{\mathbf{Z}}$ as before,

$$\tilde{f}(z, \chi) = \sum_{n=1}^{\infty} \chi(n) f_0(n) n^{-z},$$

where

$$\tilde{f}_0(z) = \sum_{n=1}^{\infty} f_0(n)n^{-z} = \prod_{ks^2>1} \zeta(ks^2 z).$$

Now, the reader who has verified the details left to him in the proof of Theorem 5.1.7. will have found that $N(f_0, x)=O(x^{1/2})$. Hence the same estimate applies to $N(|f|, x)=N(f, x)$. In this case, a variation of the proof of Lemma 5.1.8 leads to the conclusion that, in particular,

$$\tilde{f}(1) = \prod_{ks^2>1} L(ks^2, 1).$$

The theorem therefore follows with the aid of Lemma 4.4 again. □

Lastly, we single out just one further application of the 'transfer principle' 4.4. This is to the *absolute* formation structure over the category $\mathfrak{F}=\mathfrak{F}_D$ (or, strictly speaking, the semigroup $G_{\mathfrak{F}}$), i.e., the structure induced by the epimorphism $\Phi: G_{\mathfrak{F}} \to G_K$ and the absolute ideal class formation structure over G_K; here $G_{\mathfrak{F}}$ is partitioned into an 'absolute' class group $\Gamma_{\mathfrak{F}} \cong \Gamma_K$ with the same class number h_K as the field K. (For further applications, see the author [8].)

4.7. Theorem. *Let H denote any generalized ideal class in $\Gamma_{\mathfrak{F}}$. Then, as $x \to \infty$,*

$$N_H(x) = \left[h_K{}^{-1} A_K \prod_{r=2}^{\infty} \zeta_K(r) \right] x$$

$$+ \begin{cases} O(x^{1/2}) & \textit{if } [K:Q] < 3, \\ O(x^{1/2} \log x) & \textit{if } [K:Q] = 3, \\ O(x^{\eta_0}) & \textit{if } [K:Q] > 3. \end{cases}$$

Hence the formation $(G_{\mathfrak{F}}, \Gamma_{\mathfrak{F}})$ satisfies Axiom A*.

Proof. For this theorem, we need the explicit details of one of the formulae alluded to after Example 3.3. Let χ denote any character of the formation (G_K, Γ_K), and let $\Phi: G_{\mathfrak{F}} \to G_K$ be the homomorphism defined in Chapter 3, § 2. Then, in terms of the induced homomorphism Φ_*: Dir $(G_{\mathfrak{F}}) \to$ Dir (G_K) defined in Chapter 3, Proposition 2.6.3 and the formula

$$\Phi_*(\zeta_{\mathfrak{F}})(z) = \prod_{r=1}^{\infty} \zeta_{G_K}(rz)$$

imply that the L-series over $G_{\mathfrak{F}}$ associated with χ is

$$L_{\mathfrak{F}}(z, \chi\Phi) = \tilde{\zeta}_{\mathfrak{F}}(z, \chi\Phi) = [\Phi_*(\zeta_{\mathfrak{F}})]^{\sim}(z, \chi)$$

$$= \prod_{r=1}^{\infty} L_K(rz, \chi^r),$$

where $L_K(z, \psi) = \tilde{\zeta}_{G_K}(z, \psi)$. Hence

$$L_{\mathfrak{F}}(z, \chi\Phi) = L_K(z, \chi)\tilde{g}_K(z, \chi),$$

where

$$\tilde{g}_K(z, \chi) = \prod_{r=2}^{\infty} L_K(rz, \chi^r).$$

The theorem may now be deduced in essentially the same way as Theorem 4.5, and it may be interesting to note that it yields Theorem 5.1.1 as an immediate corollary. (In a similar way, the cases $m=1$ of Theorems 4.5 and 4.6 give back Corollary 5.1.2 and Theorem 5.1.7 as corollaries.) □

§ 5. Analytical properties of L-series

For the rest of this chapter, (G, Γ) *will denote an arbitrary arithmetical formation satisfying Axiom* A* as stated in § 4, and h and X will denote the class number and the character group of (G, Γ), respectively. The main object of the discussion will be to indicate in outline how the previous theory of arithmetical semigroups satisfying Axiom A may be carried over to the present systems. Applications to specific formations of interest will usually not be discussed in detail, even though once again these provide major motivation for the investigation in question; given detailed information about the constants involved in Axiom A or Axiom A* for a particular system, the reader should have little difficulty in writing down explicit statements of the immediate corollaries that apply to it. (Of course, as has been remarked before, such corollaries are only immediate for a particular semigroup G' or formation (G', Γ') on the basis of the (usually non-trivial) theorem which states that G' or (G', Γ') satisfies Axiom A or Axiom A*, and which simultaneously usually provides information about the particular constants involved.)

5.1. Proposition. *Let* $1 \neq \chi \in X$. *Then the L-series*

$$L_G(z, \chi) = \sum_{a \in G} \chi(a)|a|^{-z}$$

converges and defines an analytic function of z for all complex numbers z with $\operatorname{Re} z = \sigma > \eta$. *Further, for* $\sigma > \delta$, *the series is absolutely convergent and satisfies the Euler product formula*

$$L_G(z, \chi) = \prod_{p \in P} (1 - \chi(p)|p|^{-z})^{-1}$$

in the analytical sense. In addition, $L_G(z, \chi) \neq 0$ *for* $\sigma > \delta$, *and*

$$\sum_{a \in G[x]} \chi(a)|a|^{-z} = \begin{cases} L_G(z, \chi) + O(x^{\eta - \sigma}) & \text{for } \sigma > \eta, \\ O(\log x) & \text{for } \sigma = \eta, \\ O(x^{\eta - \sigma}) & \text{for } \sigma < \eta. \end{cases}$$

Proof. The first statement and the first estimate for $\sum_{a \in G[x]} \chi(a)|a|^{-z}$ follow immediately from Lemma 4.2.7 with $g = \chi$. Also, the absolute convergence of $L_G(z, \chi)$ follows from that of $\zeta_G(z)$, when $\sigma > \delta$. Then a minor modification of the corresponding proof for $\zeta_G(z)$ shows that $L_G(z, \chi) \neq 0$ and satisfies the Euler product formula analytically, when $\sigma > \delta$. Finally, the last two estimates are easily obtained by partial summation. □

By combining Proposition 5.1 and previous results about $\zeta_G(z)$ subject to Axiom A, or by direct use of Axiom A* and various results of Chapter 4, § 2, it is now easy to deduce a number of analytical properties of $\zeta_H(z)$ for complex z, when H is an arbitrary class in Γ. The conclusions will merely be stated:

5.2. Proposition. *Consider any class $H \in \Gamma$. Then the series*

$$\zeta_H(z) = \sum_{a \in H} |a|^{-z}$$

is absolutely and uniformly convergent, and defines an analytic function of z, for all complex z with $\operatorname{Re} z \geqq \sigma_0 > \delta$ ($\sigma_0 > \delta$ arbitrary). *Further, the function $\zeta_H(z)$ may be continued analytically into the entire region* $\operatorname{Re} z > \eta$, $z \neq \delta$, *and has a simple pole with residue $\delta A/h$ at $z=\delta$; the Laurent expansion near $z=\delta$ has the form*

$$\zeta_H(z) = \frac{\delta A/h}{(z-\delta)} + \gamma_H + \sum_{r=1}^{\infty} c_r (z-\delta)^r,$$

where

$$\gamma_H = h^{-1}\Big\{\gamma_G + \sum_{1 \neq \chi \in X} \bar{\chi}(H) L_G(\delta, \chi)\Big\}. \qquad \square$$

5.3. Proposition. *Consider any class $H \in \Gamma$. Then, as $x \to \infty$,*

$$\sum_{a \in H[x]} |a|^{-z} = \begin{cases} \zeta_H(z) + O(x^{\delta-\sigma}) & \textit{if } \operatorname{Re} z = \sigma > \delta, \\ \dfrac{\delta A}{h} \log x + \gamma_H + O(x^{\eta-\delta}) & \textit{if } z = \delta, \\ \dfrac{1}{h}\left\{\dfrac{\delta A}{(\delta-z)} x^{\delta-z} + \alpha(z) + \displaystyle\sum_{1 \neq \chi \in X} \bar{\chi}(H) L_G(z, \chi)\right\} + O(x^{\eta-\sigma}) & \\ & \textit{if } \operatorname{Re} z = \sigma > \eta,\ z \neq \delta, \\ \dfrac{\delta A}{h(\delta-z)} x^{\delta-z} + O(\log x) & \textit{if } \operatorname{Re} z = \eta, \\ \dfrac{\delta A}{h(\delta-z)} x^{\delta-z} + \dfrac{1}{h}\alpha(z) + O(x^{\eta-\sigma}) & \textit{if } \operatorname{Re} z = \sigma < \eta. \end{cases} \qquad \square$$

In view of its above properties, the constant γ_H may perhaps be called the **Euler constant** of the class $H \in \Gamma$.

§ 6. Average values of arithmetical functions over a class

The following theorem provides an extension of both Theorem 4.3.1 and Theorem 4.5.1 to the present situation. Although it is not intended to represent the most general or sharpest possible statements in this direction, it does cover quite a wide variety of the natural functions that one might wish to consider.

6.1. Theorem. *Let* $f, g \in \mathrm{Dir}\,(G)$ *be functions such that*

$$f(z) = [\zeta_G(mz)]^k g(z)$$

for some positive integers m and k.

(i) *Suppose that* $\tilde{g}(z)$ *is absolutely convergent for* $\mathrm{Re}\, z > \nu$, *where* $\nu < \delta/m$. *Then, for any class* $H \in \Gamma$, *as* $x \to \infty$,

$$N_H(f, x) = \frac{A}{h(k-1)!}\left(\frac{\delta A}{m}\right)^{k-1}[C(m, H)\tilde{g}(\delta/m) + \mathrm{o}(1)]x^{\delta/m}(\log x)^{k-1},$$

where

$$C(m, H) = \sum_{\chi \in X, \chi^m = 1} \bar{\chi}(H).$$

(ii) *Suppose further that* $N(|g|, x) = \mathrm{O}(x^{\nu})$ *as* $x \to \infty$. *Then, for any class* $H \in \Gamma$, *as* $x \to \infty$,

$$N_H(f, x) = (A/h)C(m, H)x^{\delta/m}\sum_{i=1}^{k} b_i(\log x)^{k-i}$$

$$+\begin{cases} \mathrm{O}(x^{\eta/m}) & \text{if } k = 1 \text{ and } \nu < \eta/m, \\ \mathrm{O}\big(x^{(\delta-(\delta-\eta)/k)/m}(\log x)^{k-2}\big) & \\ & \text{if } k > 1 \text{ and } \nu < (\delta-(\delta-\eta)/k)/m, \\ \mathrm{O}(x^{\nu+\varepsilon}) & \text{if } \nu \geqq (\delta-(\delta-\eta)/k)/m, \end{cases}$$

where $\varepsilon > 0$ *is arbitrary, and the* b_i *are the constants occurring in Theorem* 4.5.1.

Proof. As in the special case $\Gamma = \{1\}$ of Chapter 4, first consider the generalized divisor function $d_k(z) = [\zeta_G(z)]^k$, where $k > 1$.

6.2. Lemma. *For any class* $H \in \Gamma$,

$$N_H(d_k, x) = (A/h)x^{\delta}\sum_{i=1}^{k} c_i(\log x)^{k-i} + \mathrm{O}\big(x^{\delta-(\delta-\eta)/k}(\log x)^{k-2}\big)$$

where the c_i *are the constants appearing in Proposition* 4.3.2, *and in particular* $c_1 = (\delta A)^{k-1}/(k-1)!$.

Proof. First consider the divisor function $d=d_2$. For $1\neq\chi\in X$, substitute $f=g=\chi$ and $u=v=x^{1/2}>0$ in the second statement of Lemma 4.3.3. This gives

$$\begin{aligned}\sum_{a\in G[x]}\chi(a)d(a) &= N(\chi*\chi, x^{1/2}x^{1/2})\\ &= 2\sum_{a\in G[x^{1/2}]}\chi(a)N(\chi, x/|a|)-[N(\chi, x^{1/2})]^2\\ &= 2\sum_{a\in G[x^{1/2}]}\chi(a)O((x/|a|)^{\eta})+O(x^{\eta})\\ &= O\big(x^{\eta}\sum_{a\in G[x^{1/2}]}|a|^{-\eta}\big)+O(x^{\eta}) = O(x^{(\delta+\eta)/2})\end{aligned}$$

by Proposition 4.2.8.

Now suppose as an inductive hypothesis that, as $x\to\infty$,

$$\sum_{a\in G[x]}\chi(a)d_{k-1}(a) = O\big(x^{\delta-(\delta-\eta)/(k-1)}(\log x)^{k-3}\big) \qquad [1\neq\chi\in X];$$

the preceding conclusion shows that the induction may be started at $k=3$. Then substitute $u=x^{1/k}$, $v=x^{1-(1/k)}$ $(x\geqq 1)$ and let $f=\chi$ and $g(a)=\chi(a)d_{k-1}(a)$ $[a\in G]$ in the second statement of Lemma 4.3.3. This shows that

$$\begin{aligned}\sum_{a\in G[x]}\chi(a)d_k(a) &= N(f*g, uv)\\ &= \sum_{|a|\leqq u}\chi(a)\sum_{|b|\leqq x/|a|}\chi(c)d_{k-1}(c)+\sum_{|a|\leqq v}\chi(a)d_{k-1}(a)N(\chi, x/|a|)\\ &\quad -N(\chi, u)\sum_{|a|\leqq v}\chi(a)d_{k-1}(a)\\ &= \sum_{|a|\leqq u}\chi(a)O\big((x/|a|)^{\delta-(\delta-\eta)/(k-1)}(\log x)^{k-3}\big)\\ &\quad +\sum_{|a|\leqq v}\chi(a)d_{k-1}(a)O\big((x/|a|)^{\eta}\big)\\ &\quad +O\big(x^{\eta/k}x^{(1-(1/k))(\delta-(\delta-\eta)/(k-1))}(\log x)^{k-3}\big),\end{aligned}$$

by the inductive hypothesis. By using Proposition 4.2.8 and an estimate for $\sum_{a\in G[x]}d_{k-1}(a)|a|^{-\eta}$ occurring in the proof of Proposition 4.3.2, it may then be deduced that

$$\begin{aligned}\sum_{a\in G[x]}\chi(a)d_k(a) &= O\big(x^{\delta-(\delta-\eta)/(k-1)}(\log x)^{k-3}x^{(\delta-\eta)/k(k-1)}\big)\\ &\quad +O\big(x^{\eta}x^{(1-1/k)(\delta-\eta)}(\log x)^{k-2}\big)\\ &\quad +O\big(x^{\delta-2(\delta-\eta)/k}(\log x)^{k-3}\big)\\ &= O\big(x^{\delta-(\delta-\eta)/k}(\log x)^{k-2}\big).\end{aligned}$$

Hence the inductive statement follows for all $k\geqq 3$.

The present lemma follows by combining the last conclusion with Proposition 4.3.2 and Lemma 2.5 above. □

6.3. Corollary. *For any class $H \in \Gamma$, as $x \to \infty$,*

$$N_H(d, x) = (A/h)x^\delta(\delta A \log x + 2\gamma_G - A) + O(x^{(\delta+\eta)/2}). \qquad \square$$

In particular, for an ordinary arithmetical progression

$$H_{m,r} = \{r, m+r, 2m+r, \ldots\}$$

in $G_{\mathbf{Z}}\langle m\rangle$, this corollary yields the conclusion

$$N_{H_{m,r}}(d, x) = \sum_{\substack{n \leqq x \\ n \equiv r \,(\mathrm{mod}\, m)}} d(n)$$

$$= m^{-1}x\big(m^{-1}\varphi(m)\log x + 2\gamma_m - m^{-1}\varphi(m)\big) + O(x^{1/2}),$$

where

$$\gamma_m = \lim_{x \to \infty} \Big\{ \sum_{\substack{n \leqq x \\ (n, m) = 1}} n^{-1} - m^{-1}\varphi(m)\log x \Big\}.$$

In order to continue with the proof of Theorem 6.1, now consider:

6.4. Corollary. *Let*

$$F(z) = d_k(mz) = [\zeta_G(mz)]^k,$$

where m and k are positive integers. Then, for any class $H \in \Gamma$,

$$N_H(F, x) = \begin{cases} (A/h)C(m, H)x^{\delta/m} + O(x^{\eta/m}) & \text{if } k = 1, \\ (A/h)C(m, H)x^{\delta/m} \sum_{i=1}^{k} c_i(m^{-1}\log x)^{k-i} & \\ + O\big(x^{(\delta - (\delta-\eta)/k)/m}(\log x)^{k-2}\big) & \text{if } k > 1. \end{cases}$$

Proof. For any $\chi \in X$,

$$\sum_{a \in G[x]} \chi(a)F(a) = \sum_{|a^m| \leqq x} \chi(a^m)\, d_k(a) = \sum_{|a| \leqq x^{1/m}} \chi^m(a) d_k(a).$$

Therefore, for $k=1$, Axiom A* implies that

$$\sum_{a \in G[x]} \chi(a)F(a) = \begin{cases} A x^{\delta/m} + O(x^{\eta/m}) & \text{if } \chi^m = 1, \\ O(x^{\eta/m}) & \text{if } \chi^m \neq 1. \end{cases}$$

On the other hand, for $k>1$, an estimate for $N(F, x)$ occurring in the proof of Theorem 4.3.1 and an estimate obtained in the proof of Lemma 6.2 imply that

$$\sum_{a\in G[x]} \chi(a)F(a) = \begin{cases} Ax^{\delta/m} \sum_{i=1}^{k} c_i(m^{-1}\log x)^{k-i} + \mathrm{O}\left(x^{(\delta-(\delta-\eta)/k)/m}(\log x)^{k-2}\right) & \\ & \text{if } \chi^m = 1, \\ \mathrm{O}\left(x^{(\delta-(\delta-\eta)/k)/m}(\log x)^{k-2}\right) & \text{if } \chi^m \neq 1. \end{cases}$$

By using these estimates and Lemma 2.5, one may now deduce the corollary. □

In view of the above conclusions, part (i) of Theorem 6.1 will follow from:

6.5. Lemma. *Let $f, F, g \in \mathrm{Dir}(G)$ be arbitrary functions such that $f(z)=F(z)g(z)$. Suppose that, for any $H\in\Gamma$,*

$$N_H(F, x) = Bx^{\alpha}(\log x)^r + \mathrm{O}\left(x^{\beta}(\log x)^s\right),$$

where $\alpha>0$, $0\leqq\beta\leqq\alpha$ and $r\geqq 0$, $s\geqq 0$ are constants with the property that r and s are integers and $\beta<\alpha$ if $r=0$, while $s<r$ if $\beta=\alpha$. Suppose also that $\tilde{g}(z)$ is absolutely convergent for all complex z with $\mathrm{Re}\, z>\nu$, where $\nu<\alpha$. Then as $x\to\infty$

$$N_H(f, x) = [B\tilde{g}(\alpha)+\mathrm{o}(1)]x^{\alpha}(\log x)^r.$$

Proof. This lemma is of course a generalization of Lemma 4.3.6. Firstly, for any $\chi\in X$,

$$\sum_{a\in G[x]} \chi(a)f(a) = \sum_{|bc|\leqq x} \chi(bc)F(b)g(c) = \sum_{|c|\leqq x} \chi(c)g(c) \sum_{|b|\leqq x/|c|} \chi(b)F(b).$$

Then, for $\chi=1$, Lemma 4.3.6 implies that

$$N(f, x) = h[B\tilde{g}(\alpha)+\mathrm{o}(1)]x^{\alpha}(\log x)^r \quad \text{as } x\to\infty.$$

Also, for $\chi\neq 1$, Lemma 2.5 and the assumption about $N_H(F, x)$ imply that

$$\sum_{a\in G[x]} \chi(a)F(a) = \mathrm{O}\left(x^{\beta}(\log x)^s\right).$$

Hence

$$\sum_{a\in G[x]} \chi(a)f(a) = \sum_{|c|\leqq x} \chi(c)g(c)\mathrm{O}\left((x/|c|)^{\beta}(\log (x/|c|))^s\right).$$

Then, as in the proof of Lemma 4.3.6, in the case $r=0$ it follows that $x^{\beta}(\log x)^{s}=O(x^{\beta'})$ for some β' with $\nu<\beta'<\alpha$. Since $\tilde{g}(z)$ is absolutely convergent for $\operatorname{Re} z>\nu$, this leads to the conclusion that

$$\sum_{a\in G[x]}\chi(a)f(a)=O(x^{\beta'})$$

when $r=0$. On the other hand, if $r>0$ then the hypotheses of the lemma imply that

$$x^{\beta}(\log x)^{s}=O\big(x^{\alpha}(\log x)^{s'}\big)$$

for some non-negative $s'<r$. Then the convergence property of $\tilde{g}(z)$ implies that

$$\sum_{a\in G[x]}\chi(a)f(a)=O\big(x^{\alpha}(\log x)^{s'}O(1)\big).$$

Therefore, in all cases, the required conclusion now follows from Lemma 2.5. □

Finally, part (ii) of Theorem 6.1 is a consequence of Corollary 6.4 and:

6.6. Lemma. *Let $f,F,g\in\operatorname{Dir}(G)$ be arbitrary functions such that $f(z)=F(z)g(z)$. Suppose that for any class $H\in\Gamma$*

$$N_H(F,x)=Bx^{\alpha}\sum_{i=0}^{r}a_i(\log x)^{r-i}+O\big(x^{\beta}(\log x)^{s}\big)$$

for certain constants B, a_i, $\alpha>0$, $0\leqq\beta<\alpha$ and r, s, where r and s are non-negative integers and all the constants are independent of H. Suppose also that

$$N(|g|,x)=O(x^{\nu})\quad\textit{for some }\nu<\alpha.$$

Then

$$N_H(f,x)=Bx^{\alpha}\sum_{i=0}^{r}a_i'(\log x)^{r-i}+\begin{cases}O\big(x^{\beta}(\log x)^{s}\big) & \textit{if }\beta>\nu,\\ O(x^{\nu+\varepsilon}) & \textit{if }\beta\leqq\nu,\end{cases}$$

where $\varepsilon>0$ is arbitrary,

$$a_0'=a_0\tilde{g}(\alpha),\qquad a_1'=ra_0\tilde{g}'(\alpha)+a_1\tilde{g}(\alpha),$$

and all the constants a_i' are independent of H.

Proof. By Lemma 4.5.2, which will now be generalized,

$$N(f,x)=hBx\sum_{i=0}^{r}a_i'(\log x)^{r-i}$$
$$+\begin{cases}O\big(x^{\beta}(\log x)^{s}\big) & \text{if }\beta>\nu,\\ O(x^{\nu+\varepsilon}) & \text{if }\beta\leqq\nu,\end{cases}$$

where $\varepsilon > 0$ is arbitrary and the a_i' are constants such that a_0' and a_1' have the values stated. Also, for $1 \neq \chi \in X$,

$$\sum_{a \in G[x]} \chi(a) f(a) = \sum_{|c| \leqq x} \chi(c) g(c) \mathrm{O}\big((x/|c|)^{\beta} (\log (x/|c|)^{s}\big)$$
$$= \mathrm{O}\big(x^{\nu} (\log x)^{s}\big) + \begin{cases} \mathrm{O}\big(x^{\nu} (\log x)^{s}\big) & \text{if } \nu > \beta, \\ \mathrm{O}\big(x^{\nu} (\log x)^{s+1}\big) & \text{if } \nu = \beta, \\ \mathrm{O}\big(x^{\beta} (\log x)^{s}\big) & \text{if } \nu < \beta, \end{cases}$$

since, as in the proof of Lemma 4.5.2,

$$\sum_{|c| \leqq x} g(c) |c|^{-\beta} = \mathrm{O}(x^{\nu-\beta}) + \begin{cases} \mathrm{O}(x^{\nu-\beta}) & \text{if } \nu > \beta, \\ \mathrm{O}(\log x) & \text{if } \nu = \beta, \\ \mathrm{O}(1) & \text{if } \nu < \beta. \end{cases}$$

Therefore the required result follows from Lemma 2.5. □

As in the case of Lemma 4.5.2, the error estimates in the conclusion of the above lemma are easily sharpened when no logarithms occur in the hypotheses; this time we state the sharper conclusion explicitly:

6.7. Corollary. *Let f, F and g be as above, but with the assumed estimate for* $N_H(F, x)$ *simplified to*

$$N_H(F, x) = Bx^{\alpha} + \mathrm{O}(x^{\beta}) \quad \text{as } x \to \infty.$$

Then

$$N_H(f, x) = B\tilde{g}(\alpha) x^{\alpha} + \begin{cases} \mathrm{O}(x^{\nu}) & \text{if } \nu > \beta, \\ \mathrm{O}(x^{\beta} \log x) & \text{if } \nu = \beta, \\ \mathrm{O}(x^{\beta}) & \text{if } \nu < \beta. \end{cases}$$

In particular, if $F(z) = \zeta_G(z)$ *so that* $f(z) = \zeta_G(z) g(z)$, *then*

$$N_H(f, x) = (A/h)\tilde{g}(\delta) x^{\delta} + \begin{cases} \mathrm{O}(x^{\nu}) & \text{if } \nu > \eta, \\ \mathrm{O}(x^{\eta} \log x) & \text{if } \nu = \eta, \\ \mathrm{O}(x^{\eta}) & \text{if } \nu < \eta. \end{cases}$$

□

The last conclusion is of course closely related to that of the transfer principle 4.4, but it was convenient to have derived that principle by a direct argument earlier. It may also be remarked at this point that, if desired, one could as in Chapter 4, § 5, replace functions such as f, g in Theorem 6.1 by functions f^*, g^*, and so on, with corresponding properties in Dir ($|G|$), and then obtain corresponding estimates over the semigroup $|G|$. (The resulting slightly generalized propositions could perhaps again be indicated by an asterisk * if the need arose; the reader may perhaps like to investigate the detailed statements and proofs of such propositions, as an exercise.)

As another exercise, it may be left to the reader to apply the above general results to various special arithmetical functions of the kinds treated in Chapters 4 and 5. (Such applications are also given in the author's paper [9].)

§ 7. Abstract prime number theorem for formations

One can state and prove an abstract prime number theorem regarding the classes of formations satisfying only Axiom A* or even slightly weaker axioms; see Forman and Shapiro [1], Amitsur [1] and Müller [1, 2]. However, this leads to mild technical complications in the statement and proof which need not be considered if one is concerned mainly with applications to the specific types of formations discussed or referred to in § 4. In order to simplify matters, it is therefore convenient to introduce the following extra "technical" axiom, which turns out to be valid for the formations mentioned.

Axiom A**. $L_G(\delta, \chi) \neq 0$ *for every character* $\chi \in X$ *such that* $\chi^2 = 1$.

We may then state

7.1. Abstract Prime Number Theorem. *Suppose that the formation* (G, Γ) *satisfies both Axioms* A* *and* A**. *Let* H *denote an arbitrary class in* Γ, *and let* $\pi_H(x)$ *denote the total number of primes* $p \in H[x]$. *Then*

$$\pi_H(x) \sim x^\delta/(h\delta \log x) \quad \textit{as } x \to \infty.$$

The most immediate corollary of this theorem is the classical Prime Number Theorem for Arithmetical Progressions quoted at the beginning of the chapter. (Of course, here and in the other illustrations mentioned below the verification of Axiom A** remains to be discussed.) As further corollaries, one may deduce Landau's *Relative Prime Ideal Theorems* for ideal class formations over algebraic number fields (if the relevant theorems leading to the verification of Axioms A* and A** are taken as established).

For example, if H denotes any ideal class of the absolute ideal class formation over the number field K, then

$$\pi_H(x) \sim x/h_K \log x \quad \text{as } x \to \infty.$$

Other interesting applications of Theorem 7.1 occur in the direction of asymptotic enumeration questions concerning some of the categories discussed earlier. Here, as in earlier cases, the conclusions have an interest which is in addition to or independent of the abstract number-theoretical background. In particular, applications of this kind may be obtained by combining Theorem 7.1 with Theorems 4.5 to 4.7 which imply the validity of Axiom A* in certain cases, and results establishing Axiom A** for those cases. For example, such a procedure leads to a *Relative Simple Ring Theorem* pertaining to any generalized arithmetical progression $H_{m,r}(\mathfrak{S})$ as defined earlier. This states that

$$\pi_{H_{m,r}(\mathfrak{S})}(x) \sim x/(\varphi(m) \log x) \quad \text{as } x \to \infty.$$

Now consider any class H of the abstract formation (G, Γ) satisfying Axioms A* and A**. Our method of proof will be an extension of that used in Chapter 6. Firstly, define

$$\theta_H(x) = \sum_{p \in P \cap H[x]} \log |p|, \qquad \psi_H(x) = N_H(\Lambda, x),$$

where Λ is the von Mangoldt function. The next lemma shows that Theorem 7.1 follows from either of the conclusions

$$\theta_H(x) \sim (h\delta)^{-1} x^\delta, \qquad \psi_H(x) \sim (h\delta)^{-1} x^\delta, \qquad \text{as } x \to \infty.$$

7.2. Lemma. *As* $x \to \infty$,

$$\psi_H(x) = \theta_H(x) + O(x^{\delta/2} (\log x)^2).$$

Further, if $\theta_H(x) \sim (h\delta)^{-1} x^\delta$ *or* $\psi_H(x) \sim (h\delta)^{-1} x^\delta$ *as* $x \to \infty$, *then*

$$\pi_H(x) \sim x^\delta/(h\delta \log x) \quad \textit{as } x \to \infty.$$

Proof. For $1 < y \leqq x$,

$$\begin{aligned} \pi_H(x) &= \pi_H(y) + \sum_{\substack{p \in P \cap H \\ y < |p| \leqq x}} 1 \\ &\leqq \pi_H(y) + \sum_{\substack{p \in P \cap H \\ y < |p| \leqq x}} \frac{\log |p|}{\log y} \\ &\leqq O(y^\delta) + \frac{\theta_H(x)}{\log y}. \end{aligned}$$

If x is sufficiently large and $y=x(\log x)^{-2/\delta}$, one then obtains

$$\theta_H(x) \leqq \pi_H(x)\log x \leqq \mathrm{O}\left(\frac{x^\delta}{\log x}\right)+\theta_H(x)\log x/\log y,$$

and this leads to the second conclusion (compare the proof of Lemma 6.1.1), since

$$0 \leqq \psi_H(x)-\theta_H(x) \leqq \psi(x)-\theta(x) = \mathrm{O}\left(x^{\delta/2}(\log x)^2\right),$$

by Chapter 6, § 1. □

In view of Lemma 7.2, attention may now be restricted to $\psi_H(x)$. Firstly, note that, by the Orthogonality Lemma 2.6 and Corollary 2.6.4,

$$\begin{aligned}\sum_{a\in H} \Lambda(a)a^{-z} &= \sum_{a\in G} \Lambda(a)h^{-1} \sum_{\chi\in X} \chi(a)\bar{\chi}(H)a^{-z}\\ &= h^{-1} \sum_{\chi\in X}\bar{\chi}(H)\Lambda(z,\chi)\\ &= -h^{-1} \sum_{\chi\in X} \bar{\chi}(H)\zeta_G'(z,\chi)/\zeta_G(z,\chi).\end{aligned}$$

Writing

$$\Lambda_H(z) = \sum_{a\in H} \Lambda(a)a^{-z},$$

it follows that

$$\tilde{\Lambda}_H(z) = -h^{-1} \sum_{\chi\in X} \bar{\chi}(H)L_G'(z,\chi)/L_G(z,\chi).$$

By Lemma 4.3.5 and Proposition 5.1, this equation remains true for all complex z with Re $z>\delta$, and as in the special case $\Gamma=\{1\}$ our next aim is to consider the entire region Re $z\geqq\delta$.

7.3. Lemma. *For any real t, and $\chi\in X$,*

$$L_G(\delta+it,\chi) \neq 0.$$

Proof. Similarly to Lemma 6.1.3, which the present lemma generalizes, consider complex w with Re $w>\delta$, and (using the principal value of the logarithm) observe that the Euler product formula for $L_G(w,\chi)$ $(\chi\in X)$ now gives

$$L_G(w,\chi) = \exp\left\{\sum_{a\in G} \frac{\chi(a)\Lambda(a)}{\log|a|}|a|^{-w}\right\}.$$

Then, for real $\sigma>\delta$ and any real $t\neq 0$,

$$|\zeta_G{}^3(\sigma)L_G{}^4(\sigma+it,\chi)L_G(\sigma+2it,\chi^2)| =$$

$$= \left|\exp\left\{\sum_{a\in G}\frac{\Lambda(a)}{\log|a|}|a|^{-\sigma}\{3+4\chi(a)|a|^{-it}+\chi^2(a)|a|^{-2it}\}\right\}\right|$$

$$= \exp\left\{\sum_{a\in G}\frac{\Lambda(a)}{\log|a|}|a|^{-\sigma}\,\mathrm{Re}\,\{3+4\chi(a)|a|^{-it}+\chi^2(a)|a|^{-2it}\}\right\}$$

$$\geqq \exp 0 = 1,$$

by the inequality $\mathrm{Re}\,(3+4w+w^2)\geqq 0$ $(|w|=1)$. Hence

$$|(\sigma-\delta)\zeta_G(\sigma)|^3|(\sigma-\delta)^{-1}L_G(\sigma+it,\chi)|^4|L_G(\sigma+it,\chi^2)| \geqq (\sigma-\delta)^{-1}.$$

Just as in Lemma 6.1.3, the existence of a zero value for $L_G(\delta+it,\chi)$ would then lead to a contradiction. For $t=0$, the above argument would fail if $\chi^2=1$; but then Axiom A** gives the required conclusion. Hence the lemma follows. □

It follows now that $\tilde{\Lambda}_H(z)$ may be extended to an analytic function of z for all $z\neq\delta$ with $\mathrm{Re}\,z\geqq\delta$. Further, by Lemma 6.1.2, $\tilde{\Lambda}(z)$ has a simple pole with residue 1 at $z=\delta$, and therefore

$$\tilde{\Lambda}_H(z) = h^{-1}\tilde{\Lambda}(z)-h^{-1}\sum_{1\neq\chi\in X}\bar{\chi}(H)L_G'(z,\chi)/L_G(z,\chi)$$

has a simple pole with residue h^{-1} at $z=\delta$. Therefore Lemma 6.2.2 is applicable once more and yields the conclusion

$$\psi_H(x) = N_H(\Lambda, x) \sim (h\delta)^{-1}x^{\delta} \quad \text{as } x\to\infty.$$

This proves Theorem 7.1. □

In order to establish the specific corollaries of Theorem 7.1 referred to earlier, it remains to examine the validity of Axiom A** in various concrete situations. For this purpose, two auxiliary results will be required.

7.4. Lemma. *Let (G,Γ) denote a formation satisfying Axiom* A* *with $\delta=1$ and such that the norm mapping on G is integer-valued. Suppose that each L-series $L_G(z,\chi)$ $(\chi\in X)$ can be continued analytically into the entire region* $\mathrm{Re}\,z>\alpha$, *for some $\alpha<\frac{1}{2}$, with the exception of the point $z=1$ in the case of $\zeta_G(z)$. Then (G,Γ) satisfies Axiom* A**.

Proof. Let $\chi \in X$ be a character such that $\chi^2 = 1$, and consider the product

$$\zeta_G(z) L_G(z, \chi) = \tilde{f}(z),$$

where

$$f(a) = \sum_{d|a} \chi(d) = \prod_{\substack{p \in P \\ p|a}} (1 + \chi(p) + \dots + \chi(p^{\alpha_p})),$$

if $a = \prod_{p \in P} p^{\alpha_p}$. Then $f(a) \geqq 0$ and $f(a^2) \geqq 1$. Therefore, if σ is real and both the series converge,

$$\tilde{f}(\sigma) = \sum_{a \in G} f(a)|a|^{-\sigma} \geqq \sum_{a \in G} |a|^{-2\sigma} = \zeta_G(2\sigma)$$

$$\to \infty \quad \text{as } \sigma \to \tfrac{1}{2}+.$$

If $L_G(1, \chi) = 0$, the pole of $\zeta_G(z)$ at $z = 1$ will be "absorbed" by $L_G(z, \chi)$ in the defining product for $\tilde{f}(z)$, and so, by the hypotheses of the lemma, $\tilde{f}(z)$ will be analytic throughout the region $\operatorname{Re} z > \alpha$. Then, by the next lemma, the series for $\tilde{f}(\sigma)$ must converge for all $\sigma > \alpha$. This gives a contradiction of the preceding conclusion regarding $\sigma \to \frac{1}{2}+$. □

7.5. Lemma. *Let*

$$A(z) = \sum_{n=1}^{\infty} A_n n^{-z}$$

be an ordinary Dirichlet series with real coefficients $A_n \geqq 0$. *Suppose that this series converges whenever* $\operatorname{Re} z > \sigma_0$ *and that* $A(z)$ *is an analytic function of* z *at the point* $z = \sigma_0$. *Then the series* $\sum_{n=1}^{\infty} A_n n^{-z}$ *remains convergent whenever* $\operatorname{Re} z > \sigma_0 - \varepsilon$, *for some* $\varepsilon > 0$.

Proof. For $\varepsilon > 0$ and $\sigma_0 < \sigma < \sigma_0 + \varepsilon$,

$$A(\sigma) = \sum_{n=1}^{\infty} A_n \exp[-\sigma \log n]$$

$$= \sum_{n=1}^{\infty} A_n \exp[-(\sigma - \sigma_0 - \varepsilon) \log n] \exp[(-\sigma_0 + \varepsilon) \log n].$$

Every term $\exp[-(\sigma - \sigma_0 - \varepsilon) \log n]$ can be expanded as a power series in $w = -(\sigma - \sigma_0 - \varepsilon) \log n$. Then all the coefficients in the resulting double series are non-negative, and so the double series can be rewritten as a power series in $u = \sigma - \sigma_0 - \varepsilon$. Since $A(z)$ is analytic at $z = \sigma_0$, when $\varepsilon > 0$ is sufficiently small, this power series in u provides the Taylor expansion of $A(z)$ about σ_0. Therefore it is also convergent for $\sigma_0 - \varepsilon < \sigma \leqq \sigma_0$.

Now choose any σ in the latter range and reverse the above process so as to convert the power series in $\sigma-\sigma_0-\varepsilon$ into the given Dirichlet series for $A(\sigma)$. Then it follows that the latter series is convergent for such σ. Hence $\sum_{n=1}^{\infty} A_n n^{-z}$ is absolutely convergent for $\operatorname{Re} z>\sigma_0-\varepsilon$. □

By Propositions 4.2.13 and 5.1 above, one could choose $\alpha=\eta$ in Lemma 7.4 if $\eta<\frac{1}{2}$. Since $\eta<\frac{1}{2}$ for the classical formation $(G_{\mathbf{Z}}\langle m\rangle, \Gamma_{\mathbf{Z}}\langle m\rangle)$, it is therefore now possible to deduce the Prime Number Theorem for ordinary arithmetical progressions.

It is known that the L-series for ideal class formations over a given algebraic number field K (as well as those for ordinary arithmetical progressions) can actually be continued analytically over the entire complex plane, with the exception of the point $z=1$ for the zeta functions; see for example Hasse's *Bericht* [1]. Also, Landau [8] proved theorems which imply the validity of Axiom A* with $\delta=1$ for these formations. In this way, Theorem 7.1 may then be applied to ideal class formations, although it should be noted that here Landau [8] derived the corresponding conclusions by direct arguments (which also yield error-term estimates).

Next consider the formation structure over the category $\mathscr{A}\langle m\rangle$ discussed earlier. Here

$$L_{\mathscr{A}\langle m\rangle}(z, \chi|\ |) = \prod_{k=1}^{\infty} L(kz, \chi^k) = L(z, \chi)L(2z, \chi^2)F_0(z, \chi),$$

where

$$F_0(z, \chi) = \prod_{k=3}^{\infty} L(kz, \chi^k).$$

Then

$$F_0(z, 1) = \zeta_{\mathscr{A}'}(z),$$

where $\mathscr{A}'$ denotes the category $\mathfrak{F}^{\langle k\rangle}$ discussed in Chapter 5, § 1, in the case when $\mathfrak{F}=\mathscr{A}$ and $\langle k\rangle=(3, 4, 5, \ldots)$. A slight modification of the proof of Lemma 5.1.4 and of the discussion of uniform convergence in the proof of Proposition 4.2.1 enable one to deduce that the product representation for $F_0(z, \chi)$ holds analytically for $\operatorname{Re} z>\frac{1}{3}$, and that $F_0(z, \chi)$ is an analytic function of z in this region. Since $L(z, \chi)$ and $L(2z, \chi^2)$ can also be continued analytically into the region $\operatorname{Re} z>\frac{1}{3}$, it follows that in the present case one may choose $\alpha=\frac{1}{3}$ for the purposes of Lemma 7.4. (By considering successive factorizations

$$L_{\mathscr{A}\langle m\rangle}(z, \chi|\ |) = L(z, \chi)L(2z, \chi^2)\ldots L(Nz, \chi^N) \prod_{k=N+1}^{\infty} L(kz, \chi^k),$$

it may be deduced similarly that $\alpha=N^{-1}$ is a suitable choice, for any $N>2$; hence $\alpha=0$ is also a suitable choice.)

In this situation, one therefore obtains the conclusion

$$\pi_{H_{m,r}(\mathscr{A})}(x) \sim x/(\varphi(m)\log x) \quad \text{as } x \to \infty,$$

where $H_{m,r}(\mathscr{A})$ is a generalized arithmetical progression of abelian groups as defined earlier.

A similar argument may be applied to the generalized arithmetical progressions over the category $\mathfrak{S}\langle m\rangle$. Here

$$L_{\mathfrak{S}\langle m\rangle}(z, \chi|\ |) = L_{\mathscr{A}\langle m\rangle}(z, \chi|\ |)F(z, \chi),$$

where

$$F(z, \chi) = \prod_{k=1}^{\infty} \prod_{s=2}^{\infty} L(ks^2 z, \chi^{ks^2}).$$

Then $F(z, 1)=\zeta_{\mathfrak{S}'}(z)$, where $\mathfrak{S}'$ is the subcategory of $\mathfrak{S}$ referred to in the proof of Lemma 5.1.8. By now slightly modifying the discussion of $F_0(z, \chi)$ above, it may be deduced that $F(z, \chi)$ is an analytic function of z when $\operatorname{Re} z>\frac{1}{4}$ (and that the above product representation is valid analytically in this region). Therefore here one may substitute $\alpha=\frac{1}{4}$ in Lemma 7.4, and this leads to the 'Relative Simple Ring Theorem' stated at the beginning of the section.

Formation structures over the category $\mathfrak{F}$, such as the absolute formation $(G_{\mathfrak{F}}, \Gamma_{\mathfrak{F}})$, may be dealt with similarly on the basis of previous statements about ideal class formations.

The above discussion completes our treatment of some basic aspects of the theory of arithmetical formations. As an exercise, the reader might perhaps care to extend some of the other results of Chapter 6 to the context of this section. (A treatment of such results, based on slightly different preliminary theorems, is given in the author's paper [9].)

Lastly, it is worth noting that it can be of interest to derive error-term estimates in relation to Theorem 7.1. Sharp, and also more general, abstract results of this type have been obtained by Müller [1, 2], while corresponding conclusions for various specific arithmetical categories, which are sometimes still sharper, appear in the author's paper [8]. (The latter paper uses an elementary 'transfer principle' to deduce results from previously known theorems about ordinary arithmetical progressions and ideal classes in number fields, as given by Landau [8, 11], Mitsui [1] and Walfisz [1] for example.)

Selected bibliography for Chapter 9

Sections 1 *and* 2: Amitsur [1], Forman and Shapiro [1], Hayes [1], Knopfmacher [8, 9].

Sections 3 *and* 4: Duttlinger [2], Knopfmacher [8, 9], Landau [6, 8, 9], Richert [1], Šapiro–Pyateckiĭ [1].

Sections 5 *and* 7: Amitsur [1], Bredihin [6, 7], Fogels [1], Forman and Shapiro [1], Hecke [2], Knopfmacher [8, 9], Landau [8], Müller [1, 2], Rémond [1].

Section 6: Knopfmacher [9].

APPENDIX 1

SOME UNSOLVED QUESTIONS

This final section contains a selection of questions or projects relating to the subject-matter of this book, that were apparently still open at the time (June, 1974) of completion of the manuscript. The questions are not classified, or graded in terms of difficulty; some may be trivial or only of a fairly routine nature for a reasonably competent and interested mathematician, but at least a few seem to be more difficult or even out of range at present. Certain problems are at least partly of a non-specific conceptual type, and so the nature of their solutions will to some extent depend on the interpretations of individual research workers. Lastly, it should be emphasized that the selected problems recorded below were intended to have some bearing on *abstract* analytic number theory or the application of classical methods of analytic number theory to *other* parts of mathematics; with minor exceptions, they do not include well-known unsolved questions of classical analytic or algebraic number theory.

1. Develop an analytical theory of 'arithmetical morphisms' which would embrace such things as, for example:

(i) the examples of arithmetical 'isomorphisms' and 'monomorphisms' that occur (implicitly) in the present monograph and in the author's papers [3—8],

(ii) the transfer principle of Chapter 9, § 4, and the 'elementary transfer principles' of the author's paper [8],

(iii) zeta functions and L-series of morphisms,

(iv) 'abstract prime number theorems' for morphisms subject to suitable hypotheses.

If possible, relate the concept of arithmetical morphism to that of a (suitable) functor between arithmetical categories.

2. In the theory of finite graphs and certain closely related fields, and in the study of polynomials in several indeterminates over a finite field, one comes across natural examples of arithmetical semigroups whose norm or degree mappings are really derived from prior vector-valued norm or degree mappings. For example, the total degree of a polynomial is the sum of its degrees in the individual variables, while finite graphs are often assumed to be "coloured" in two or more different ways or are classified in terms of both the number of vertices and the number of edges in a given graph. Develop a theory of 'higher-dimensional' arithmetical semigroups which would include such and other cases. (For some initial facts about graphs or polynomials, see especially Harary and Palmer [1] and S. D. Cohen [2, 3].)

3. In various mathematical fields, one finds interesting natural examples of "infinite" or "quasi" arithmetical semigroups, i.e., systems satisfying all the initial hypotheses about an arithmetical semigroup except that concerning the finiteness of $N_G(x)$. For example, polynomial rings in one or more indeterminates over an infinite field give rise to such semigroups, and so do many categories of mathematical objects for which a form of the Krull–Schmidt theorem is valid. By perhaps using forms of integration over suitable measure spaces, is it possible in at least some cases to introduce additional measures of "size" for the elements of such semigroups, in such a way that it becomes nevertheless possible to develop interesting quantitative studies of "averages" and "densities" for the systems in question? (Amongst the systems one might perhaps investigate initially are the polynomial rings $F[t]$, where F is one of the fields $\mathbf{Q}$, $\mathbf{R}$ or $\mathbf{C}$.)

4. There are so many interesting enumeration questions regarding natural categories of mathematical objects that occur in various parts of mathematics, particularly abstract algebra, and appear to be virtually out of range at present, that it would be futile to list more than a minute sample. Usually, even though it is clear that some particular category is an arithmetical category, present knowledge is insufficient for an effective calculation of the terms of the category's Euler product formula — not to mention the derivation of asymptotic formulae. As specific examples, consider the categories $\mathscr{G}$ and $\mathscr{R}$ of all finite groups, or all finite rings, respectively. Find asymptotic formulae for $N_{\mathscr{G}}(x)$ and $N_{\mathscr{R}}(x)$ in terms of elementary functions like exponentials and logarithms. Do the same for nilpotent finite groups and nilpotent finite rings. In particular, for $\mathscr{G}$ and the last two cases, do the relevant enumerating functions for objects of cardinal n possess finite asymptotic mean-values? Similar questions may be posed for commutative finite rings.

In connection with these particular examples, the cases of rings and of nilpotent groups should be more amenable than that of arbitrary finite groups. For, finite rings have unique decompositions into direct sums of rings of prime-power order while nilpotent finite groups decompose uniquely into direct products of p-groups, and some asymptotic enumeration theorems are already available for rings and groups of prime-power order; see Higman [1], Sims [1], Kruse and Price [1, 2] and the author [2, 4, 5, 7]. (The reader who wishes to pursue this further should note that for finite p-rings the asymptotic enumeration theorems available at present provide only asymptotic inequalities; it was pointed out in the author's papers [4, 7] that there is a gap in the proof leading to certain apparently sharper results for p-rings in Kruse and Price [1, 2].) A general upper bound for the number of non-isomorphic finite groups of a given order is provided by Gallagher [1]; see also Dickenson [1], Doornhof and Spitznagel [1], and Neumann [1], for certain further asymptotic results about finite groups.

5. There are various other open problems relating to the asymptotic enumeration theorems for finite p-groups and p-rings referred to in the previous question. Firstly, sharpen the asymptotic inequalities that were mentioned to genuine asymptotic formulae. Then note that even when this has been achieved, the resulting formulae will to begin with provide only logarithmic estimates; the same is true for the Higman–Sims theorem about finite p-groups. Consequently, it would be of interest to find still sharper asymptotic estimates in all these cases. (There are a number of additional asymptotic inequalities, providing partial asymptotic enumeration theorems for a number of categories besides those mentioned above, in the author's papers [4, 7]; similar questions may be posed for those cases as well.)

Next, given the asymptotic enumeration results for the arithmetical categories just referred to, can these be used to deduce abstract prime number theorems for those categories? In other words, can the given results be used for the purpose of deducing asymptotic enumeration theorems for the indecomposable objects in the relevant arithmetical categories? Such results, and a fuller arithmetical theory along the general lines of Chapter 8, § 5, for example, should be obtainable as corollaries of suitable general 'Tauberian' and 'Abelian' theorems. Establish such theorems.

6. Derive general Tauberian and Abelian theorems relating to Axiom C of Chapter 8, which are sharper than those discussed in Chapter 8 and hence would lead to sharper asymptotic enumeration theorems, as well as to sharper or more complete results along the lines of Chapter 8, § 5 in the context of Axiom C.

7. As far as possible, sharpen the asymptotic estimates for the various special enumerative functions discussed in Chapter 8, as well as for certain similar generalized partition functions considered in the author's papers [3, 5, 7] along the general lines laid down by Hardy and Ramanujan [2] and Rademacher [1, 2, 4] for the classical partition function $p(n)$ in particular.

8. The author's note [12] contains some asymptotic enumeration results for finite abelian Π-groups, semisimple finite Π-rings and related types of objects. Derive these results directly from the known ones for finite abelian p-groups, semisimple finite p-rings and other corresponding objects. (This is related to that part of Question 4 above which suggested that the known facts about finite p-groups or p-rings should lead to asymptotic enumeration theorems about general finite nilpotent groups, or about finite rings of various kinds.)

9. In Chapter 4, § 4 and Chapter 5, § 1 a number of special arithmetical functions, such as $a(n)$, $S(n)$ and β, for example, were shown to possess finite asymptotic k^{th} moments for every $k=1, 2, \ldots$. If f denotes one of these particular functions, investigate the asymptotic behaviour of the asymptotic k^{th} moment $m(f^k)$ as $k\to\infty$.

The author's note [14] provides a somewhat indirect proof that, when $G=G_{\mathrm{Z}}$ and f is one of the functions just referred to, $m(f^k)$ coincides with the kth moment of the asymptotic distribution function of f. Obtain sufficient conditions for functions on a general arithmetical semigroup G satisfying Axiom A which would simultaneously lead to more direct proofs of the results just mentioned and yield similar conclusions for such functions as β, d/d_* and d_*/d in the more general case.

10. Find the "true" maximum orders of magnitude of the various special functions whose indicators were determined in Chapter 5, § 2, and in the author's papers [8, 10]. Determine the indicator $\tau(K)$ of K when K is a non-abelian algebraic number field, and under the same situation find the indicators of the categories $\mathfrak{F}$ and $\mathfrak{S}_D$; see the theorems about these categories in the author's paper [10], and also consider similar questions for the 'relative' indicators discussed there. In this direction, compare also some of the results of Heppner [1, 2], and Schwarz and Wirsing [1].

11. Regarding the well-known Riemann hypothesis and its generalizations within classical analytic and algebraic number theory, one might ask whether there exists an arithmetical semigroup G satisfying Axiom A whose norm mapping is *integer-valued* and is such that $\zeta_G(z)$ has zeros arbitrarily close to the line $\operatorname{Re} z=\delta$? In this connection, compare the results of Malliavin [1] and R. S. Hall [1] and also the discussion in § 12 of Bateman and Diamond [1]. For discussions of similar types of questions, concerned especially with possible deductions subject to hypotheses of a weaker kind than Axiom A, and with delicate breaking-points in the analogy between ordinary and "generalized" integers, see in addition Amitsur [1], Beurling [1], Bredihin [1—8], Diamond [1—4], R. S. Hall [2, 3], Rémond [1], Ryavec [1, 2] and Segal [1] (amongst others). (The emphasis of most of these articles is on obtaining analytical or asymptotic conclusions under the weakest possible hypotheses, without reference to possible applications outside pure analysis or classical number theory.)

12. As far as possible, sharpen the asymptotic estimates for $N_{\mathfrak{F}}(x)$ and $N_{\mathfrak{S}_D}(x)$ given in Chapter 5 and in the author's paper [7] along the general lines available for $N_{\mathscr{A}}(x)$ as a result of the work of Kendall and Rankin [1], Richert [1], P. G. Schmidt [1], Schwarz [1] and Srinivasan [1]; regarding $\mathfrak{S}_{\mathbf{Z}}$, see the author [5, 8] and Duttlinger [1]. In connection with $\mathfrak{F}$ and $\mathfrak{S}_D$ when K is an arbitrary algebraic number field, it should be noted that this question is closely related to that concerning best possible asymptotic estimates for $N_K(x)$, regarding which particular reference may be made to Landau [9], Narkiewicz [1] and Joris [1]. Similar questions may be asked in relation to enumeration over the generalized ideal classes in $\mathfrak{F}$ and $\mathfrak{S}_D$, which were discussed in Chapter 9 and in the author's paper [8]. Regarding $\mathscr{A}$ and $\mathfrak{S}_{\mathbf{Z}}$ in this respect, reference may be made to Richert [1], Šapiro–Pyateckiĭ [1] and the author [8]. Again, for a general number field K, such questions for $\mathfrak{F}$ and $\mathfrak{S}_D$ are related to similar ones for K; in that case, see especially Landau [8] and Joris [2] in addition to the other references above. See also Duttlinger [2].

13. Establish asymptotic enumeration theorems for the "generalized arithmetical progressions" in such categories as $\mathscr{A}$ and $\mathfrak{S}=\mathfrak{S}_{\mathbf{Z}}$, which consist of all the objects of cardinal $\equiv k \pmod q$, in the case when k and q are *not* coprime. Concerning $\mathscr{A}$, see the reviewer's comments in *Mathematical Reviews* regarding the paper of Šapiro–Pjateckiĭ [1] and also Duttlinger [2].

14. An asymptotic formation-type theory over the polynomial ring $D = \mathrm{GF}[q, t]$ has been developed by Hayes [1]. Derive a similar theory for the corresponding arithmetical categories $\mathfrak{F}(D)$ and $\mathfrak{S}(D)$ discussed in the author's paper [7]. As a special case, this would include asymptotic arithmetical theories for the entire categories in each case. Here a starting point would be provided by the results about D discussed in Chapter 3, and by Carlitz [1, 2], S. D. Cohen [5] and Shader [1].*

15. Novoselov [1] has developed a genuinely probabilistic theory of ordinary arithmetical functions over $G_{\mathbf{Z}}$ with the aid of a certain type of compactification of the ring $\mathbf{Z}$ of all rational integers, and the Haar measure on this compactification. Develop a similar theory for functions on an arbitrary arithmetical semigroup G satisfying Axiom A, with the aid of the almost even compactification G^* and its probability measure ν, if possible.

16. Let f denote an arithmetical function over an arithmetical semigroup G satisfying Axiom A. Obtain convenient "number-theoretical" necessary and sufficient conditions that f should be uniformly almost even, or almost even (B^λ) for a given λ. (See the references to papers given at the end of Chapter 7.)**

17. (Suggested to the author by A. M. Sinclair.) Given an arithmetical formation (G, Γ), can one in some canonical way associate with it an arithmetical semigroup $G_\#$ in such a way that:

(i) $G_\#$ satisfies Axiom A if and only if (G, Γ) satisfies Axiom A* and (perhaps) Axiom A**,

(ii) asymptotic results like the abstract prime number theorem for (G, Γ) subject to Axioms A* and A** become simple corollaries of the corresponding results for $G_\#$ subject to Axiom A?

For some indirect hints at this possibility, compare the relationship between the prime number theorem for $(G_{\mathbf{Z}}\langle 4\rangle, \Gamma_{\mathbf{Z}}\langle 4\rangle)$ and that for $G_{\mathbf{Z}[\sqrt{-1}]}$ outlined in Chapter 5, § 2, and compare also Landau [12] and Shapiro [2].

18. (Suggested to the author by B. Volkmann.) Develop a "comparative prime number theory" for the various formation structures relative to such categories as $\mathscr{A}$, $\mathfrak{S}$, and so on, along the general lines of papers like those of Knapowski and Turán [1, 2] and Grosswald [1, 2] in particular.

* See the author [16].

** See Schwarz and Spilker [2].

APPENDIX 2

VALUES OF $p(n)$ AND $s(n)$

n	$p(n)$	$s(n)$	n	$p(n)$	$s(n)$
1	1	1	26	2 436	5 523
2	2	2	27	3 010	7 000
3	3	3	28	3 718	8 922
4	5	6	29	4 565	11 235
5	7	8	30	5 604	14 196
6	11	13	31	6 842	17 777
7	15	18	32	8 349	22 336
8	22	29	33	10 143	27 825
9	30	40	34	12 310	34 720
10	42	58	35	14 883	43 037
11	56	79	36	17 977	53 446
12	77	115	37	21 637	65 942
13	101	154	38	26 015	81 423
14	135	213	39	31 185	100 033
15	176	284	40	37 338	122 991
16	231	391	41	44 583	150 481
17	297	514	42	53 174	184 149
18	385	690	43	63 261	224 449
19	490	900	44	75 175	273 614
20	627	1197	45	89 134	332 291
21	792	1549	46	105 558	403 445
22	1002	2025	47	124 754	488 273
23	1255	2600	48	147 273	590 824
24	1575	3377	49	173 525	712 728
25	1958	4306	50	204 226	859 381

LIST OF SPECIAL SYMBOLS

(IN ORDER OF APPEARANCE)

BIBLIOGRAPHY

(Books are indicated by means of asterisks.)

AHERN, P. R.

[1] The asymptotic distribution of prime ideals in ideal classes. *Proc. Koninkl. Nederl. Akad. Wetensch.* (A) **67** (1964) 10–14.

[2] Elementary methods in the theory of primes. *Trans. Am. Math. Soc.* **118** (1965) 221–242.

ALBERT, A. A.

*[1] *Structure of Algebras*, Am. Math. Soc. Colloq. Publ. **24** (Am. Math. Soc., Providence, R. I, 1961).

AMICE, Y. and J. FRESNEL

[1] Fonctions zêta p-adiques des corps de nombres abèliens réels. *Acta Arith.* **20** (1972) 353–384.

AMITSUR, S. A.

[1] Arithmetic linear transformations and abstract prime number theorems. *Can. J. Math.* **13** (1961) 83–109, **21** (1969) 1–5.

[2] On a lemma in elementary proofs of the prime number theorem. *Bull. Res. Council Israel* (F) **10** (1962) 101–108.

ANDERSON, D. R. and T. M. APOSTOL

[1] The evaluation of Ramanujan's sum and generalizations. *Duke Math. J.* **20** (1953) 211–216.

ARCHIBALD, R. G.

[1] Highly composite ideals. *Trans. Roy. Soc. Canada* **30** (III) (1936) 41–47.

ARTIN, E.

[1] Quadratische Körper im Gebiete der höreren Kongruenzen, I–II. *Math. Z.* **19** (1924) 153–246.

AULUCK, F. C. and C. B. HASELGROVE

[1] On Ingham's Tauberian theorem for partitions. *Proc. Cambridge Philos. Soc.* **48** (1952) 566–570.

BAĬBULATOV R. S.

[1] The Erdös—Wintner theorem for normed semigroups. *Dokl. Akad. Nauk. UzSSR* (8) (1967) 3–6 (in Russian); *Math. Rev.* **46** (1973) #5273.

BASS, H.

*[1] *Algebraic K-Theory* (Benjamin, New York, 1968).

BATEMAN, P. T.

[1] The distribution of values of the Euler function. *Acta Arith.* **21** (1972) 329–345.

BATEMAN, P. T. and H. G. DIAMOND

[1] Asymptotic distribution of Beurling's generalized prime numbers. In: W. LeVeque, ed., *Studies in Number Theory,* MAA Stud. in Math. Vol. **6** (Prentice-Hall, Englewood Cliffs, N. J., 1969).

BELGY, J. N.

[1] Fonctions Arithmetiques. Thesis, Univ. of Clermont-Ferrand (1970).

BENDER, E. A.

[1] Central and local limit theorems applied to asymptotic enumeration. *J. Combin. Theory* (A) **15** (1973) 91–111.

[2] Asymptotic methods in enumeration. *SIAM Rev.* **16** (1974) 485–515, **18** (1976) 292.

BENDER, E. A. and J. R. GOLDMAN

[1] The enumerative uses of generating functions. *Indiana Univ. Math. J.* **20** (1971) 753–766.

BESICOVITCH, A. S.

*[1] *Almost Periodic Functions* (Dover, New York, 1954).

BEURLING, A.

[1] Analyse de la loi asymptotique de la distribution des nombres premiers généralisés, I. *Acta Math.* **68** (1937) 255–291.

BOLLMAN, D. and H. RAMÍREZ

[1] On the enumeration of matrices over finite commutative rings. *Am. Math. Monthly* **76** (1969) 1019–1023.

BOMBIERI, E.

[1] Sull'analogo della formula di Selberg nei corpi di funzioni. *Rend. Accad. Naz. Lincei* (8) **35** (1963) 252–257.

BRAUER, R.

[1] On the zeta functions of algebraic number fields, I–II. *Am. J. Math.* **69** (1947) 243–250, **72** (1950) 739–746.

[2] A note on zeta-functions of algebraic number fields. *Acta Arith.* **24** (1973) 325–327.

BREDIHIN, B. M.

[1] Free numerical semigroups with power densities. *Dokl. Akad. Nauk. SSSR* **118** (1958) 855–857 (in Russian); *Math. Rev.* **20** (1959) #5175.

[2] Natural densities of some semigroups of numbers. *Ukrain. Mat. Ž.* **11** (1959) 137–145 (in Russian); *Math. Rev.* **22** (1961) #3708.

[3] Elementary solutions of inverse problems on bases of free semigroups. *Mat. Sb.* N. S. **50** (92) (1960) 221–232 (in Russian); *Math. Rev.* **26** (1963) #246.

[4] The remainder term in the asymptotic formula for $\nu_G(x)$ *Izv. Vysš. Učebn. Zaved. Mat.* **6** (19) (1960) 40–49 (in Russian); *Math. Rev.* **26** (1963) #92.

[5] Free numerical semigroups with power densities. *Mat. Sb.* N. S. **46** (88) (1958) 143–158 (in Russian); *Math. Rev.* **21** (1960) #87.

[6] On characters of numerical semigroups with a sufficiently sparse base. *Dokl. Akad. Nauk. SSSR.* N. S. **90** (1953) 707–710 (in Russian); *Math. Rev.* **15** (1954) 105.

[7] On sum functions of characters of numerical semigroups. *Dokl. Akad. Nauk. SSSR* N. S. **94** (1954) 609–621 (in Russian); *Math. Rev.* **15** (1954) 940.

[8] On power densities of certain subsets of free semigroups. *Izv. Vysš. Učebn. Zaved. Mat.* **3** (4) (1958) 24–30 (in Russian); *Math. Rev.* **26** (1963) #6150.

[9] Distribution of generating elements in ordered semigroups with a regular normalization. *Proc. First Sci. Conf. Math. Dept. Volg. Region,* pp. 4–8 (in Russian); *Math. Rev.* **33** (1967) #1294.

BRIGGS, W. E.

[1] Prime-like sequences generated by a sieve process. *Duke Math. J.* **30** (1963) 297–311.

BRIGHAM, N. A.

[1] A general asymptotic formula for partition functions. *Proc. Am. Math. Soc.* **1** (1950) 181–191.

BROUGHAN, K. A.

[1] On the number of structures of reflexive and transitive relations. *Can. J. Math.* **25** (1973) 1269–1273.

BURCKEL, R. B.

*[1] *Weakly Almost Periodic Functions on Semigroups* (Gordon and Breach, New York, 1970).

BUSCHMAN, R. G.

[1] Number-theoretic functions for generalized integers. *Jñānabha* (A) **3** (1973) 7–14.

CARLITZ, L.

[1] The arithmetic of polynomials in a Galois field. *Am. J. Math.* **54** (1932) 39–50.

[2] On polynomials in a Galois field. *Bull. Am. Math. Soc.* **38** (1932) 736–744.

[3] The distribution of irreducible polynomials in several indeterminates. *Illinois J. Math.* **7** (1963) 371–375.

[4] The distribution of irreducible polynomials in several indeterminates, II. *Can. J. Math.* **17** (1965) 261–266.

[5] Rings of arithmetic functions. *Pacific J. Math.* **14** (1964) 1165–1171.

CARMICHAEL, R. D.

[1] Expansions of arithmetical functions in infinite series. *Proc. London Math. Soc.* (2) **34** (1932) 1–26.

CASHWELL, E. D. and C. J. EVERETT

[1] The ring of number-theoretic functions. *Pacific J. Math.* **9** (1959) 975–985.

[2] Formal power series. *Pacific J. Math.* **13** (1963) 45–64.

CASSELS, J. W. S. and A. FRÖHLICH (Editors)

*[1] *Algebraic Number Theory* (Academic Press, New York, 1967).

CHANDRASEKHARAN, K.

*[1] *Introduction to Analytic Number Theory* (Springer, Berlin, 1968).

*[2] *Arithmetical Functions* (Springer, Berlin, 1970).

CHANDRASEKHARAN, K. and R. NARASIMHAN

[1] The approximate functional equation for a class of zeta-functions. *Math. Ann.* **152** (1963) 30–64.

CIBUL'SKITE, D.

[1] The distribution of generating elements in free numerical semigroups, I–III. *Litovsk. Mat. Sb.* **10** (1970) 397–415, 593–610, 835–844 (in Russian); *Math. Rev.* **46** (1973) #8989.

COHEN, E.

[1] On the average number of direct factors of a finite abelian group. *Acta Arith.* **6** (1960) 159–173.

[2] On the normal number of irreducible factors of a finite abelian group. *Ricerche Mat.* **9** (1960) 203–212.

[3] Arithmetical functions of finite abelian groups. *Math. Ann.* **142** (1961) 165–182.

[4] On the inversion of even functions of finite abelian groups (mod H). *J. Reine Angew. Math. 207* (1961) 192–202.

[5] Almost even functions of finite abelian groups. *Acta Arith.* **7** (1962) 311–323.

[6] An extension of Ramanujan's sum, I–III. *Duke Math. J.* **16** (1949) 85–90, **22** (1955) 543–550, **23** (1956) 623–630.

[7] Rings of arithmetic functions, I–II. *Duke Math. J.* **19** (1952) 115–129, **21** (1954) 9–28.

[8] A class of arithmetical functions. *Proc. Nat. Acad. Sci. U.S.A.* **41** (1955) 939–944.

[9] Representations of even functions (mod r), I–III. *Duke Math. J.* **25** (1958) 401–421, **26** (1959) 165–182, 491–500.

[10] Trigonometric sums in elementary number theory. *Am. Math. Monthly* **66** (1959) 105–117.

[11] A class of residue systems (mod r) and related arithmetical functions, I–II. *Pacific J. Math.* **9** (1959) 13–23, 667–679.

[12] Arithmetical inversion formulas. *Can. J. Math.* **12** (1960) 399–409.

[13] Arithmetical functions associated with the unitary divisors of an integer. *Math. Z.* **74** (1960) 66–80.

[14] Unitary functions (mod r). *Duke Math. J.* **28** (1961) 475–485.

[15] Fourier expansions of arithmetical functions. *Bull. Am. Math. Soc.* **67** (1961) 145–147.

[16] The elementary arithmetical functions. *Scripta Math.* **25** (1960) 221–227.

[17] Arithmetical functions associated with arbitrary sets of integers. *Acta Arith.* **5** (1959) 407–415.

COHEN, S. D.

[1] The distribution of irreducible polynomials in several indeterminates over a finite field. *Proc. Edinburgh Math. Soc.* **16** (2) (1968) 1–17.

[2] Some arithmetical functions in finite fields. *Glasgow Math. J.* **11** (1969) 21–36.

[3] Further arithmetical functions in finite fields. *Proc. Edinburgh Math. Soc.* **16** (2) (1969) 349–363.

[4] The distribution of polynomials over finite fields, I–II. *Acta Arith.* **17** (1970) 255–271, **20** (1972) 53–62.

[5] Uniform distribution of polynomials over finite fields. *J. London Math. Soc.* (2) **6** (1972) 93–102.

COHN, H.

*[1] *A Second Course in Number Theory* (Wiley, New York, 1962).

CORDUNEANU, C.

*[1] *Almost Periodic Functions* (Interscience, New York, 1968).

DANILOV, A. N.

[1] On the sequences of values of additive arithmetical functions defined on the set of ideals of a field K of degree n over the field of rational numbers. *Leningrad. Gos. Ped. Inst. Učen. Zap.* **274** (1965) 59–70 (in Russian); *Math. Rev.* **33** (1967) #5595.

DAVENPORT, H.

*[1] *Multiplicative Number Theory* (Markham, Chicago, Ill., 1967).

DAVISON, T. M. K.

[1] A Tauberian theorem and analogues of the prime number theorem. *Can. J. Math.* **20** (1968) 362–367.

DE KROON, J. P. M.

[1] The asymptotic behaviour of additive functions in algebraic number theory. *Compositio Math.* **17** (1965) 207–261.

DELANGE, H.

[1] Théorèmes taubériens et applications arithmétiques. *Mem. Soc. Roy. Sci. Liège* (4) **16** (I) (1955) 5–87.

[2] Sur la distribution des entiers ayant certaines propriétés. *Ann. Sci. École Norm. Sup.* **73** (1956) 15–74.

DELSARTE, J.

[1] Essai sur l'application de la théorie des fonctions presque-périodiques à l'arithmétique. *Ann. Sci. École Norm. Sup.* **62** (1945) 185–204.

DÉNES, J., P. ERDÖS and P. TURÁN

[1] On some statistical properties of the alternating group of degree *n*. *Enseignement Math.* **15** (1969) 89–99.

DEURING, M.

*[1] *Algebren*, 2nd ed. (Springer, Berlin, 1968).

*[2] *Lectures on the Theory of Algebraic Functions of One Variable* (Springer, Berlin, 1973).

DIAMOND, H. G.

[1] The prime number theorem for Beurling's generalized numbers. *J. Number Theory* **1** (1969) 200–207.

[2] Asymptotic distribution of Beurling's generalized integers. *Illinois J. Math.* **14** (1970) 12–28.

[3] A set of generalized numbers showing Beurling's theorem to be sharp. *Illinois J. Math.* **14** (1970) 29–34.

[4] Chebyshev estimates for Beurling generalized primes. *Proc. Am. Math. Soc.* **39** (1973) 503–508.

DICKENSON, G. A.

[1] On the enumeration of certain classes of soluble groups. *Quart. J. Math.* (2) **20** (1969) 383–394.

DIXON, J. D.

[1] The probability of generating the symmetric group. *Math. Z.* **110** (1969) 199–205.

DOORNHOF, L.

[1] Simple groups are scarce. *Proc. Am. Math. Soc.* **19** (1968) 692–696.

DOORNHOF, L. and E. L. SPITZNAGEL

[1] Density of finite simple group orders. *Math. Z.* **106** (1968) 175–177.

DOUBILET, P., G.-C. ROTA and R. STANLEY

[1] On the foundations of combinatorial theory, VI: The idea of generating function. In: *Proc. 6th Berkeley Symp. on Mathematical Statistics and Probability*, Vol. 2 (1970) 267–318.

DRESSLER, R.

[1] A density which counts multiplicity. *Pacific J. Math.* **34** (1970) 371–378.

DUGUNDJI, J.

*[1] *Topology* (Allyn and Bacon, Boston, Mass., 1966).

DUTTLINGER, J.

[1] Eine Bemerkung zu einer asymptotischen Formel von Herrn Knopfmacher. *J. Reine Angew. Math.* **266** (1974) 104–106.

[2] Über die Anzahl Abelscher Gruppen gegebener Ordnung. *J. Reine Angew. Math.* **273** (1975) 61–76.

EDWARDS, H. M.

*[1] *Riemann's Zeta Function* (Academic Press, New York, 1974).

EHLICH, H.

[1] Über die elementaren Beweise der Primzahlsätze. *J. Reine Angew. Math.* **203** (1960) 143–153.

EICHLER, M.

*[1] *Introduction to the Theory of Algebraic Numbers and Functions* (Academic Press, New York, 1966).

ERDÖS, P.

*[1] *The Art of Counting—Selected Writings* (M.I.T. Press, Cambridge, Mass., 1973).

ERDÖS, P. and A. RÉNYI

[1] Probabilistic methods in group theory. *J. Analyse Math.* **14** (1965) 127–138.

[2] On random graphs, I. *Publ. Math. Debrecen* **6** (1959) 290–297.

[3] On the mean value of nonnegative multiplicative number-theoretical functions. *Mich. Math. J.* **12** (1965) 321–338.

[4] On the evolution of random graphs. *Magyar Tud. Akad. Mat. Fiz. Oszt. Közl.* **5** (1960) 17–61.

ERDÖS, P. and J. SPENCER

*[1] *Probabilistic Methods in Combinatorics* (Academic Press, New York, 1974).

ERDÖS, P. and G. SZEKERES

[1] Über die Anzahl der Abelschen Gruppen gegebener Ordnung und über ein verwandtes zahlentheoretisches Problem. *Acta Sci. Math. (Szeged)* **7** (1935) 95–102.

ERDÖS, P. and P. TURÁN

[1] On some problems of a statistical group theory, I–VII. *Z. Warscheinlichkeitstheorie Verw. Geb.* **4** (1965) 175–186, *Acta Math. Acad. Sci. Hung.* **18** (1967) 151–163, 309–320, **19** (1968) 413–435, *Per. Math. Hung.* **1** (1971) 5–13, **2** (1972) 149–163.

ERDÖS, P. and A. WINTNER

[1] Additive functions and almost periodicity (B^2). *Am. J. Math.* **62** (1940) 635–645.

ERNÉ, M.

[1] Struktur- and Anzahlformeln für Topologien auf endlichen Mengen. *Manuscripta Math.* **11** (1974) 221–259.

FAĬNLEĬB, A. S.

[1] Summation of the values of multiplicative functions that are defined on a normed semigroup. *Izv. Akad. Nauk. SSSR* 9 (1965) 49–55 (in Russian); *Math. Rev.* **33** (1967) #2605.

FLUCH, W.

[1] Über einen Satz von Hardy–Ramanujan. *Monatsh. Math.* **73** (1969) 31–35.

FOGELS, E.

[1] On the abstract theory of primes, I–III. *Acta Arith.* **10** (1964) 137–182, 333–358, **11** (1966) 292–331.

[2] On the distribution of prime ideals. *Acta Arith.* **7** (1962) 255–269.

FORMAN, W. and H. N. SHAPIRO

[1] Abstract prime number theorems. *Commun. Pure Appl. Math.* **7** (1954) 587–619.

FREDMAN, M. L.

[1] The distribution of absolutely irreducible polynomials in several indeterminates. *Proc. Am. Math. Soc.* **31** (1972) 387–390.

FRESNEL, J.

[1] Nombres de Bernoulli et fonctions *L* *p*-adiques. *Ann. Inst. Fourier (Grenoble)* **17** (1967) 281–333.

FRIEDLANDER, J.

[1] On the number of ideals free from large prime divisors. *J. Reine Angew. Math.* **255** (1972) 1–7.

GALAMBOS, J.

[1] Distribution of arithmetical functions: a survey. *Ann. Inst. H. Poincaré* **B6** (1970) 281–305.

GALLAGHER, P. X.

[1] Counting finite groups of given order. *Math. Z.* **102** (1967) 236–237.

[2] The large sieve and probabilistic Galois theory. In: *Proc. Symp. Pure Math.* **24** (Am. Math. Soc., Providence, R. I., 1973) 91–101.

GILLETT, J. R.

[1] On the largest prime divisors of ideals in fields of degree *n*. *Duke Math. J.* **37** (1970) 589–600.

GODEMENT, R. and H. JACQUET

*[1] *Zeta Functions of Simple Algebras* (Springer, Berlin, 1972).

GOLDBERG, M. and J. W. MOON

[1] On the maximum order of the group of a tournament. *Can. Math. Bull.* **9** (1966) 563–569.

GOLDSTEIN, L. J.

[1] A generalization of the Siegel–Walfisz theorem. *Trans. Am. Math. Soc.* **149** (1970) 417–429.

*[2] *Analytic Number Theory* (Prentice-Hall, Englewood Cliffs, N. J., 1971).

GOLOMB, S. W.

[1] Sets of primes with intermediate density. *Math. Scand.* **3** (1955) 264–274.

GREENBERG, L. and M. NEWMAN

[1] Some results on solvable groups. *Arch. Math.* **21** (1970) 349–352.

GROSSWALD, E.

[1] Oscillation theorems of arithmetical functions. *Trans. Am. Math. Soc.* **126** (1967) 1–28.

[2] Oscillation theorems. In: *Lecture Notes in Maths.* **251** (Springer, Berlin, 1972) 141–168.

GUINAND, A. P.

[1] Concordance and the harmonic analysis of sequences. *Acta Math.* **101** (1959) 235–271.

GUSTAFSON, W. H.

[1] What is the probability that two group elements commute? *Am. Math. Monthly* **80** (1973) 1031–1034.

GUTMAN, H.

[1] Anwendung Tauberschen Sätze und Laurentschen Reihen in der zahlentheoretischen Asymptotik. *Verhandl. Naturforsch. Ges. Basel* **69** (1959) 119–144.

[2] Beitrag zur Zahlentheorie der verallgemeinerten Primzahlen. *Verhandl. Naturforsch. Ges. Basel* **70** (1959) 167–192.

HABERLAND, K.

[1] Über die Anzahl der Erweiterungen eines algebraischen Zahlkörpers mit einer gegebenen abelschen Gruppe als Galoisgruppe. *Acta Arith.* **26** (1975) 153–158.

HALL, M., jr.

*[1] *The Theory of Groups* (MacMillan, New York, 1959).

HALL, R. R.

[1] On a theorem of Erdös and Rényi concerning abelian groups. *J. London Math. Soc.* (2) **5** (1972) 143–153.

HALL, R. R. and A. SUDBURY

[1] On a conjecture of Erdös and Rényi concerning abelian groups. *J. London Math. Soc.* (2) **6** (1972) 177–189.

HALL, R. S.

[1] Theorems about Beurling's Generalized Primes and the Associated Zeta Function. Thesis, Univ. of Illinois, Urbana, Ill. (1967).

[2] The prime number theorem for generalized primes. *J. Number Theory* **4** (1972) 313–320.

[3] Beurling generalized prime number systems in which the Chebyshev inequalities fail. *Proc. Am. Math. Soc.* **40** (1973) 79–82.

HARARY, F.

[1] The number of linear, directed, rooted, and connected graphs. *Trans. Am. Math. Soc.* **78** (1955) 445–463.

HARARY, F. and E. PALMER

*[1] *Graphical Enumeration* (Academic Press, New York, 1973).

HARDY, G. H.

[1] Note on Ramanujan's trigonometrical function $c_q(n)$ and certain series of arithmetical functions. *Proc. Cambridge Philos. Soc.* **20** (1921) 263–271.

HARDY, G. H., J. E. LITTLEWOOD and G. PÓLYA

*[1] *Inequalities* (Cambridge Univ. Press, London, 1952).

HARDY, G. H. and S. RAMANUJAN

[1] Asymptotic formulae concerning the distribution of integers of various types. *Proc. London Math. Soc.* (2) **16** (1917) 112–132.

[2] Asymptotic formulae in combinatory analysis. *Proc. London Math. Soc.* (2) **17** (1918) 75–115.

HARDY, G. H. and E. M. WRIGHT

*[1] *An Introduction to the Theory of Numbers* (Oxford Univ. Press, London, 1960).

HARRIS, B.

[1] The asymptotic distribution of the order of elements in symmetric semigroups. *J. Combin. Theory* **A15** (1973) 66–74.

HARRISON, D. K.

[1] Finite and infinite primes for rings and fields. *Mem. Am. Math. Soc.* **68** (1966).

HARRISON, M.

[1] A census of finite automata. *Can. J. Math.* **17** (1965) 100–113.

HARTMAN, P. and A. WINTNER

[1] On the almost periodicity of additive number-theoretical functions. *Am. J. Math.* **62** (1940) 753–758.

[2] Additive functions and almost periodicity. *Duke Math. J.* **9** (1942) 112–119.

HASSE, H.

*[1] *Bericht über neuere Untersuchungen und Probleme aus der Theorie der algebraischen Zahlkörper* (Physica, Würzburg, 1970).

*[2] *Vorlesungen über Zahlentheorie* (Springer, Berlin, 1964).

HAWKINS, D.

[1] The random sieve. *Math. Mag.* **31** (1958) 1–3.

[2] Random sieves, II. *J. Number Theory* **6** (1974) 192–200.

HAYES, D. R.

[1] The distribution of irreducibles in GF$[q, x]$. *Trans. Am. Math. Soc.* **117** (1965) 101–127.

HECKE, E.

[1] Über die Zetafunktion beliebiger algebraischer Zahlkörper. *Nachr. K. Ges. Wiss. Göttingen, Math.-Phys. Kl.* (1917) 77–89.

[2] Über die L-Funktionen und den Dirichletschen Primzahlsatz für einen beliebigen Zahlkörper. *Nachr. K. Ges. Wiss. Göttingen, Math.-Phys. Kl.* (1917) 299–318.

*[3] *Vorlesungen über die Theorie der algebraischen Zahlen* (Chelsea, Bronx, N. Y., 1970).

HELGASON, S.

*[1] *Differential Geometry and Symmetric Spaces* (Academic Press, New York, 1962).

HELMBERG, G.

[1] Abstract theory of uniform distribution. *Compositio Math.* **16** (1964) 72–82.

HEPPNER, E.

[1] Die maximale Ordnung primzahl-unabhängiger multiplikativer Funktionen. *Arch. Math.* **24** (1973) 63–66.

[2] Primzahl-unabhängige multiplikative Funktionen. Thesis, J. W. Goethe University, Frankfurt/Main (1973).

HERSTEIN, I. N.

*[1] *Noncommutative Rings* (Wiley, New York, 1968).

HEWITT, E. and K. STROMBERG

*[1] *Real and Abstract Analysis* (Springer, Berlin, 1965).

HEY, K.

[1] Analytische Zahlentheorie in Systemen hyperkomplexer Zahlen. Thesis, University of Hamburg (1929).

HIGMAN, G.

[1] Enumerating p-groups. I: Inequalities; II: Problems whose solution is PORC. *Proc. London Math. Soc.* (3) **10** (1960) 24–30, 566–582.

HORADAM, E. M.

[1] Arithmetical functions of generalized primes. *Am. Math. Monthly* **68** (1961) 626–629.

[2] Arithmetical functions associated with the unitary divisors of a generalized integer. *Am. Math. Monthly* **69** (1962) 196–199.

[3] The order of arithmetical functions of generalized integers. *Am. Math. Monthly* **70** (1963) 506–512.

[4] The Euler φ function for generalized integers. *Proc. Am. Math. Soc.* **14** (1963) 754–762.

[5] A calculus of convolutions for generalized integers. *Proc. Koninkl. Nederl. Akad. Wetensch.* A **66** (1963) 695–698.

[6] Selberg's inequality for generalized integers using the calculus of convolutions. *Proc. Koninkl. Nederl. Akad. Wetensch.* A **66** (1963) 699–704.

[7] The number of unitary divisors of a generalized integer. *Am. Math. Monthly* **71** (1964) 893–895.

[8] Ramanujan's sum for generalized integers. *Duke Math. J.* **31** (1964) 697–702, **33** (1966) 705–708.

[9] Unitary divisor functions of a generalized integer. *Portugal. Math.* **24** (1965) 131–143.

[10] The average order of the number of generalized integers representable as a product of a prime and a square. *J. Reine Angew. Math.* **217** (1965) 64–68.

[11] Exponential functions for arithmetical semigroups. *J. Reine Angew. Math.* **222** (1966) 14–19.

[12] A sum of a certain divisor function for arithmetical semigroups. *Pacific J. Math.* **22** (1967) 407–412.

[13] On the number of pairs of generalised integers with l. c. m. not exceeding x. *Am. Math. Monthly* **74** (1967) 811–812.

[14] Ramanujan's sum and Nagell's totient function for arithmetical semigroups. *Math. Scand.* **22** (1968) 269–281.

[15] Ramanujan's sum and its applications to the enumerative functions of certain sets of elements of an arithmetical semigroup. *J. Math. Sci.* **3** (1968) 47–70.

[16] Normal order for divisor functions of generalised integers. *Portugal. Math.* **27** (1968) 201–207.

[17] Normal order and the Euler Φ function for generalised integers. *An. Fac. Ci. Univ. Porto* **51** (1968) 211–219.

HUXLEY, M. N.

*[1] *The Distribution of Prime Numbers.* (Clarendon, Oxford, 1972).

HYSLOP, J. M.

*[1] *Infinite Series* (Oliver and Boyd, Edinburgh, 1965).

INDLEKOFER, K.-H.

[1] A remark on solvable groups. *Arch. Math.* **24** (1973) 57–58.

INGHAM, A. E.

*[1] *The Distribution of Prime Numbers* (Cambridge Univ. Press, London, 1932).

[2] A Tauberian theorem for partitions. *Ann. Math.* **42** (1941) 1075–1090.

IWASAWA, K.

*[1] *Lectures on p-adic L-functions.* Ann. Math. Studies **74** (Princeton Univ. Press, Princeton, N. J., 1972).

JACOBSON, N.

*[1] *Theory of Rings* (Am. Math. Soc., Providence, R. I., 1943).

*[2] *Lie Algebras* (Interscience, New York, 1962).

JENNER, W. E.

[1] Zeta functions of non-maximal orders in rational semisimple algebras. *Duke Math. J.* **30** (1963) 541–544.

JÓNSSON, B. and A. TARSKI

*[1] *Direct Decompositions of Finite Algebraic Systems* (Notre Dame Univ. Press, Notre Dame, Ind., 1947).

JORIS, H.

[1] Ω-Sätze für zwei arithmetische Funktionen. *Comment. Math. Helv.* **47** (1972) 220–248.

[2] Ω-Sätze für gewisse multiplikative arithmetische Funktionen. *Comment. Math. Helv.* **48** (1973) 409–435.

JUŠKIS, Z.

[1] Limit theorems for additive functions defined on ordered semigroups with a regular norm. *Litovsk. Mat. Sb.* **4** (1964) 565–603 (in Russian); *Math. Rev.* **30** (1965) #4741.

[2] An asymptotic decomposition of distribution laws for certain functions defined on ordered semigroups with regular normalization. *Litovsk. Mat. Sb.* **5** (1965) 167–183 (in Russian); *Math. Rev.* **34** (1967) #5795.

[3] A theorem of P. Erdös and A. Wintner on ordered semigroups with regular norm. *Litovsk Mat. Sb.* **4** (1964) 429–450 (in Russian); *Math. Rev.* **30** (1965) #1197.

KAC, M.

*[1] *Statistical Independence in Probability, Analysis and Number Theory* (Wiley, New York, 1959).

[2] Almost periodicity and the representation of integers as sums of squares. *Am. J. Math.* **62** (1940) 122–126.

KAC, M., E. R. VAN KAMPEN and A. WINTNER

[1] Ramanujan sums and almost periodic functions. *Am. J. Math.* **62** (1940) 107–114.

KALNIŃ, I. M.

[1] Analogues of elementary theorems on primes in the semigroup of real numbers. *Rīgas Politehn. Inst. Zinātn. Raksti.* **10** (1) (1963) 151–165 (in Russian); *Math. Rev.* **33** (1967) #7315.

KAPLANSKY, I.

[1] Modules over Dedekind rings and valuation rings. *Trans. Am. Math. Soc.* **72** (1952) 327–340.

KENDALL, D. G. and R. A. RANKIN

[1] On the number of abelian groups of a given order. *Quart. J. Math.* **18** (1947) 197–208.

KNAPOWSKI, S. and P. TURÁN

[1] Comparative prime-number theory, I–VIII. *Acta Math. Acad. Sci. Hung.* **13** (1962) 229–364, **14** (1963) 31–78, 241–268.

[2] Further developments in the comparative prime-number theory, I–VII. *Acta Arith.* **9** (1964) 23–40, **10** (1964) 293–313, **11** (1965) 115–127, 147–161, 192–202, **12** (1966) 85–96, **21** (1972) 193–201.

KNOPFMACHER, J.

[1] Note on finite topological spaces. *J. Austral. Math. Soc.* **9** (1969) 252–256.

[2] Arithmetical properties of finite rings and algebras, and analytic number theory. *Bull. Am. Math. Soc.* **76** (1970) 830–833.

[3] Asymptotic enumeration of manifolds and Lie groups. *Math. Ann.* **190** (1970) 129–134.

[4] Finite modules and algebras over Dedekind domains, and analytic number theory. *Bull. Am. Math. Soc.* **78** (1972) 193–196.

[5] Arithmetical properties of finite rings and algebras, and analytic number theory, I. *J. Reine Angew. Math.* **252** (1972) 16–43.

[6] Arithmetical properties of finite rings and algebras, and analytic number theory, II: Categories and analytic number theory. *J. Reine Angew. Math.* **254** (1972) 74–99.

[7] Arithmetical properties of finite rings and algebras, and analytic number theory, III: Finite modules and algebras over Dedekind domains. *J. Reine Angew. Math.* **259** (1973) 157–170.

[8] Arithmetical properties of finite rings and algebras, and analytic number theory, IV: Relative asymptotic enumeration and *L*-series. *J. Reine Angew. Math.* **270** (1974) 97–114.

[9] Arithmetical properties of finite rings and algebras, and analytic number theory, V: Categories and relative analytic number theory. *J. Reine Angew. Math.*, **271** (1974) 95–121.

[10] Arithmetical properties of finite rings and algebras, and analytic number theory, VI: Maximum orders of magnitude. *J. Reine Angew. Math.* **277** (1975) 45–62.

[11] A prime-divisor function. *Proc. Am. Math. Soc.* **40** (1973) 373–377.

[12] On the asymptotic enumeration of finite rings and modules, *Arch. Math.* **26** (1975) 615–619.

[13] Arithmetical properties of finite graphs and polynomials, *J. Combin. Theory.* **B20** (1976) 205–215.

[14] On the moments of certain arithmetical functions arising in algebra. Appendix to [15] below.

[15] Fourier analysis of arithmetical functions, *Ann. Mat. Pura Appl.* **109** (1976) 177–201.

[16] Analytic arithmetic of algebraic function fields. *Lect. Notes in Pure & Appl. Math.* **50** (M. Dekker, New York, 1979).

KNOPFMACHER, J. and J. N. RIDLEY

[1] Prime-independent arithmetical functions. *Ann. Mat. Pura Appl.* **101** (1974) 153–169.

KNOPP, K.

*[1] *Theorie und Anwendung der unendlichen Reihen* (Springer, Berlin, 1964).

KNUTH, D. E.

[1] The asymptotic number of geometries. *J. Combin. Theory* **16** (1974) 398–400.

KOHLBECKER, E. E.

[1] Weak asymptotic properties of partitions. *Trans. Am. Math. Soc.* **88** (1958) 346–365.

Kornblum, H.
[1] Über die Primfunktionen in einer arithmetische Progression. *Math. Z.* **5** (1919) 100–111.
Krätzel, E.
[1] Die maximale Ordnung der Anzahl der wesentlich verschiedenen Abelschen Gruppen n-ter Ordnung. *Quart. J. Math.* (2) **21** (1970) 273–275.
[2] Zahlen k-ter art. *Am. J. Math.* **94** (1972) 309–328.
Kruse, R. L. and D. T. Price
[1] Enumerating finite rings. *J. London Math. Soc.* (2) **1** (1970) 149–159.
*[2] *Nilpotent Rings* (Gordon and Breach, New York, 1969).
Kruskal, M. D.
[1] The expected number of components under a random mapping function. *Am. Math. Monthly* **61** (1954) 392–397.
Kubilius, J.
*[1] *Probabilistic Methods in the Theory of Numbers.* Am. Math. Soc. Transl. of Math. Monographs, Vol. **11** (1964).
Kubota, T. and H. W. Leopoldt
[1] Eine p-adische Theorie der Zetawerte, I. *J. Reine Angew. Math.* **214/215** (1964) 328–339.
Kurosh, A. G.
*[1] *Lectures on General Algebra* (Chelsea, Bronx, N. Y., 1963).

Landau, E.
[1] Über die zu einem algebraischen Zahlkörper gehörige Zetafunktion und die Ausdehnung der Tschebyscheff'schen Primzahltheorie auf das Problem der Verteilung der Primideale. *J. Reine Angew. Math.* **125** (1903) 64–188.
[2] Neuer Beweis des Primzahlsatzes und Beweis des Primidealsatzes. *Math. Ann.* **56** (1903) 645–670.
[3] Über die zahlentheoretische Funktion $\mu(k)$. *Sitz ber. Kaiserl. Akad. Wissensch. Wien* **112** (1903) 537–570.
[4] Über die Maximalordnung der Permutationen gegebenen Grades. *Arch. Math. Phys.* **5** (1903) 92–103.
[5] Über den Zusammenhang einiger neueren Sätze der analytischen Zahlentheorie. *Sitz ber. Kaiserl. Akad. Wissensch. Wien* **115** (1906) 589–632.
[6] Über die Verteilung der Primideale in den Idealklassen eines algebraischen Zahlkörpers. *Math. Ann.* **63** (1907) 145–204.
[7] Über eine idealtheoretische Funktion. *Trans. Am. Math. Soc.* **13** (1912) 1–21.
[8] Über Ideale und Primideale in Idealklassen. *Math. Z.* **2** (1918) 52–154.
*[9] *Einführung in die elementare und analytische Theorie der algebraischen Zahlen und der Ideale* (Chelsea, Bronx, N. Y., 1949).
*[10] *Handbuch der Lehre von der Verteilung der Primzahlen* (Chelsea, Bronx, N. Y., 1953).
*[11] *Vorlesungen über Zahlentheorie* (Chelsea, Bronx, N. Y., 1969).
[12] Über die Primzahlen einer arithmetischen Progression. *Sitzber. Kaiserl. Akad. Wissensch. Wien* **112** (1903) 493–509.
Lang, S.
*[1] *Algebra* (Addison–Wesley, Reading, Mass., 1967).

*[2] *Algebraic Number Theory* (Addison–Wesley, Reading, Mass., 1970).
[3] On the zeta function of number fields. *Invent. Math.* **12** (1971) 337–345.
[4] Sur les séries L d'une variété algébrique. *Bull. Soc. Math. France* **84** (1956) 385–407.

LARDON, R.
[1] Évaluations asymptotiques concernant la fonction Ω dans certaines semi-groupes normés à factorisation unique. *C. R. Acad. Sci. Paris* A **273** (1971) 76–79.

LEVEQUE, W. J.
*[1] *Topics in Number Theory* (Addison–Wesley, Reading, Mass., 1956).

LLOYD, E. K.
[1] Pólya's theorem in combinatorial analysis applied to enumerate multiplicative partitions. *J. London Math. Soc.* **43** (1968) 224–230.

LOOMIS, L. H.
*[1] *An Introduction to Abstract Harmonic Analysis* (Van Nostrand, New York, 1953).

LU, C.-P.
[1] On the unique factorization theorem in the ring of number theoretic functions. *Illinois J. Math.* **9** (1965) 40–46.

LUKACS, E.
*[1] *Characteristic Functions,* 2nd ed. (Griffin, London, 1970).

MAAK, W.
*[1] *Fastperiodische Funktionen* (Springer, Berlin, 1950).

MACBEATH, A. M.
[1] Combinatorial analysis with values in a semigroup. In: A. O. L. Atkin and B. J. Birch eds., *Computers in Number Theory* (Academic Press, New York, 1971).

MCCARTHY, P. J.
[1] The generation of arithmetical identities. *J. Reine Angew. Math.* **203** (1960) 55–63.

MALLIAVIN, P.
[1] Sur le reste de la loi asymptotique de répartition des nombres premiers généralisés de Beurling. *Acta Math.* **106** (1961) 281–298.

MARTIN, N. F. G.
[1] A class of zeta functions for ergodic theory. *J. Math. Anal. Appl.* **38** (1972) 735–745.

MAZUR, B. and P. SWINNERTON-DYER
[1] Arithmetic of Weil curves. *Invent. Math.* **25** (1974) 1–61.

MEINARDUS, G.
[1] Asymptotische Aussagen über Partitionen. *Math. Z.* **59** (1954) 388–398.

MIECH, R. J.
[1] On a conjecture of Erdös and Rényi. *Illinois J. Math.* **11** (1967) 114–127.

MITSUI, T.
[1] On the prime ideal theorem. *J. Math. Soc. Japan* **20** (1968) 233–247.

MONSKY, P.
*[1] *p-adic Analysis and Zeta Functions.* Lectures in Maths., Kyoto Univ. (1970).

MONTGOMERY, H. L.
*[1] *Topics in Multiplicative Number Theory* (Springer, Berlin, 1971).

MOON, J. W.
*[1] *Topics on Tournaments* (Holt, Rinehart and Winston, New York, 1968).
[2] On the maximum degree in a random tree. *Mich Math. J.* **15** (1968) 429–432.
[3] Climbing random trees. *Aequationes Math.* **5** (1970) 68–74.

[4] A problem on random trees. *J. Combin. Theory.* **10** (1971) 201–205.
[5] Counting labelled trees. *Math. Coll. Univ. Cape Town* **6** (1970–71) 56–76.

MOON, J. W. and L. MOSER
[1] Almost all tournaments are irreducible. *Can. Math. Bull.* **5** (1962) 61–65.
[2] Almost all (0,1) matrices are primitive. *Studia Sci. Math. Hung.* **1** (1966) 153–156.

MÜLLER, H.
[1] Über abstrakte Primzahlsätze mit Restglied. Thesis, Freie Universität, Berlin (1970).
[2] Ein Beitrag zur abstrakten Primzahltheorie. *J. Reine Angew. Math.* **259** (1973) 171–182.
[3] Zur Gleichverteilung von Beurlingschen Zahlen. *Abh. Math. Sem. Hamburg* **43** (1975) 186–191.
[4] Asymptotische Verteilung von Beurlingschen Zahlen. *J. Reine Angew. Math.* **289** (1977) 181–187.
[5] Über *k*-freie Elemente in Nebenklassen verallgemeinerter ganze Zahlen. *J. Reine Angew. Math.,* **273** (1975) 138–143.

NARKIEWICZ, W.
*[1] *Elementary and Analytic Theory of Algebraic Numbers* (PWN, Polish Scientific Publishers, Warsaw, 1974).

NEUDECKER, W. and D. WILLIAMS
[1] The 'Riemann hypothesis' for the Hawkins random sieve. *Compositio Math.* **29** (1974) 197–200.

NEUMANN, P. M.
[1] An enumeration theorem for finite groups. *Quart. J. Math.* (2) **20** (1969) 395–401.

NOVOSELOV, E. V.
[1] A new method in probabilistic number theory. *Am. Math. Soc. Transl.* (2) **52** (1966) 217–275.

NYMAN, B.
[1] A general prime number theorem. *Acta Math.* **81** (1949) 299–307.

NYMAN, J. E.
[1] On the probability that *k* positive integers are relatively prime. *J. Number Theory* **4** (1972) 469–473.

OBERSCHELP, W.
[1] Kombinatorische Anzahlbestimmungen in Relationen. *Math. Ann.* **174** (1967) 53–78.

O'NEIL, P. E.
[1] Asymptotics and random matrices with row-sum and column-sum restrictions. *Bull. Am. Math. Soc.* **75** (1969) 1276–1282.

PARAMESWARAN, S.
[1] Partition functions whose logarithms are slowly oscillating. *Trans. Am. Math. Soc.* **100** (1961) 217–240.

PAREIGIS, B.
*[1] *Categories and Functors* (Academic Press, New York, 1970).

PASSMAN, D. S.
[1] Isomorphic groups and group rings. *Pacific J. Math.* **15** (1965) 561–583.

PHILIBERT, G.

[1] Une formule fondamentale sur des fonctions multiplicatives définies sur certains semi-groupes et son application à un théorème d'équirépartition asymptotique. *C. R. Acad. Sci. Paris* A **274** (1972) 1764–1767.

PIFF, M. J. and D. J. A. WELSH

[1] The number of combinatorial geometries. *Bull. London Math. Soc.* **3** (1971) 55–56.

POLLARD, H.

*[1] *The Theory of Algebraic Numbers* (Wiley, New York, 1950).

PÓLYA, G.

[1] Kombinatorische Anzahlbestimmungen für Gruppen, Graphen, und chemische Verbindungen. *Acta Math.* **68** (1937) 145–254.

POPKEN, T.

[1] On convolutions in number theory. *Proc. Koninkl. Nederl. Akad. Wetensch.* (A) **58** (1955) 10–15.

[2] Algebraic dependence of arithmetic functions. *Ibid.* (A) **65** (1962) 155–168.

[3] Algebraic dependence of certain zêta functions. *Ibid.* (A) **69** (1966) 1–5.

[4] A measure for the differential-transcendence of the zêta-function of Riemann. In: P. Turán, ed., *Number Theory and Analysis* (Plenum, New York, 1969) 245–255.

PRACHAR, K.

[1] Verallgemeinerung eines Satzes von Hardy und Ramanujan auf algebraische Zahlkörper. *Monatsh. Math.* **56** (1952) 229–232.

*[2] *Primzahlverteilung* (Springer, Berlin, 1957).

RADEMACHER, H.

[1] On the partition function $p(n)$. *Proc. London Math. Soc.* (2) **43** (1937) 241–254.

[2] Fourier expansions of modular forms and problems of partitions. *Bull. Am. Math. Soc.* **46** (1940) 59–73.

[3] On the number of certain types of polyhedra. *Illinois J. Math.* **9** (1965) 361–380.

*[4] *Topics in Analytic Number Theory* (Springer, Berlin, 1973).

RADFORD, D. E.

[1] On the convergence of the zeta function of a commutative ring. *Duke Math. J.* **38** (1971) 521–526.

RAMANUJAN, S.

[1] Highly composite numbers. *Proc. London Math. Soc.* (2) **14** (1915) 347–409.

[2] On certain trigonometrical sums and their applications in the theory of numbers. *Trans. Cambridge Philos. Soc.* **22** (1918) 259–276.

REICHARDT, H.

[1] Der Primdivisorsatz für algebraische Funktionenkörper über einem endlichen Konstantenkörper. *Math. Z.* **40** (1936) 713–719.

RÉMOND, P.

[1] Étude asymptotique de certaines partitions dans certains semi-groupes. *Ann. Sci. École Norm. Sup.* **83** (1966) 343–410.

RICHERT, H.-E.

[1] Über die Anzahl Abelscher Gruppen gegebener Ordnung, I–II. *Math. Z.* **56** (1952) 21–32, **58** (1953) 71–84.

RIDDELL, R. J. and G. E. UHLENBECK

[1] On the theory of the virial development of the equation of state of monoatomic gases. *J. Chem. Phys.* **21** (1953) 2056–2064.

RIEGER, G. J.

[1] Über die Anzahl der Teiler der Ideale in einem algebraischen Zahlkörper. *Arch. Math.* **8** (1957) 162–165.

[2] Verallgemeinerung der Selbergschen Formel auf Idealklassen mod $\mathfrak{f}$ in algebraischen Zahlkörpern. *Math. Z.* **69** (1958) 183–194.

[3] Über die Anzahl der Ideale in einer Idealklasse mod $\mathfrak{f}$ eines algebraischen Zahlkörpers. *Math. Ann.* **135** (1958) 444–466.

[4] Verallgemeinerung der Siebmethode von A. Selberg auf algebraische Zahlkörper, I–III. *J. Reine Angew. Math.* **199** (1958) 208–214, **201** (1959) 157–171, **208** (1961) 79–90.

[5] Ein weiterer Beweis der Selbergschen Formel für Idealklassen mod $\mathfrak{f}$ in algebraischen Zahlkörpern. *Math. Ann.* **134** (1958) 403–407.

[6] Eine Selbergsche Identität für algebraische Zahlen. *Math. Ann.* **145** (1962) 77–80.

[7] Über die Anzahl der Primfaktoren algebraischer Zahlen und das Gauss'sche Fehlergesetz. *Math. Nachr.* **24** (1962) 77–89.

[8] Ramanujansche Summen in algebraischen Zahlkörpern. *Math. Nachr.* **22** (1960) 371–377.

[9] Über die Verteilung gewisser Idealmengen in einem algebraischen Zahlkörper. *Arch. Math.* **8** (1957) 401–404.

ROTA, G.-C.

[1] On the foundations of combinatorial theory, I: Theory of Möbius functions. *Z. Warscheinlichkeitstheorie Verw. Geb.* **2** (1964) 340–368.

RUDIN, W.

*[1] *Fourier Analysis on Groups* (Interscience, New York, 1962).

RYAVEC, C.

[1] Euler products associated with Beurling's generalized prime number systems. In: *Proc. Symp. Pure Math.* **24** (Am. Math. Soc., Providence, R. I., 1973) 263–266.

[2] The analytic continuation of Euler products with applications to asymptotic formulae. *Illinois J. Math.* **17** (1973) 608–618.

ŠAPIRO-PJATECKIĬ, I. I.

[1] On an asymptotic formula for the number of Abelian groups whose order does not exceed *n*. *Mat. Sb.* N. S. **26** (68) (1950) 479–486 (in Russian); *Math. Rev.* **12** (1951) 316.

SCHEID, H.

[1] Arithmetische Funktionen über Halbordnungen, I–II. *J. Reine Angew. Math.* **231** (1968) 192–214, **232** (1968) 207–220.

[2] Einige Ringe zahlentheoretischer Funktionen. *J. Reine Angew. Math.* **237** (1969) 1–11.

[3] Über ordnungstheoretische Funktionen. *J. Reine Angew. Math.* **238** (1969) 1–13.

SCHMIDT, F. K.

[1] Analytische Zahlentheorie in Körpern der Characteristik *p*. *Math. Z.* **33** (1931) 1–32.

SCHMIDT, P. G.

[1] Zur Anzahl Abelscher Gruppen gegebener Ordnung, I–II. *J. Reine Angew. Math.* **229** (1968) 34–42, *Acta Arith.* **13** (1968) 405–417.

SCHOENBERG, I. J.

[1] On asymptotic distributions of arithmetical functions. *Trans. Am. Math. Soc.* **93** (1936) 315–330.

SCHWARZ, W.

[1] Über die Anzahl Abelscher Gruppen gegebener Ordnung, I–II. *Math. Z.* **92** (1966) 314–320, *J. Reine Angew. Math.* **228** (1967) 133–138.

[2] Schwache asymptotische Eigenschaften von Partitionen. *J. Reine Angew. Math.* **232** (1968) 1–16.

[3] Asymptotische Formeln für Partitionen. *J. Reine Angew. Math.* **234** (1969) 172–178.

*[4] *Einführung in Methoden und Ergebnisse der Primzahltheorie* (Bibliogr. Inst. Mannheim, 1970.)

[5] Einige Bemerkungen über periodische zahlentheoretische Funktionen. *Math. Nachr.* **31** (1966) 125–136.

[6] Ramanujan-Entwicklungen stark multiplikativer zahlentheoretischer Funktionen, I–II. *Acta Arith.* **22** (1973) 329–338, and forthcoming.

[7] Ramanujan-Entwicklungen stark multiplikativer Funktionen. *J. Reine Angew. Math.* **262/3** (1973) 66–73.

[8] Die Ramanujan-Entwicklung reellwertiger multiplikativer Funktionen vom Betrage kleiner oder gleich Eins. *J. Reine Angew. Math.*, **271** (1974) 171–176.

SCHWARZ, W. and J. SPILKER

[1] Eine Anwendung des Approximationssatzes von Weierstrass–Stone auf Ramanujan-Summen. *Nieuw Arch. Wisk.* (3) **19** (1971) 198–209.

[2] Mean values and Ramanujan expansions of almost even arithmetical functions. In: *Proc. 1974 Colloqu. on Number Theory, Debrecen, Colloq. Math. Soc. J. Bolyai* **13** (1974) 315–357.

SCHWARZ, W. and E. WIRSING

[1] The maximal number of non-isomorphic abelian groups of order *n*. *Arch. Math.* **24** (1973) 59–62.

SEGAL, S. L.

[1] Prime number theorem analogues without primes. *J. Reine Angew. Math.* **265** (1974) 1–22.

SEMADENI, Z.

*[1] *Banach Spaces of Continuous Functions* (PWN, Polish Scientific Publishers, Warsaw, 1971).

SERRE, J.-P.

[1] Zeta and *L* functions. In: O. Schilling, ed., *Arithmetical Algebraic Geometry.* (Harper and Row, New York, 1965).

[2] Formes modulaires et fonctions zêta *p*-adiques. In: *Lecture Notes in Maths.* **350** (Springer, Berlin, 1973) 191–268.

SHADER, L.

[1] Arithmetic functions associated with unitary divisors in GF[q,x], I–II. *Ann. Mat. Pura Appl.* **86** (1970) 79–98.

[2] The unitary Brauer–Rademacher identity. *Rend. Accad. Lincei* (8) **48** (1970) 403–404.

SHAPIRO, H. N.

[1] An elementary proof of the prime ideal theorem. *Commun Pure Appl. Math.* **2** (1949) 309–323.

[2] On primes in arithmetic progressions, I–II. *Ann. Math.* **52** (1950) 217–243.

[3] Tauberian theorems and elementary prime number theory. *Commun Pure Appl. Math.* **12** (1959) 579–610.

[4] On the convolution ring of arithmetic functions. *Commun Pure Appl. Math.* **25** (1972) 287–336.

SIMS, C.

[1] Enumerating p-groups. *Proc. London Math. Soc.* (3) **15** (1965) 151–166.

SMALE, S.

[1] Differentiable dynamical systems. *Bull. Am. Math. Soc.* **73** (1967) 747–817.

SPRINDŽUK, V. G.

[1] 'Almost every' algebraic number field has a large class number. *Acta Arith.* **25** (1974) 411–413.

SRINIVASAN, B. R.

[1] On the number of Abelian groups of a given order. *Acta Arith.* **23** (1973) 195–205.

STAŚ, W. and K. WIERTELAK

[1] Some estimates in the theory of Dedekind zeta-functions. *Acta Arith.* **23** (1973) 127–135.

SURYANARAYANA, D. and R. SITA RAMA CHANDRA RAO

[1] The number of square-full divisors of an integer. *Proc. Am. Math. Soc.* **34** (1972) 79–80.

SURYANARAYANA, D. and V. SIVA RAMA PRASAD

[1] The number of pairs of generalized integers with L.C.M. $\leq x$. *J. Austral. Math. Soc.* **13** (1972) 411–416.

[2] The number of k-ary divisors of a generalized integer. *Portugal. Math.* **33** (1974) 85–92.

TATE, J.

[1] The arithmetic of elliptic curves. *Invent. Math.* **23** (1974) 179–206.

TITCHMARSH, E. C.

*[1] *The Theory of the Riemann Zeta-Function* (Clarendon, Oxford 1951).

*[2] *The Theory of Functions* (Oxford Univ. Press, London, 1952).

TITS, J.

*[1] *Tabellen zu den einfachen Lie-Gruppen und ihren Darstellungen* (Springer, Berlin, 1967).

URAZBAEV, B. M.

[1] On the distribution of certain absolutely abelian fields. *Izv. Akad. Nauk Kazah. SSR Ser. Fiz.–Mat. Nauk* **1** (1966) 10–13 (in Russian); *Math. Rev.* **34** (1967) 5806. (Regarding related articles by this author, see also *Math. Rev.* **15** (1954) 403, 937 **17** (1956) 829, **30** (1965) 3075, **34** (1967) 2558, **35** (1968) 1577.)

VAN KAMPEN, E. R.

[1] On uniformly almost periodic multiplicative and additive functions. *Am. J. Math.* **62** (1940) 627–634.

VAN KAMPEN, E. R. and A. WINTNER

[1] On the almost periodic behavior of multiplicative number-theoretical functions. *Am. J. Math.* **62** (1940) 613–626.

VAUGHAN, R. C.

[1] The problime number theorem. *Bull. London Math. Soc.* **6** (1974) 337–340.

WALFISZ, A.

*[1] *Weylsche Exponentialsummen in der neueren Zahlentheorie* (VEB Deutscher Verlag der Wissensch., Berlin, 1953).

WALKUP, D.

[1] The number of plane trees. *Mathematika* **19** (1973) 200–204.

WEGMANN, H.

[1] Beiträge zur Zahlentheorie auf freien Halbgruppen, I–II. *J. Reine Angew. Math.* **221** (1965) 20–43, 150–159.

WEIL, A.

*[1] *Basic Number Theory* (Springer, Berlin, 1968).

WILLIAMS, D.

[1] A study of a diffusion process motivated by the sieve of Eratosthenes. *Bull. London Math. Soc.* **6** (1974) 155–164.

WILLIAMS, R. F.

[1] Zeta functions in global analysis. In: *Proc. Symp. Pure Math.* **14** (Am. Math. Soc., Providence, R. I, 1970) 335–339.

WINTNER, A.

*[1] *The Theory of Measure in Arithmetical Semigroups* (Waverly, Baltimore, Md., 1944).

*[2] *Eratosthenian Averages* (Waverly, Baltimore, Md., 1943).

[3] On a statistics of the Ramanujan sums. *Am. J. Math.* **64** (1942) 106–114.

[4] Number-theoretical almost-periodicities. *Am. J. Math.* **67** (1945) 173–193.

WIRSING, E.

[1] Das asymptotische Verhalten von Summen über multiplikative Funktionen, I–II. *Math. Ann.* **143** (1961) 75–102, *Acta Math. Acad. Sci. Hung.* **18** (1967) 411–467.

WOLF, J. A.

*[1] *Spaces of Constant Curvature* (McGraw-Hill, New York, 1967).

WRIGHT, E. M.

[1] A relationship between two sequences, I–III. *Proc. London Math. Soc.* (3) **17** (1967) 296–304, 547–552, *J. Lond. Math. Soc.* **43** (1968) 720–724.

[2] Asymptotic relations between enumerative functions in graph theory. *Proc. London Math. Soc.* (3) **20** (1970) 558–572.

[3] Asymptotic enumeration of connected graphs. *Proc. Roy. Soc. Edinburgh* A **68** (1970) 298–308.

[4] Counting coloured graphs. *Can. J. Math.* **13** (1961) 683–693.

[5] Graphs on unlabelled nodes with a given number of edges. *Acta Math.* **126** (1971) 1–9.

[6] The probability of connectedness of an unlabelled graph can be less for more edges. *Proc. Am. Math. Soc.* **35** (1972) 21–25.

[7] The number of unlabelled graphs with many nodes and edges. *Bull. Am. Math. Soc.* **78** (1972) 1032–1034.

[8] The probability of connectedness of a large unlabelled graph. *Bull. Am. Math. Soc.* **79** (1973) 767–769.

[9] Graphs on unlabelled nodes with a large number of edges. *Proc. London Math. Soc.* (3) **28** (1974) 577–594.

WUNDERLICH, M. C.

[1] A probabilistic setting for prime number theory. *Acta Arith.* **26** (1974) 59–81.

[2] The prime number theorem for random sequences, *J. Number Theory,* to appear.

ZARISKI, O. and P. SAMUEL

*[1] *Commutative Algebra,* 2 vols. (Van Nostrand, New York, 1958, 1960).

ADDITIONAL BIBLIOGRAPHY

BABAI, L.

[1] The probability of generating the symmetric group. *J. Combin. Th.* **A52** (1989) 148–153.

BALOG, A.

[1] Statistical theorems about the embedding of abelian groups into symmetrical ones. *Acta Math. Acad. Sci. Hung.* **39** (1982) 117–124.

BAUERMEISTER, H.

[1] Funktionen mit formalem Eulerprodukt. *Manuscr. Math.* **26** (1978) 83–121.

BELL, E. T.

*[1] *Algebraic Arithmetic* (Am. Math. Soc., Providence, 1927).

BENDER, E. A.

[1] An asymptotic expansion for the coefficients of some formal power series. *J. London Math. Soc.* (2) **9** (1975) 451–458.

BENDER, E. A. and L. B. RICHMOND

[1] Central and local limit theorems applied to asymptotic enumeration, I–III. *J. Combin. Th.* **A15** (1973) 91–111, **A34** (1983) 255–265, and **A35** (1983) 263–278.

[2] The asymptotic enumeration of rooted convex polyhedra. *J. Combin. Th.* **B36** (1984) 276–283.

[3] An asymptotic expansion for the coefficients of some power series, II. *Discrete Math.* **50** (1984) 135–141.

BERTRAM, E. A.

[1] A density theorem on the number of conjugacy classes in finite groups. *Pacific J. Math.* **55** (1974) 329–333.

BOGNÁR, K.

[1] On a problem of statistical group theory. *Studia Sci. Math. Hung.* **5** (1970) 129–136.

BOLLOBAS, B.

*[1] *Extremal Graph Theory* (Academic Press, New York, 1978).

*[2] *Random Graphs* (Academic Press, New York, 1985).

BOREL, J.-P.

[1] Quelques résultats d'équi-répartition liés aux nombres premiers généralisés de Beurling. *Acta Arith.* **38** (1980) 255–272.

[2] Sur le prolongement des fonctions ζ associées à un système de nombres premiers généralisés de Beurling. *Acta Arith.* **43** (1984) 273–282.

BORWEIN, J. M. and B. RICHMOND

[1] How many matrices have roots? *Can. J. Math.* **36** (1984) 286–299.

BOVEY, J. D.
[1] An approximate probability distribution for the order of elements of the symmetric group. *Bull. London Math. Soc.* **12** (1980) 41–46.
[2] The probability that some power of a permutation has small degree. *Ibid.* **12** (1980) 47–51.

BOVEY, J. and A. WILLIAMSON
[1] The probability of generating the symmetric group. *Bull. London Math. Soc.* **10** (1978) 91–96.

BUSHNELL, C. J. and I. REINER
[1] Zeta functions of arithmetic orders and Solomon's conjectures. *Math. Z.* **173** (1980) 135–161.
[2] L-functions of arithmetic orders and asymptotic distribution of ideals. *J. Reine Angew. Math.* **327** (1981) 156–183.
[3] Functional equations for L-functions of arithmetic orders. *Ibid.* **329** (1981) 88–124.
[4] The prime ideal theorem in non-commutative arithmetic. *Math. Z.* **181** (1982) 143–170.
[5] Zeta functions and composition factors for arithmetic orders. *Math. Z.* **194** (1987) 415–428.

CAHN, R. S.
[1] Lattice points and Lie groups, I–II. *Trans. Am. Math. Soc.* **183** (1973) 119–137.
[2] The asymptotic expansion of the zeta-function of a compact semisimple Lie group. *Proc. Am. Math. Soc.* **54** (1976) 459–462.

CAHN, R. S. and J. A. WOLF
[1] Zeta-functions and their asymptotic expansions for compact symmetric spaces of rank one. *Comment. Math. Helv.* **51** (1976) 1–22.

CARLITZ, L.
[1] Arithmetic functions in an unusual setting, I–II. *Am. Math. Monthly* **73** (1966) 582–590, and *Duke Math. J.* **34** (1967) 757–760.

COHEN, S. D.
[1] The function field abstract prime number theorem. *Math. Proc. Cambr. Phil. Soc.* **106** (1989) 7–12.

CONNELL, I. G.
[1] A number theory problem concerning finite groups and rings. *Can. Math. Bull.* **7** (1964) 23–34.

DAS, S. K.
[1] On the enumeration of finite maximal connected topologies. *J. Combin. Th.* **B15** (1973) 184–199, and **B21** (1976) 285.

DELSARTE, P.
[1] A generalization of the Legendre symbol for finite abelian groups. *Discrete Math.* **27** (1979) 187–192.

DHAR, D.
[1] Asymptotic enumeration of partially ordered sets. *Pacific J. Math.* **90** (1980) 299–305.

DIACONIS, P., F. MOSTELLER and H. ONISHI

[1] Second-order terms for the variance and covariance of the number of prime factors. *J. Numb. Th.* **9** (1977) 187–202.

DIACONIS, P. and C. STEIN

[1] Some Tauberian theorems related to coin tossing. *Ann. of Prob.* **6** (1978) 483–490.

DIAMOND, H. G.

[1] When do Beurling generalized integers have a density? *J. Reine Angew. Math.* **295** (1977) 22–39.

DÜR, A.

*[1] *Mobius Functions, Incidence Algebras and Power Series Representations. Lect. Notes in Math.* **1202** (Springer, Berlin, 1986).

ELLIOTT, P. D. T. A.

[1] A mean value theorem for multiplicative functions. *Proc. London Math. Soc.* (3) **3** (1975) 418–438.

*[2] *Probabilistic Number Theory,* Vols. I–II (Springer, Berlin, 1979/80).

ERDÖS, P. and R. R. HALL

[1] Some new results in probabilistic group theory. *Comment. Math. Helv.* **53** (1978) 448–458.

[2] Probabilistic methods in group theory, II. *Houston J. Math.* **2** (1976) 173–180.

ERDÖS, P. and E. JABOTINSKY

[1] On sequences generated by a sieving process. *Proc. K. Nederl. Akad. Wet.* **A61** (1958) 115–128.

ERDÖS, P. and M. E. MAYS

[1] On nilpotent but not abelian groups and abelian but not cyclic groups. *J. Numb. Th.* **28** (1988) 363–368.

ERDÖS, P., M. RAM MURTY and V. KUMAR MURTY

[1] On the enumeration of finite groups. *J. Numb. Th.* **25** (1987) 360–378.

ERDÖS, P. and E. G. STRAUSS

[1] How abelian is a finite group? *Lin. Multilin. Alg.* **3** (1976) 307–312.

FINE, N. J. and I. N. HERSTEIN

[1] The probability that a matrix be nilpotent. *Illinois J. Math.* **2** (1958) 449–454.

FINE, T. and J. GILL

[1] The enumeration of comparative probability relations. *Ann. of Prob.* **4** (1976) 667–673.

FLAJOLET, P. and M. SORIA

[1] Gaussian limiting distributions for the number of components in combinatorial structures. *J. Combin. Th.* **A53** (1990) 165–182.

FOGELS, E.

[1] On the distribution of analogues of primes. (In Russian) *Dokl. Akad. Nauk SSSR* **146** (1962) 318–321.

[2] On the zeros of L-functions. *Acta Arith.* **11** (1965) 67–96.

FOSTER, D. and D. WILLIAMS

[1] The Hawkins sieve and Brownian motion. *Compos. Math.* **37** (1978) 279–289.

FOTINO, I. P.

[1] Generalized convolution ring of arithmetic functions. *Pacific J. Math.* **61** (1975) 103–116.

FRANKS, J. M.

[1] Morse inequalities for zeta functions. *Ann. of Math.* **102** (1975) 143–157.

FREDMAN, M. L.

[1] Arithmetical convolution products and generalizations. *Duke Math. J.* **37** (1970) 231–242.

[2] Congruence formulas obtained by counting irreducibles. *Pacific J. Math.* **35** (1970) 613–624.

GALKIN, V. M.

[1] Zeta-functions of certain one-dimensional rings. *Math. of USSR–Izv.* **7** (1973) 1–17.

GALLIAN, J. A. and J. VAN BUSKIRK

[1] The number of homomorphisms from $\mathbf{Z}_m$ into $\mathbf{Z}_n$. *Am. Math. Monthly* **91** (1984) 196–197.

GERSTENHABER, M.

[1] On the number of nilpotent matrices with coefficients in a finite field. *Illinois J. Math.* **5** (1961) 330–333.

GIOIA, A. A.

[1] The K-product of arithmetic functions. *Can. J. Math.* **17** (1965) 970–976.

GOLDSMITH, D. L.

[1] A generalized convolution for arithmetic functions. *Duke Math. J.* **38** (1971) 279–283.

GOSS, D.

[1] v-adic Zeta functions, L-series and measures for function fields. *Invent. Math.* **55** (1979) 107–116.

[2] The algebraist's upper half-plane. *Bull. Am. Math. Soc.* **2** (1980) 391–415.

[3] The arithmetic of function fields, 2: the 'cyclotomic' theory. *J. Alg.* **81** (1983) 107–149.

[4] On a new type of L-function for algebraic curves over finite fields. *Pacific J. Math.* **105** (1983) 143–181.

GREEN, T. A.

[1] Asymptotic enumeration of generalized Latin rectangles. *J. Combin. Th.* **A51** (1989) 149–160.

GREENE, C.

[1] On the Möbius algebra of a partially ordered set. *Adv. in Math.* **10** (1973) 177–187.

GREGER, K.

[1] Random sieves and the prime number theorem. *Nordisk Mat. Tidskr.* **21** (1973) 57–66, 127.

GROSSWALD, E. and F. J. SCHNITZER

[1] A class of modified ζ and L-functions. *Pacific J. Math.* **74** (1974) 357–364.

GRUNEWALD, F. J., D. SEGAL and G. C. SMITH

[1] Subgroups of finite index in nilpotent groups. *Invent. Math.* **93** (1988) 185–223.

GRYTCZUK, A.

[1] The Halberstam-Richert type theorems for submultiplicative functions on arithmetical semigroups. *Demonstratio Math.* **19** (1986/7) 629–636.

HAFNER, J. L.

[1] The distribution and average order of the coefficients of Dedekind ζ-functions. *J. Numb. Th.* **17** (1983) 183–190.

HALÁSZ, G.

[1] Über die Mittelwerte multiplikativer zahlentheoretischer Funktionen. *Acta Math. Acad. Sci. Hung.* **19** (1968) 365–403.

HALL, R. R.

[1] Extensions of a theorem of Erdös-Renyi in probabilistic group theory. *Houston J. Math.* **3** (1977) 225–234.

HAWKINS, D. and W. E. BRIGGS

[1] The lucky number theorem. *Math. Mag.* **31** (1957/8) 81–84 and 277–280.

HEJHAL, D. A.

[1] The Selberg trace formula and the Riemann zeta function. *Duke Math. J.* **43** (1976) 441–482.

*[2] *The Selberg Trace Formula for* PSL (2, R), I–II. *Lect. Notes in Math.* **548, 1001** (Springer, Berlin, 1976, 1983).

HEPPNER, E.

[1] Über die Dichte der Ordnungen einfacher Gruppen. *Math. Z.* **149** (1976) 17–18.

[2] Über die Anzahl der natürlichen Zahlen *n* kleiner oder gleich *x* für die jede Gruppe der Ordnung *n* auflösbar ist. *Arch. Math.* **32** (1979) 548–550.

HEYDE, C. C.

[1] On asymptotic behavior for the Hawkins random sieve. *Proc. Am. Math. Soc.* **56** (1976) 277–280.

[2] A loglog improvement to the Riemann hypothesis for the Hawkins random sieve. *Ann. of Prob.* **6** (1978) 870–875.

HORADAM, E. M.

[1] Solved, semi-solved and unsolved problems in generalized integers: a survey. *Fib. Quart.* **16** (1978) 370–381.

ILANI, I.

[1] Counting finite index subgroups and the P. Hall enumeration principle. *Israel J. Math.* **68** (1989) 18–26.

INDLEKOFER, K.-H.

[1] Some remarks on almost-even and almost-periodic functions. *Arch. Math.* **37** (1981) 353–358.

[2] Properties of uniformly summable multiplicative functions. *Periodica Math. Hung.* **17** (1986) 143–161.

[3] Cesàro means of additive functions. *Analysis* **6** (1986) 1–24.

[4] Gleichgradige Summierbarkeit. To appear.

[5] The abstract prime number theorem for function fields. Preprint, 1989.

INDLEKOFER, K.-H. and J. KNOPFMACHER

[1] Direct factors of additive arithmetical semigroups. Preprint, 1989.

INDLEKOFER, K.-H., E. MANSTAVICIUS and R. WARLIMONT

[1] On a certain class of infinite products with an application to arithmetical semigroups. Preprint, 1989.

IVIĆ, A.

[1] An asymptotic formula for elements of a semigroup of integers. *Mat. Vesnik* **10**(25) (1973) 255–257.

[2] On an arithmetical semigroup connected with quadratic residues. *C. R. Acad. Bulgare Sci.* **29** (1976) 1257–1259.

[3] The distribution of values of the enumerating function of non-isomorphic abelian groups of finite order. *Arch. Math.* **30** (1978) 274–279.

[4] On the number of finite non-isomorphic abelian groups in short intervals. *Math. Nachr.* **101** (1981) 257–271.

[5] On the number of abelian groups of a given order and on certain related multiplicative functions. *J. Numb. Th.* **16** (1983) 119–137.

*[6] *The Riemann Zeta-Function* (Wiley, New York, 1985).

[7] The number of finite non-isomorphic abelian groups in mean square. *Hardy-Ramanujan J.* **9** (1986) 17–23.

IVIĆ, A. and W. SCHWARZ

[1] Remarks on some number-theoretical functional equations. *Aequ. Math.* **20** (1980) 80–89.

KATZ, N. M.

[1] An overview of Deligne's proof of the Riemann hypothesis for varieties over finite fields. *Proc. Symp. Pure Math.* **28** (1976) 275–305.

KINGSLEY, R. A.

[1] On the number of matrices of fixed rank and size over a finite field. *Utilitas Math.* (1975) 287–292.

KLEITMAN, D. J. and B. L. ROTHSCHILD

[1] The number of finite topologies. *Proc. Am. Math. Soc.* **25** (1970) 276–282.

[2] Asymptotic enumeration of partial orders on a finite set. *Trans. Am. Math. Soc.* **205** (1975) 205–220.

KLEITMAN, D. J., B. L. ROTHSCHILD and J. H. SPENCER

[1] The number of semigroups of order *n*. *Proc. Am. Math. Soc.* **55** (1976) 227–232.

KLEITMAN, D. J. and K. J. WINSTON

[1] The asymptotic number of lattices. *Ann. Discr. Math.* **6** (1980) 243–249.

KLOTZ, W. and L. LUCHT

[1] Endliche Verbände. *J. Reine Angew. Math.* **247** (1971) 58–68.

KNOPFMACHER, A. and J. KNOPFMACHER

[1] The exact length of the Euclidean algorithm in $\mathbf{F}_q[X]$. *Mathematika* **35** (1988) 297–304.

[2] Counting polynomials with a given number of zeros in a finite field. *Linear and Mult. Algebra* **26** (1990) 287–292.

[3] The distribution of values of polynomials over a finite field. *Linear Algebra and Appl.* (1990), in press.

KNOPFMACHER, J.

[1] Finite modules and algebras over rings of algebraic functions. *Bull. London Math. Soc.* **8** (1976) 289–293.

[2] An abstract prime number theorem relating to algebraic function fields. *Arch. Math.* **24** (1977) 271–279.

[3] Generalized Euler constants. *Proc. Edin. Math. Soc.* **21** (1978) 25–32.

[4] Solomon's zeta function and enumeration of lattices over orders. *Analysis* **5** (1985) 29–42.

[5] Solomon's zeta function over algebraic function fields. *Manuscr. Math.* **53** (1985) 101–106.

[6] Direct factors of polynomial rings over finite fields. *J. Combin. Th.* **A40** (1985) 429–434.

[7] Harmonic analysis of arithmetical functions. Proc. Symposium on Harmonic Analysis in Number Theory, RIMS, Kyoto University, 1987. *RIMS Kôkyûroku* **631** (1987) 110–122.

KNOPFMACHER, J. and W. SCHWARZ

[1] Binomial expected values of arithmetical functions, I–II. *J. Reine Angew. Math.* **328** (1981) 84–98, and *Colloq. Math. Soc. J. Bolyai* **34** (1984) 863–904.

KNOPFMACHER, J. and P. G. SLATTERY

[1] Ramanujan expansions over additive arithmetical semigroups. *Bull. Soc. Math. Belgique* **B41** (1989) 109–127.

KOLCHIN, V. F.

*[1] *Random Mappings* (Springer, Berlin, 1986).

KOLESNIK, G.

[1] On the number of Abelian groups of a given order. *J. Reine Angew. Math.* **329** (1981) 164–175.

KONUSBEKOV, K. K. and G. A. POPOV

[1] Tauberian theorem on semigroups. *Dokl. Akad. Nauk SSSR,* No. 7 (1977), 10–11.

KORSHUNOV, A. D.

[1] Random graphs with a large number of vertices. *Russian Math. Surv.* **40,** No. 1 (1985) 121–198.

KOVACS, A.

[1] On the probability that the product of k $n \times n$ matrices over a finite field will be zero. *J. Combin. Th.* **A45** (1987) 290–299.

[2] Some enumeration problems for matrices over a finite field. *Lin. Alg. Appl.* **94** (1987) 223–236.

[3] Products of matrices over a finite field. *Illinois J. Math.* **32** (1988) 211–221.

KRÄTZEL, E.

[1] Die Werteverteilung der Anzahl der nicht-isomorphen Abelschen Gruppen endlicher Ordnung in kurzen Intervallen. *Math. Nachr.* **98** (1980) 135–144.

[2] The number of non-isomorphic abelian groups of a given order. *Forschs. Ergeb. Fr-Schiller Univ. Jena* (1982).

[3] On the average number of direct factors of a finite abelian group. *Acta Arith.* **51** (1988) 369–379.

KRISHNAMURTHY, V.

[1] Counting of finite topologies and a dissection of Stirling numbers of the second kind. *Bull. Austral. Math. Soc.* **12** (1975) 111–124.

KRYŽIUS, Z.

[1] Additive arithmetic functions on semigroups and the preservation of weak convergence of measures. *Lith. Math. J.* **25** (1985) 35–43.

[2] Almost even functions on semigroups. *Ibid.* **25** (1985) 128–136.

[3] Limit periodic arithmetic functions. *Ibid.* **25** (1985) 243–250.

KUNG, J. P. S., M. RAM MURTY and G. C. ROTA
[1] On the Rédei zeta function. *J. Numb. Th.* **12** (1980) 421–436.

LEHMER, D. H.
[1] Euler constants for arithmetical progressions. *Acta Arith.* **27** (1975) 125–142.

LYNCH, J. F.
[1] An asymptotic formula for the number of classes of sets of n indistinguishable elements. *J. Combin. Th.* **A19** (1975) 109–112.

MALOLETKIN, G. N.
[1] Zeta functions of a semisimple algebra over the field of rational numbers. *Sov. Math. Dokl.* **12** (1971) 1678–1681.

MANNING, A.
[1] Axiom A diffeomorphisms have rational zeta functions. *Bull. London Math. Soc.* **3** (1971) 215–220.

MARCU, D.
[1] An application of vector spaces to the calculation of the number of connected components of a finite directed graph. *Stud. Cerc. Mat.* **28** (1976) 199–203.

MARTIN, U.
[1] Almost all p-groups have automorphism group a p-group. *Bull. Am. Math. Soc.* **15** (1986) 78–82.

MATSUOKA, Y.
[1] Generalized Euler constants. In: *Number Theory and Combinatorics,* pp. 279–295 (World Sci. Publ., Tokyo, 1985).

MAUCLAIRE, J.-L.
*[1] *Intégration et Théorie des Nombres* (Hermann, Paris, 1986).

MAYS, M. E.
[1] Counting abelian, nilpotent, solvable and supersolvable group orders. *Arch. Math.* **31** (1978) 536–538.
[2] Groups of square-free order are scarce. *Pacific J. Math.* **91** (1980) 373–375.
[3] Counting p-nilpotent group orders. *West Virginia Acad. Sci.* (1979/80) 92–94.
[4] A lower bound on the number of finite simple groups. *Int. J. Math. and Math. Sci.* **3** (1980) 797–799.

McDONALD, D. A.
[1] Arithmetical functions on a locally finite, locally semi-modular, local lattice. *J. Reine Angew. Math.* **278/9** (1975) 37–47.

McHALE, D.
[1] Commutativity in finite rings. *Am. Math. Monthly* **83** (1976) 30–32.

McIVER, A. and P. M. NEUMANN
[1] Enumerating finite groups. *Quart. J. Math.* (2) **38** (1987) 473–488.

MILLER, W.
[1] The maximum order of an element of a finite symmetric group. *Am. Math. Monthly* **94** (1987) 497–506.

MOON, J. W.
*[1] *Counting Labelled Trees* (Can. Math. Congress, Montreal, 1970).

Müller, H.

[1] Zur Gleichverteilung von Beurlingschen Zahlen. *Abh. Math. Sem. Hamburg* **43** (1975) 186–191, and **49** (1979) 252.

[2] Eine Bemerkung zur Zahlentheorie auf freien Halbgruppen. *Mitt. Math. Gesellsch. Hamburg* **10** (1978) 459–469.

[3] Über eine Klasse modifizierter ζ-und L-Funktionen. *Arch. Math.* **36** (1981) 157–161.

[4] On generalized zeta-functions at negative integers. *Illinois J. Math.* **32** (1988) 222–229.

Narkiewicz, W.

[1] On a class of arithmetical convolutions. *Colloq. Math.* **10** (1963) 81–94.

Narlikar, M. J. and S. Srinivasan

[1] On orders solely of abelian groups, II. *Bull. London Math. Soc.* **20** (1988) 211–216.

Neudecker, W.

[1] On twin 'primes' and gaps between successive 'primes' for the Hawkins sieve. *Math. Proc. Cambr. Phil. Soc.* **77** (1975) 365–368.

Newman, M.

[1] Asymptotic formulas related to products of cyclic groups. *Math. Comp.* **30** (1976) 836–846.

Nicolas, J.-L.

[1] Sur l'ordre maximum d'un élément dans le groupe S_n des permutations. *Acta Arith.* **14** (1968) 315–332.

[2] Ordre maximal d'un élément du groupe S_n des permutations et "highly composite numbers." *Bull. Soc. Math. France* **97** (1969) 129–191.

[3] Ordre maximal d'un groupe de permutations. *C. R. Acad. Sci. Paris* **A270** (1970) 1473–1476.

[4] Sur les entiers N pour lesquels il y a beaucoup de groupes abéliens d'ordre N. *Ann. Inst. Fourier* (Grenoble) **28** (1978) 1–16.

[5] Grandes valeurs d'une certaine classe de fonctions arithmétiques. *Studia Sci. Math. Hung.* **15** (1980) 71–77.

[6] Petites valeurs de la fonction d'Euler. *J. Numb. Th.* **17** (1983) 375–388.

Nicolas, J.-L. and P. Erdős

[1] Probabilistic and combinatory methods in number theory. *Bull. Sci. Math.* **100** (1976) 301–320.

Nymann, J. E. and W. J. Leahey

[1] On the probability that integers chosen according to the binomial distribution are relatively prime. *Acta Arith.* **31** (1976) 205–211.

[2] On the probability that an integer chosen according to the binomial distribution be k-free. *Rocky Mtn. J. Math.* **7** (1977) 769–774.

Palmer, E. M.

*[1] *Graphical Evolution* (Wiley, New York, 1985).

Parry, W.

[1] An analogue of the prime number theorem for shifts of finite type and their suspensions. *Israel J. Math.* **45** (1983) 41–52.

PARRY, W. and M. POLLICOTT

[1] An analogue of the prime number theorem for closed orbits of Axiom A flows. *Ann. of Math.* **118** (1983) 573–591.

PAUL, E. M.

[1] Density in the light of probability theory, I–III. *Sankyha (Indian J. Statistics)* **A24** (1962) 103–114 and 209–212, and **A25** (1963) 273–280.

[2] Some properties of additive arithmetical functions. *Ibid.* **A29** (1967) 279–282.

POMERANCE, C.

[1] On the average number of groups of square-free order. *Proc. Am. Math. Soc.* **99** (1987) 223–231.

PORUBSKÝ, S.

[1] On exponents in arithmetical semigroups. *Monatsh. Math.* **84** (1977) 49–53.

[2] On the density of certain sets in arithmetical semigroups. *Czech. Math. J.* **29** (104) (1979) 148–152.

[3] Density for generalized integers. *Colloq. Math. Soc. J. Bolyai.* **34** (1984) 1295–1315.

PORUBSKÝ, S. and W. SCHWARZ

[1] Sums of additive functions on arithmetical semigroups. *Colloq. Math.* **53** (1987) 137–146.

RAGGI-CÁRDENAS, A.

[1] Zeta functions of two-sided ideals in arithmetic orders. *Math. Z.* **192** (1986) 353–382.

RAM MURTY, M. and V. KUMAR MURTY

[1] On the density of various classes of groups. *J. Numb. Th.* **17** (1983) 29–36.

[2] On the number of groups of a given order. *Ibid.* **18** (1984) 178–191.

[3] On groups of square-free order. *Math. Ann.* **267** (1984) 299–309.

RAM MURTY, M. and S. SRINIVASAN

[1] On the number of groups of square-free order. *Can. J. Math.* **30** (1987) 412–420.

REARICK, D.

[1] Operators on algebras of arithmetic functions. *Duke Math. J.* **35** (1968) 761–766.

REICH, A.

[1] Universelle Werteverteilung von Eulerprodukten. *Nachr. Akad. Wiss. Göttingen, Math.-Phys. Kl.* II (1977), No. 1.

REINER, I.

[1] On the number of matrices with given characteristic polynomial. *Illinois J. Math.* **5** (1961) 324–329.

RÉNYI, A.

[1] On random generating elements of a finite Boolean algebra. *Acta Sci. Math.* **22** (1961) 75–81.

RICHMAN, D. J. and H. SCHNEIDER

[1] Primes in the semigroup of non-negative matrices. *Lin. Multilin. Alg.* **2** (1974) 135–140.

RICHMOND, L. B. and M. V. SUBBARAO

[1] On certain weighted partitions and finite semisimple rings. *Proc. Am. Math. Soc.* **64** (1977) 13–19.

RICHMOND, L. B. and N. C. WORMALD

[1] The asymptotic number of convex polyhedra. *Trans. Am. Math. Soc.* **273** (1982) 721–735.

RIDLEY, J. N. and D. B. SEARS

[1] Asymptotic and analytic formulae for enumerating modules. *Arch. Math.* **32** (1979) 149–154.

ROBIN, G.

[1] Grandes valeurs de la fonction somme des diviseurs et hypothèse de Riemann. *J. Math. Pures et Appl.* **63** (1984) 187–213.

[2] Grandes valeurs de la fonction somme des diviseurs dans les progressions arithmétiques. *Ibid.* **66** (1987) 337–349.

ROTHAUS, O. S.

[1] On the number of irreducible representations of degree at most n of a Lie group. *Proc. Am. Math. Soc.* **51** (1975) 217–220.

RUELLE, D.

[1] Zeta-functions for expanding maps and Anasov flows. *Invent. Math.* **34** (1976) 231–242.

RUZSA, I. Z.

[1] Semigroup-valued multiplicative functions. *Acta Arith.* **42** (1982/3) 79–90.

SARNAK, P.

[1] The arithmetic and geometry of some hyperbolic 3-manifolds. *Acta Math.* **151** (1983) 253–295.

SCHEID, H.

[1] Funktionen über lokal endlichen Halbordnungen, I–II. *Monatsh. Math.* **74** (1970) 336–347, and **75** (1971) 44–56.

SCHROHE, E.

[1] Über eine Klasse modifizierter Zetafunktionen. *Arch. Math.* **43** (1984) 427–433.

SCHWARZ, W.

[1] Aus der Theorie der zahlentheoretischen Funktionen. *Jber. Deutsch. Math.-Ver.* **78** (1976) 147–167.

[2] Fourier-Ramanujan-Entwicklungen zahlentheoretischer Funktionen und Anwendungen. *Festschrift J. W. Goethe-Universität, Frankfurt/Main* (1981) 399–415.

SHERMAN, G.

[1] What is the probability an automorphism fixes a group element? *Am. Math. Monthly* **82** (1975) 261–264.

SHIU, P.

[1] The maximum orders of multiplicative functions. *Quart. J. Math.* (Oxford) (2) **31** (1980) 247–252.

SLATTERY, P. G.

[1] Ramanujan expansions of arithmetical functions over arithmetical semigroups. Ph.D. Thesis, University of the Witwatersrand, Johannesburg, 1989.

SMITH, D. A.

[1] Incidence functions as generalized arithmetic functions, I–III. *Duke Math. J.* **34** (1967) 617–634, **36** (1969) 15–30 and 353–368.

[2] Multiplication operators on incidence algebras. *Indiana Univ. Math. J.* **20** (1970) 369–383.

[3] Generalized arithmetic function algebras. *Lect. Notes Math.* **251,** 205–245 (Springer, Berlin, 1972).

SOLOMON, L.

[1] Zeta functions and integral representations. *Adv. in Math.* **26** (1977) 306–326.

SRINIVASAN, S.

[1] On orders solely of abelian groups. *Glasgow Math. J.* **29** (1987) 105–108.

STEIN, C. M.

[1] Asymptotic enumeration of Latin rectangles. *J. Combin. Th.* **A25** (1978) 38–49.

STONG, R.

[1] Some asymptotic results on finite vector spaces. *Adv. in Appl. Math.* **9** (1988) 167–199.

SUBBARAO, M. V.

[1] On some arithmetic convolutions. *Lect. Notes Math.* **251,** 247–271. (Springer, Berlin, 1972).

SURYANARAYANA, D. and R. S. R. CHANDRA RAO

[1] On the true maximum order of magnitude of a class of arithmetical functions. *Math. J. Okayama Univ.* **17,** No. 2 (1975) 95–101.

SZALAY, M.

[1] On the maximal order in S_n and S_n^*. *Acta Arith.* **37** (1980) 321–331.

SZALAY, M. and P. TURÁN

[1] On some problems of the statistical theory of partitions with applications to characters of the symmetric group, I–III. *Acta Math. Acad. Sci. Hung.* **29** (1977) 361–392 and **32** (1978) 129–156.

TULJAGANOVA, M. I.

[1] Distribution of values of Euler's function defined on a normalized semigroup. (In Russian) *Izv. Akad. Nauk UzSSR, Fiz.-Mat. Nauk.* **12,** No. 5 (1968) 33–37.

[2] Asymptotic formulae for sums of multiplicative functions that are determined on a normed semigroup. (In Russian) *Ibid.* **13,** No. 4 (1969) 30–35.

[3] Some asymptotic formulas for the sums of multiplicative functions that are defined on a normed semigroup. (In Russian) *Ibid.* **20,** No. 6 (1976) 21–24 and 87.

TUTTE, W. T.

[1] On the enumeration of convex polyhedra. *J. Combin. Th.* **B28** (1980) 105–126.

VERSHIK, A. and A. SCHMIDT

[1] Limit measures arising in the asymptotic theory of symmetric groups, I–II. *Th. Prob. and Appl.* **22** (1977) 70–85, and **23** (1978) 36–49.

WARLIMONT, R.

[1] Über die Anzahl der Losungen von $x^n = 1$ in der symmetrischen Gruppe S_n. *Arch. Math.* **30** (1978) 591–594.

[2] On the set of natural numbers which only yield orders of abelian groups. *J. Numb. Th.* **20** (1985) 354–362.

[3] Multiplicative functions into finite abelian groups. *Arch. Math.* **50** (1988) 534–537.

WARREN, R. H.

[1] The number of topologies. *Houston J. Math.* **8** (1982) 297–301.

WILLE, D.

[1] Asymptotische Abschätzung einer Anzahlfunktion in endlichen Relationssystemen. *Monatsh. Math.* **78** (1974) 174–180.

[2] Asymptotic formulas for the number of oriented graphs. *J. Combin. Th.* **B21** (1976) 270–274.

WILSON, R. M.

[1] Nonisomorphic Steiner triple systems. *Math. Z.* **135** (1973/4) 303–313.

WIRSING, E.

[1] Das asymptotische Verhalten von Summen über multiplikative Funktionen, I–II. *Math. Ann.* **143** (1961) 75–100, and *Acta Math. Acad. Sci. Hung.* **18** (1967) 411–467.

WRIGHT, E. M.

[1] The number of irreducible tournaments. *Glasgow Math. J.* **11** (1970) 97–101.

[2] Asymptotic enumeration of connected graphs. *Proc. Roy. Soc. Edin.* **68** (1968/9) 298–308.

[3] The probability of connectedness of a large unlabelled graph. *J. London Math. Soc.* (2) **11** (1975) 13–16.

[4] The asymptotic enumeration of unlabelled graphs. *Utilitas Math.* (1976) 665–677.

[5] The evolution of unlabelled graphs. *J. London Math. Soc.* (2) **14** (1976) 554–558.

[6] The proportion of unlabelled graphs which are Hamiltonian. *Bull. London Math. Soc.* **8** (1976) 241–244.

[7] The number of connected sparsely-edge graphs, I–III. *J. Graph Th.* **1** (1977) 317–330, **2** (1978) 299–305 and **4** (1980) 393–407.

WUNDERLICH, M. C.

[1] Sieve-generated sequences. *Can. J. Math.* **18** (1966) 291–299.

[2] Sieve-generated sequences with a limiting distribution property. *J. London Math. Soc.* **43** (1968) 339–346.

[3] A general class of sieve-generated sequences. *Acta Arith.* **16** (1969) 41–56.

WUNDERLICH, M. C. and W. E. BRIGGS

[1] Second and third term approximations of sieve-generated sequences. *Illinois J. Math.* **10** (1966) 694–700.

YEUNG, C. N.

[1] An asymptotic formula for the number of K-free groups of order $\leq x$. *J. Nat. Sci. and Math.* **11** (1971) 243–256.

ZHANG, W.-B.

[1] Chebyshev type estimates for Beurling generalized prime numbers. *Proc. Am. Math. Soc.* **101** (1987) 205–212.

[2] Density and O-density of Beurling generalized integers. *J. Numb. Th.* **30** (1988) 120–139.

[3] A generalization of Halasz's theorem to Beurling's generalized integers. *Illinois J. Math.* **31** (1987) 645–664.

INDEX

A CATALOG OF SELECTED

DOVER BOOKS

IN ALL FIELDS OF INTEREST

A CATALOG OF SELECTED DOVER BOOKS IN ALL FIELDS OF INTEREST

DRAWINGS OF REMBRANDT, edited by Seymour Slive. Updated Lippmann, Hofstede de Groot edition, with definitive scholarly apparatus. All portraits, biblical sketches, landscapes, nudes. Oriental figures, classical studies, together with selection of work by followers. 550 illustrations. Total of 630pp. 9⅛ × 12¼.
21485-0, 21486-9 Pa., Two-vol. set $25.00

GHOST AND HORROR STORIES OF AMBROSE BIERCE, Ambrose Bierce. 24 tales vividly imagined, strangely prophetic, and decades ahead of their time in technical skill: "The Damned Thing," "An Inhabitant of Carcosa," "The Eyes of the Panther," "Moxon's Master," and 20 more. 199pp. 5⅜ × 8½. 20767-6 Pa. $3.95

ETHICAL WRITINGS OF MAIMONIDES, Maimonides. Most significant ethical works of great medieval sage, newly translated for utmost precision, readability. Laws Concerning Character Traits, Eight Chapters, more. 192pp. 5⅜ × 8½.
24522-5 Pa. $4.50

THE EXPLORATION OF THE COLORADO RIVER AND ITS CANYONS, J. W. Powell. Full text of Powell's 1,000-mile expedition down the fabled Colorado in 1869. Superb account of terrain, geology, vegetation, Indians, famine, mutiny, treacherous rapids, mighty canyons, during exploration of last unknown part of continental U.S. 400pp. 5⅜ × 8½. 20094-9 Pa. $6.95

HISTORY OF PHILOSOPHY, Julián Marías. Clearest one-volume history on the market. Every major philosopher and dozens of others, to Existentialism and later. 505pp. 5⅜ × 8½. 21739-6 Pa. $8.50

ALL ABOUT LIGHTNING, Martin A. Uman. Highly readable non-technical survey of nature and causes of lightning, thunderstorms, ball lightning, St. Elmo's Fire, much more. Illustrated. 192pp. 5⅜ × 8½. 25237-X Pa. $5.95

SAILING ALONE AROUND THE WORLD, Captain Joshua Slocum. First man to sail around the world, alone, in small boat. One of great feats of seamanship told in delightful manner. 67 illustrations. 294pp. 5⅜ × 8½. 20326-3 Pa. $4.95

LETTERS AND NOTES ON THE MANNERS, CUSTOMS AND CONDITIONS OF THE NORTH AMERICAN INDIANS, George Catlin. Classic account of life among Plains Indians: ceremonies, hunt, warfare, etc. 312 plates. 572pp. of text. 6⅛ × 9¼. 22118-0, 22119-9 Pa. Two-vol. set $15.90

ALASKA: The Harriman Expedition, 1899, John Burroughs, John Muir, et al. Informative, engrossing accounts of two-month, 9,000-mile expedition. Native peoples, wildlife, forests, geography, salmon industry, glaciers, more. Profusely illustrated. 240 black-and-white line drawings. 124 black-and-white photographs. 3 maps. Index. 576pp. 5⅜ × 8½. 25109-8 Pa. $11.95